BOSTON STUDIES IN THE PHILOSOPHY OF SCIENCE

VOLUME XXIII

THE UNDERSTANDING OF NATURE

SYNTHESE LIBRARY

MONOGRAPHS ON EPISTEMOLOGY,

LOGIC, METHODOLOGY, PHILOSOPHY OF SCIENCE,

SOCIOLOGY OF SCIENCE AND OF KNOWLEDGE,

AND ON THE MATHEMATICAL METHODS OF

SOCIAL AND BEHAVIORAL SCIENCES

VOLUME 66

BOSTON STUDIES IN THE PHILOSOPHY OF SCIENCE

EDITED BY ROBERT S. COHEN AND MARX W. WARTOFSKY

VOLUME XXIII

MARJORIE GRENE

University of California, Davis

THE UNDERSTANDING

OF NATURE

Essays in the Philosophy of Biology

D. REIDEL PUBLISHING COMPANY

DORDRECHT-HOLLAND/BOSTON-U.S.A.

Library of Congress Catalog Card Number 74–76477

Cloth edition: ISBN 90 277 0462 7
Paperback edition: ISBN 90 277 0463 5

Published by D. Reidel Publishing Company,
P.O. Box 17, Dordrecht, Holland

Sold and distributed in the U.S.A., Canada, and Mexico
by D. Reidel Publishing Company, Inc.
306 Dartmouth Street, Boston,
Mass. 02116, U.S.A.

Printed in Belgium

EDITORIAL PREFACE

No student or colleague of Marjorie Grene will miss her incisive presence in these papers on the study and nature of living nature, and we believe the new reader will quickly join the stimulating discussion and critique which Professor Grene steadily provokes. For years she has worked with equally sure knowledge in the classical domain of philosophy and in modern epistemological inquiry, equally philosopher of science and metaphysician. Moreover, she has the deeply sensible notion that she should be a critically intelligent learner as much as an imaginatively original thinker, and as a result she has brought insightful expository readings of other philosophers and scientists to her own work.

We were most fortunate that Marjorie Grene was willing to spend a full semester of a recent leave here in Boston, and we have on other occasions sought her participation in our colloquia and elsewhere. Now we have the pleasure of including among the Boston Studies in the Philosophy of Science this generous selection from Grene's philosophical inquiries into the understanding of the natural world, and of the men and women in it.

Boston University Center for the　　　　　R. S. COHEN
Philosophy and History of Science　　　　M. W. WARTOFSKY
April 1974

PREFACE

This collection spans – spottily – years from 1946 ('On Some Distinctions between Men and Brutes') to 1974 ('On the Nature of Natural Necessity'). It would be a bad lookout if one's views had not changed (or developed) in such a lengthy period. In my case, the most obvious alteration, which needs some accounting for, is the change in my reflections on the conceptual structure of evolutionary theory (from 'Two Evolutionary Theories' in 1958 to 'Explanation and Evolution' in 1973). Two prefatory remarks need to be made in this connection. First, as a philosopher trying to analyze the conceptual puzzles inherent in a scientific theory – and/or, in 'Two Evolutionary Theories', the nature of scientific controversy – I was *not* (as some of my critics appear to have believed) trying to tell biologists what happened in the course of evolution. That is not a philosopher's business. I was trying to sort out some of the ambiguities inherent in a theory as comprehensive as Darwinism (or the synthetic theory) has aimed at being, particularly ambiguities in the concepts of 'improvement' or 'progress' or 'adaptation'. Secondly, since my return in 1965 from the isolation first of an Irish farmstead and then of British provincial universities to the more communicative milieu of a University of California campus, I have learned a good deal from talks with colleagues and students, so that, I believe, I now understand better where the philosophical problems in evolutionary theory lie. As I first argued in 'Aristotle and Modern Biology' (1972), it is wiser, or so I have come to believe, to restrict evolutionary theory to what it can cleanly and clearly handle: changes in relative gene frequencies, and to abandon hope for a 'scientific' theory of emergence. That living systems can, and should, be studied at a number of levels – in terms of what Howard Pattee calls 'Hierarchy Theory' – is an important methodological and epistemological thesis. But systems analysis and evolutionary theory are, I believe, forced to coalesce, if at all, either through *ad hoc* principles or through the kinds of conceptual confusions I had earlier – and mistakenly – considered to be characteristic of evolutionary theory as such. In this view, admittedly, I am in disagreement with some eminent evolutionists, even with some of those from whom I have learned most. But again, I must emphasize, this is a disagreement in philosophical interpretation, not *in* science, but *about* it. For although, admittedly, conceptual analysis and empirical investigation are not totally dissociable from one another, as many philosophers used to suppose, their interrelations are both subtle and limited. On the one hand, philosophy reflects on the conceptual structure of scientific statements or theories; on the other hand, philosophical reflection, since it is situated, like every human activity, within the human world, is also influenced by the outcome of scientific research. The latter influence is a difficult one to specify. Scientists can't tell us how to solve our problems – which are always (almost always?) meta-problems. But on the other hand we certainly cannot tell them (except when they try to philosophize) how to solve theirs. I hope, therefore, that these papers will be taken for what they were meant to be: reflections by a student of philosophy about epistemological problems arising in some areas of human knowledge. The last few essays are perhaps more ambitious; the last in particular (as well as the paper on perception) suggests a direction for a 'philosophical anthropology', not so much a philosophy of mind as a philosophy of man (or man-in-nature) which would permit new and more fruitful attacks on some traditional philosophical questions. The theory of evolution would form an essential ingredient of such an anthropology, though not, I believe, its comprehensive framework.

University of California, Davis MARJORIE GRENE
January, 1974

TABLE OF CONTENTS

ACKNOWLEDGEMENTS

Most of the papers in this volume have been published elsewhere, as follows:

Chapter I: *Philosophy* **38** (1963), 149–159.

Chapter III: *The New Scholasticism* **41** (1967), 94–123.

Chapter IV: in *Interpretations of Life and Mind* (ed. by M. Grene), Routledge and Kegan Paul, London; Humanities Press, New York, pp. 14–37.

Chapter V: *Journal of the History of Ideas* **33** (1972), 395–424.

Chapter VII: *British Journal for the Philosophy of Science* **9** (1958), 110–127 and 185–193.

Chapter VIII: *B.J.P.S.* **12** (1961), 25–42.

Chapter IX: *Cambridge Review*, February 1964, pp. 269–273.

Chapter X: in *Towards a Theoretical Biology. II: Sketches* (ed. by C. H. Waddington), Edinburgh University Press, Edinburgh, 1969, pp. 61–69.

Chapter XI: in *Proceedings of the Colloquium 'Connaissance scientifique et Philosophie'*, held at Brussels on May 16 and 17, 1974 in honor of the bicentennial of the Académie Royale des Sciences, des Lettres et des Beaux-Arts de Belgique.

Chapter XII: Review of Waddington's 'The Ethical Animal', *B.J.P.S.* **13** (1962), 173–176.

Chapter XV: *Ethics* **57** (1947), 121–127.

Chapter XVI: from my *Approaches to a Philosophical Biology*, Basic Books, New York, 1969, pp. 3–54.

Chapter XVII: *Review of Metaphysics* **21** (1967), 94–123 (under the title 'Straus's Phenomenological Psychology').

Chapter XVIII: *Review of Metaphysics* **20** (1966), 250–277 (under the title 'Positionality in the Philosophy of Helmuth Plessner').

Chapter XIX: *Philosophical Forum* **3** (1972), 157–172 (also published in *Philosophia Naturalis* **14** (1973), 25–38).

I am grateful to all the above for their permission to reprint – including a few cases where, technically, no permission was necessary.

CHAPTER I

CAUSES

In an essay on 'The Notion of Cause' reprinted in *Mysticism and Logic* (and which constituted the presidential address to the Aristotelian Society in the year 1912) Russell argued 'that the word "cause" is so inextricably bound up with misleading associations as to make its complete extrusion from the philosophical vocabulary desirable.'[1] His argument here to the effect that 'cause' is not a central concept in science, as philosophers have thought it, is reminiscent of Norman Campbell's statement in *Physics: The Elements* (1920) and in *What is Science?* (1921). 'Cause', we are told by Campbell, and again by Russell in *Our Knowledge of the External World* (1914), is not a concept that forms part of any 'developed' science, and so whatever its place in medicine (as: A protozoan belonging to the class Sporozoa causes malaria) or history (as: Discuss the causes of the Civil War!) or everyday life (as: The fog caused many accidents), the idea that something called 'causal laws' are important in the exact sciences and *a fortiori* in all knowledge worthy of the name, appears to be one of those idols of the theatre to which philosophers persistently and mistakenly pay homage.

Yet Russell himself, two years later, and again thirty-six years later, had by no means succeeded in extruding the word 'cause' from his philosophical vocabulary. In *Our Knowledge of the External World* he still talked about 'causal laws', and in *Human Knowledge: Its Scope and Limits* (1948) he arrived at the statement of five postulates needed to validate the scientific method, three of which explicitly and the other two implicitly involve a reference to cause. In fact, the term 'cause' is introduced much earlier in the argument as an undefined and indispensable term, in the preliminary definition of truth: 'A sentence of the form "this is *A*" is called "true" when it is caused by what "*A*" means.'[2]

This is puzzling. Let me push the puzzle a little further back in time and ask: if we consider the history of the influence of the ideal of scientific knowledge on philosophical thinking, where does the concept of 'cause' come into this story? In what sense, if any, did the new way of knowing of the seventeenth century entail a characteristic concept of cause, or did it really mean that in so far as people now thought 'scientifically' they stopped thinking in terms of cause at all? What partly worries me about this situation, of course, is that the concept of 'cause' bulks so large in the arguments of Hume and Kant, both of whom thought they were leaning heavily on Newtonian science as the paradigm of knowledge. What I propose to do here (this is only a very fragmentary beginning of what ought to be done) is to look at a few statements by its practitioners about the new 'experimental philosophy' in order to see where 'cause' comes into them, then to consider what happens to 'cause' in Hume and Kant, and finally to come back to Russell in *Human Knowledge*. Thus I want to exclude the meanings of cause in human affairs, i.e., in any context involving voluntary action, as well as the use of cause in the biological sciences.

On September 27th, 1661, Henry Oldenburg wrote to Spinoza:

In our Philosophical Society, we indulge, as far as our powers allow, in diligently making experiments and observations, and we spend much time in preparing a History of the Mechanical Arts, feeling certain that the forms and qualities of things can best be explained by the principles of Mechanics, and that all the effects of Nature are produced by motion, figure, texture, and the varying combinations of these, and that there is no need to have recourse to inexplicable forms and occult qualities, as to a refuge for ignorance.[3]

Two parts of this statement are relevant to our inquiry here: 'that the forms and qualities of things can best be *explained* by the principles of Mechanics', and 'that all the effects of nature are *produced* by motion, figure, texture, and the varying combinations of these'. To say that 'forms and qualities of things' are *explained* by the principles of mechanics means that these forms and qualities can be deduced as theorems from the principles of mechanics taken as axioms. In experimental philosophy, as Newton was to say twenty years later in the *Principia*, we first have ' "particular propositions" inferred from the phenomena, and afterwards rendered general by induction,' and these become *explanations* when further phenomena (forms and qualities of things) are derived by demonstrative reasoning from the generalised statements. Newton describes in his Preface how this method applies to the *Principia*:

By the propositions mathematically demonstrated in the former books, we in the third derive from the celestial phaenomena the forces of gravity with which bodies tend to the sun and several planets. Then from these forces, by other propositions which are also mathematical, we deduce the motions of the planets, the comets, the moon, and the sea.

Now this is of course the model to which philosophers ever since, including Hume and Kant, have been looking as to the fulfilment of experimental reasoning, and it is this very structure of scientific explanation to which Russell and Campbell are referring when they say that science has nothing to do with 'causes'. For the relation of explanation to the phenomena is one of ground to consequent, not one of cause to effect. In terms of traditional logic or of common sense, cause and effect seem to be concepts referring to the relation between events in a time sequence. As Miss Stebbing says in her *A Modern Introduction to Logic*, 'Whatever view we adopt we must admit that there would be no significance in the assertion of causation unless we at least meant to assert that *whenever* a given occurrence happens, then some other given occurrence happens.'[4] But the relation of a theory to the phenomena explained by it is an if-then relation, not a when-then, or before and after, relation. The theory is not an event in time, but a general proposition, a mechanical principle, in Oldenburg's language, or as it turned out in classical mechanics a differential equation, from which statements about phenomena can be rigorously deduced. There are the phenomena and there is a formula and the phenomena, or rather statements about the phenomena, are deducible from the formula. This is simply the deductive part of the 'hypothetico-deductive' method.

What about the other half of Oldenburg's statement? 'All the effects of nature are *produced* by motion, figure, texture, and the varying combinations of these.' In other words, all the effects of nature are caused by figure, motion, etc. This is indeed a statement about causes, but is it a statement about occurrences of which we could say that whenever they happen (now) something else happens (thereafter)? It seems rather to be referring to something somehow underlying the phenomena which has in it the 'power' of producing the phenomena. The phenomena, which are our perceptions of the physical world, are produced by the fundamental properties of bodies, in Boyle's words, to use a rather more inclusive list than Oldenburg's, 'the determinate size, figure, motion and connection, and suchlike mechanical affections of bodies.'[5] Thus in an eclipse of the sun, for instance, the phenomenon is the occurrence of darkness when it is normally daylight, but the cause is to be found in the motions of the earth, moon and sun relatively to one another. True, motion is usually considered just as plainly a phenomenon as darkness – but what we are doing is to generalise one kind of phenomenon or one aspect of the phenomena and to declare that motions, figures, etc., and the combinations of these *cause* not only particular motions but phenomena which are not, as they appear to us, obviously motions at all. This looks like the procedure described by Newton in his third rule of reasoning (*Principia*, Book III):

The qualities of bodies, which admit neither intension nor remission of degrees, and which are found to belong to all bodies within the reach of our experiments, are to be esteemed the universal qualities of all bodies whatsoever.

Now this could be just a device for explanation as Hobbes had made it: we invent a language for dealing with sensations and we must invent a 'mechanical' language – a language whose basic terms refer to bodies in motion – if we are to succeed in finding our way among sensations, since any other language would be contradicted by sensation itself. This is of course the ancient Lucretian argument for atomism as well as the modern argument of the Vienna circle in their 'physicalist' period. But the language of Oldenburg and of his contemporaries seems to imply more than the relation of an artificially constructed axiom system to its theorems. It implies that there really *are* causes, we may even say metaphysical causes, of which the phenomena really are the effects. The effects of nature, darkness, light, heat, cold and so on, are produced by certain mechanical realities, by the shape, size, densities, etc., of the bodies which really exist as against the phenomena which merely appear. So we have, along with the logical relation of explication, some kind of metaphysical relation of cause to effect. True, Oldenburg by no means intended to separate these two extrapolations from the phenomena – any more than did his friend the very illustrious Mr Boyle, of whose experiments he was writing to Spinoza. As Boyle puts it: 'To explicate a phenomenon, it is not enough to ascribe it to one general efficient, but we must intelligibly show the particular manner, how that general cause produces the proposed effect.'[6] Here explanation and the discovery of causes are clearly identified – as they are also in Newton's first two rules of reasoning:

Rule I. We are to admit no more causes of natural things than such as are both true, and sufficient to explain their appearances.

Rule II. Therefore to the same natural effects we must, as far as possible, assign the same causes. As to respiration in a man and in a beast; the descent of stones in Europe and in America; the light of our culinary fire and of the sun; the reflection of light in the earth, and in the planets.

In short, principles are statements of causes and phenomena are effects and it is because causes produce effects that principles succeed in explaining phenomena. Again, however, this does not appear to be a case of causes and effects in the sense of before and after. The universal properties of bodies do not come first and the phenomena follow after them. This is once more a deductive relation, not one of temporal sequence; and what the language of 'causation' adds to the purely formal statement, if C, then E, is the metaphenomenal if not metaphysical assertion that C – in the case of the Oldenburg-Royal Society programme, the assertion that Nature is composed of hard, impenetrable bodies, etc. In other words, explanation succeeds in explaining because things *are* the way the explanation says they are. If an explanation is true, not only does the description of the phenomena follow logically from it, but the phenomena themselves are the effects of the state of affairs which the explanation asserts. Physical explanation becomes causal insofar as it is metaphysical: not through linking phenomena necessarily or invariably to one another in a time sequence but through tying all the phenomena together as consequences of things' being really of a certain sort.

Yet the causes of phenomena, Newton says, must be derived *from* the phenomena and may be stated in experimental philosophy only insofar as they are so derived. He refuses to speculate about the cause of gravity because he has not so far been able to discover it from the phenomena. But can we really derive any *causes* from the phenomena? The phenomena suggest explanations from which statements about further phenomena are derivable – in what phenomenal, non-metaphysical sense do these constitute statements of *causes*? If we really stick to phenomena, to the sequence of sense impressions strung along after one another and eliminate any reference to unseen, underlying causes, what predications of cause can we honestly make? This is, it seems to me, one of the few questions in the history of philosophy that has received a definitive answer, and that answer is: none. For Hume has demonstrated conclusively that all the items of our experience are separable and that therefore no two of them can be connected by a necessary relation such as that of cause and effect is supposed to be. Nor for that matter can we, from the sequence of sense impressions alone, derive probable relations, for among the infinite possible variations of the data of experience there is no *a priori* reason why any two should be more probably related to one another than the next two. No modern 'refutation' of Hume has, so far as I know, met this argument. Take Jeffreys' *Scientific Inference* as one example among many. He supposes that he has constructed a formal model of scientific inference based on a calculus of probability. In his theory, he says,

the classical view of causality is inverted. Instead of saying that every event has a cause, we recognise that observations vary and regard scientific method as a procedure for analysing the variation. Our starting point is to consider all variation as random; then successive significance tests warrant the treatment of more and more of it as predictable.[7]

Yet if variations were really random we could never begin to expect anything. This is apparent from the starting point of Jeffreys' calculus, for we have to start, he says, by assuming that 'all considered propositions shall have positive probabilities on H', where 'H is some general datum that is included in all experience; it might for instance be the rules of pure mathematics.'[8] But what among other things characterises the rules of pure mathematics is precisely that on them all the data of experience are *equi*probable. No one datum could have any positive probability on H. So we cannot begin calculating probabilities at all. And that is just what Hume has told us. In short, far from 'admitting no more causes of natural things than such as are both true and sufficient to explain their appearances', if we really stick to appearances we can admit no causes at all. Russell seems to be right, then, when he tells us to extrude the word 'cause' from our philosophical vocabulary: if we are rigorous phenomenalists we can proceed only by what in *Human Knowledge* he calls 'animal inference'. In Russell's terms, we simply follow those habits which do not lead us to rude shocks and stop following those that do. Or in Hume's more subjective formulation:

All probable reasoning is nothing but a species of sensation. It is not solely in poetry and music we must follow our taste and sentiment, but likewise in philosophy. When I am convinced of any principle, it is only an idea which strikes more strongly upon me. When I give a preference to one set of arguments above another, I do nothing but decide from my feeling concerning the superiority of their influence.[9]

If we stick religiously to the phenomena, therefore, there is no sense in talking about phenomenal causes, for we can find none, only habits; and *a fortiori* we cannot talk about metaphysical causes, like Oldenburg's 'produced by', for in terms of phenomena this makes no sense. Be it noticed, however, that we can give no explanations either. Perhaps we do just follow our noses a lot of the time in a 'burnt child shuns the fire' sort of way, but if that produces knowledge it is knowledge of only a very rudimentary sort. When you halter-break a foal you must always give it oats every time you lead it into the stall. This produces, so far as one can tell, a purely external and mechanical association: halter, oats, halter, oats, halter, idea of oats, nice oats, nice halter, or not too bad anyway – but the foal is very unlikely to produce any differential equations from which, say, even the idea of a bran mash would follow, let alone an understanding of the whole context within which the halter-oats procedure makes sense. Such basic associative mechanisms may of course be elaborated into much more complicated ones, as in the McCulloch-Pitts model of the nervous system, or in Hebb's account in the *Organization of Behavior*, but by becoming more complicated these come no closer to explaining our explanations of anything, for they never tell us how we understand anything as distinct from uncomprehendingly expecting it.

In any case, it may be questioned, I think, whether even our expectations, let alone our explanations, really do proceed always in this purely associative way. Robinson Crusoe, we say, saw a footprint in the sand, and since he was conditioned by past experience to expect a footprint of such a shape to have been caused by a man's foot and since he had found men's feet in the past always attached to men, a vivid idea of a man leapt to his mind, and he said, 'Aha! There is another man besides myself on this island.' But how many of us, unless we spent all our holidays at the seashore, have observed so many footprints on the sand that we are driven simply by force of habit from the impression *footprint* to the lively idea *man stepping on sand*? Besides, even if Man Friday had walked on his hands, Crusoe would probably have inferred from the resultant hand prints that some person had made these marks, although prints left on sands by acrobats doing hand stands are not exactly common in most people's experience. In other words, impressions, whether many or few, generate expectations of this sort when they serve us as clues to some kind of context in which we believe they make sense. Or conversely, it takes an act of comprehension by somebody to put impressions together so that they generate an explanation from which in turn an intelligent expectation can follow. Explanation is something the mind does, not something that happens to it.

That was the essence of Kant's answer to Hume. In the passive flow of my sensations there are indeed no necessary connections; but this flow of what Kant called the 'inner sense' – the stream of consciousness – would not be possible even as such, he argued, let alone as the bearer of my experience of a world of objects, were it not bound together into a unity by the concepts through which I interpret it. The raw material of sensation comes to me, willy nilly, but it is solid, objective experience of things out there only insofar as it is ordered and unified by the rules and maxims through which I apprehend it as such.

Now in this argument of Kant's, which did, it seems to me, establish a very important truth about the relation between the knowing person and the objectivity of the world he knows, the concept (in Kant's terms, the category) of cause, and the principle that 'every change has a cause' again figure very conspicuously. Let us look at one of the passages in which Kant gives us an example of the use of the category of cause. 'When I perceive the freezing of water', he says, 'I am apprehending two states (fluidity and solidity), which stand to one another in a relation of time.'[10] Just the temporal sequence of the two perceptions, call them for short this water percept and this ice percept, these two, Kant goes on to explain, follow one another in subjective time, Kant's inner sense. For instance, I might remember Lake Wingra, near the house where I lived as a child in Wisconsin, and the next moment I might remember the lagoon there where I used to skate in winter; thus I should have, in time, an image of water, then of ice – but I shouldn't say, Oh look, the lake has frozen over. But if I went out and watered the cows in my Irish farmyard one evening and came out next morning and found the trough frozen over, I might remark on this, pointing out that the water on the surface had turned to ice. Now, Kant explains, if I subtract the subjective temporal sequence characteristic of all my moments of awareness, what distinguishes my apprehension of an objective change, my awareness that in this case it is not just my memory that has strayed, but the water that has in actual fact frozen, what distinguishes the latter case, is, in Kant's words, 'the category of cause which I apply to my sense perceptions and according to which I determine everything that happens in time.'[11] Now I can see that the difference between a succession of memory images and a pair of actual perceptions of real situations has to do with the difference between the way in which these two pairs of data do or do not fit into the objective framework by which I normally orient myself in the physical world – and I agree with Kant that the activity of the mind has a great deal to do with the establishment and maintenance of such a framework. But what *cause* is really doing in Kant's example is another question. Cause, he says elsewhere, is 'a special kind of synthesis in which to one thing A another altogether different thing B is added according to a rule.'[12] Now, here we have, he says, two states, A fluidity and B solidity – is A the cause of B? Surely not. Is B the cause of A? Even less so. What is the cause of the trough's freezing over? Admittedly, if there were no water in it, it would not freeze over. But the cause of the change from water A to ice B was not the water itself but the cold. So in connecting percepts A and B through the concept of causality I am invoking another concept C – cold or absence of heat – at any rate a drop in temperature – and depending on my sophistication or otherwise I may be invoking also a theory of heat and of the effect on the particles of a substance like water of a lowered temperature. This may also involve a theory of the structure of matter as between amorphous and crystalline states and so on. Whatever my explanation, however, assigning the *cause* of percept B involves something more than reference back to the percept A which preceded it. It involves at least a reference to an event C which also preceded it, and to an explicative relation alleged to hold between B and AC. C must be related in some explicative context to A in such a way as to produce B. But then given A and the concept or theory or hypothesis C or $C, D, E, \ldots$ etc. (perhaps the whole universe?) we can deduce B; or given $C, D, E, \ldots$ we can deduce A and B. Then again, however, we have not really a straight causal relation in time but a relation of explication between concepts or principles C, D, E, etc., on the one hand and percepts A and B on the other. If we have rescued the validity of our judgment about the physical world from the threat of Humean scepticism, it is not cause and effect in the phenomena that we have reinstated, but the relation of principles as sources of order to the phenomena so ordered. Kant would perhaps have clarified matters if he could prophetically have followed Russell's advice and have refrained from speaking of cause at all.

But what has happened to the other side of Oldenburg's programme from which we started: the assertion that the forms and qualities of things as we perceive them are produced by certain properties of the real world? Such relationships would not be 'cause' in Kant's sense at all, nor could we according to him specify anything at all about those properties of things in themselves which bring it about that the phenomena are in fact intelligible to us in terms of the rules through which we order them. All that is left of things-in-themselves for knowledge is the bare object $= X$ which we know affects our senses somehow, though we cannot so much as understand it to be an object except through reference to the forms through which we perceive and understand it. Yet Kant's account of explanation is not meant to make it artificial as in Hobbes. What holds it together? Nothing, I believe in Kant's theory of knowledge, but only his underlying faith in the unity of God's creation, a justification which will not do, in its Kantian form, for most modern epistemologists.

What happens then? Must we revert to Hume, who allows us neither cause nor explanation? That brings me back to Russell in *Human Knowledge*. We have, according to Russell, no alternative to Hume's scepticism unless we postulate certain propositions as premises for our knowledge, and all these postulates (which look suspiciously like Kantian principles) are, as I said at the start, explicitly or implicitly statements about *cause*. Apparently, therefore, an account of scientific explanation does after all involve reference to causes, as we saw at the beginning Russell agreed with Campbell that it does not. Russell does not define the term cause, although he discusses and rejects various commonsense beliefs about it. He says instead that science elaborates various forms of causal law. This would seem to indicate (1) that to think about knowledge at all we have to know what we mean by 'cause' – cause seems to be a primitive idea in thinking about knowledge in general or science in particular; and that either (2) 'causes' still play a crucial role in the statements of science itself or (at least) (3) they figure in some prescientific but essential form in the presuppositions of science. On (2) it seems to me Russell is not entirely clear about the position he is taking, but he clearly does assert (3) and this seems to entail (1), or at least (1) is strongly suggested by the prominence of the term cause in the discussion of his postulates. What is this primitive and underlying meaning of cause? We cannot define it, apparently, but let us look at one of the postulates, one especially relevant to the present discussion – and see how the concept of 'cause' operates in it. This is what Russell calls the *structural* postulate, i.e. 'the postulate of the common *causal* origin of similar structures ranged round a centre.'[13] This relation of common causal origin of similar structures is again subdivided into two fundamental kinds. 'What I am suggesting', Russell writes, in the chapter on Structure and Causal Laws,

is that we are not merely to seek simple laws such as A causes B, but are to enunciate a principle of the following sort: given two identical structures, it is probable that they have a causal connection of one of two kinds. The first kind consists in having a *common causal ancestor*. This is illustrated by the different visual sensations of a number of people looking at a given object, and by the different auditory sensations of a number of people hearing a given speech. The second kind arises where two structures are composed of similar ingredients and there exists a *causal law* leading such ingredients to arrange themselves in a certain pattern. The most obvious examples of this kind are atoms, molecules and crystals.[14]

Now in both these cases it seems to me we do have statements of *metaphysical* causation very like Oldenburg's statement about the 'effects of nature' being produced by motions, etc., though complicated by the greater complexity of modern physics and of our beliefs about what the things are like that *cause* the phenomena to appear as they do. And we can see from Russell's discussion something about what is involved in such an assertion of metaphysical cause. First, a statement of metaphysical cause is necessary *at least* to validate the objectivity of our perceptions. Our perceptions can be organised into a system of knowledge only insofar as we believe that there are *things* (or events or structures) causing those perceptions in an orderly and reliable way. And more than this, we must be convinced, if we are to know anything about nature, that the effects we perceive, in their inter-relationships, are in fact effects of structurally corresponding relationships in the things which are causing our perceptions. We believe not only that our perceptions are caused in a regular way by things but that the general regularities in our perceptions are caused by general regularities – structures – in the things – and these regularities in turn may be described as depending on causal laws', i.e. something in the nature of bodies which makes them display the structures which, trusting in our perceptions, we find they do display. This is essentially Oldenburg's programme in a more sophisticated form. It says: 'We feel certain that the forms and qualities of things can best be explained by the principles of physics (not just mechanics any more, of course), and that all our perceptions of events in nature are produced by real events in Nature (this is true even of hallucinations, since they are produced by events in the CNS, which is part of nature) and these events in turn are caused by real regularities in nature resembling those regularities which the principles of physics state.' This means that for Russell as for Oldenburg explanations explain not because we arbitrarily postulate this or that as premise for our systems of explanations, but because we *believe* that things are the way they have to be to make our explanations explain. Without this causal relation between perceptions and events, and between the regularities existing in the real world, our explanations would not explain at all. Thus if science does not consist in tracking down causes in the sense of necessary when-then relations stateable in the form 'phenomenon A causes phenomenon B', scientific discovery and scientific knowledge do depend on beliefs about a *metaphysical* relation of cause and effect which alone validates our otherwise inexplicable habit of inferring explanations from phenomena as logically we have no conceivable right to do. Not that these fundamental causal beliefs are necessarily true. Any or all of them may well be mistaken, but the measure of our confidence in their truth is nevertheless the measure of the degree to which we succeed in explaining anything at all.

NOTES

[1] *Mysticism and Logic*, Pelican ed., 1953, p. 171.
[2] *Human Knowledge*, p. 134.
[3] A. Wolf, *Correspondence of Spinoza*, 1928, p. 80.
[4] Second edition, 1933, p. 276.
[5] M. Boas, *Robert Boyle*, p. 93.
[6] *Works*, Vol. V, p. 245.
[7] Sir H. Jeffreys, *Scientific Inference* (2d. ed.), 1957, p. 78.
[8] *Ibid.*, p. 31, p. 28.
[9] *Treatise*, Bk. I, Pt. III, Section 8.
[10] *Kr. r. V.*, B 162–3.
[11] *Ibid.*, B 163.
[12] *Ibid.*, B 122.
[13] *Human Knowledge*, p. 506.
[14] *Ibid.*, p. 486.

CHAPTER II

THREE ASPECTS OF PERCEPTION

I

Knowledge is not perception, Socrates is supposed to have proved to Theaetetus, since knowledge must be both incorrigible and (of the) real. But perception, insofar as it is incorrigible, is not of the real: it is just what seems to me now; and insofar as it is, or claims to be, of the real, it is not incorrigible: our senses, as purveyors of information, may err. So knowledge is not perception. So at least runs the traditional interpretation, which I am following here for the purpose of my argument. For the point is that even if we admit the defeat of the thesis that knowledge is perception, we still want to claim that perception is one kind of, and indeed the primordial, and fundamental, type of knowledge. For the senses are our only source of informational input from the world; without them we could know nothing. Perhaps we may drop one of Plato's criteria (as we are taking them to be), claim that knowledge is of the real, but not incorrigible, and so admit perception to the class of cognitive activities. Perception is the most fundamental type of knowledge, on which others build, and they are all fallible: there should be, on principle, no difficulty in the case. Still, recurrent puzzles about perception have their good grounds, and I would like to try to sort out once more, very simply, some of the features that produce these puzzles.

II

First, perception is involuntary. Supposing I see yellow marigolds. I can't help seeing them. True, it is I who have placed myself opposite them. If I turn round, I see, not the marigolds, but the dahlias on the other side of the garden. But given my present bodily condition, the state of the light, and so on, the marigolds are what I have to see. Similarly, though admittedly I may look at the house from top to bottom or from bottom to top, it is a house I see and not a castle or a bird's nest. Given the appropriate circumstances, seeing and hearing and smelling and even tasting and feeling are not anything I can choose to do or not do. Sights, sounds, smells, tastes, feels impose themselves upon me, whether I will or no.

What does one think about perception if one reflects on its involuntary character? First, one takes it to be passive. Thus the Cartesian 'idea', which is both an act of the mind and a mental content, is taken, in the empiricist tradition, as content only. 'Ideas' are sensed or imagined bits which are 'in' the mind, not achievements of mind, but its 'affections'. For Hume, indeed, that is all mind is: an aggregate of percepts or their paler remembered, and imagined, copies.

If perception is thrust upon us, however, secondly, it seems to be thrust upon us by something somehow outside ourselves, or at least by something extraneous to the perception itself. Thus the involuntariness of perception gives rise, on reflection, to the so-called causal theory. I see the marigolds because they're there. When Berkeley and Hume insist that the 'vulgar' consider their percepts to be real objects, it is not really an identity but rather this sort of direct causality that is involved. Hume himself sets a general causal theory at the head of the *Treatise*: impressions are those vivid presentations caused in us by some unknown – and, as it turns out, unknowable – objects. He tries to put this external source aside, as the proper subject more of anatomists than of philosophers, but it is there in the background, and even makes its way occasionally back into his argument, where it has no right to be. Kant, too, accepts this general principle when he presents perception as our power of being affected by things, in contrast to the active organization of experience by the rule-giving work of the 'understanding'. Like Hume, he insists that we can say nothing about these external causes: things in themselves are ineffable and unknown. Yet they must *be*, since it is plain that our sensed experience is thrust on us by *some*thing.

Perception so interpreted, then, is the effect term of a cause and effect relation. In other words, it is the second term of a two-term, asymmetrical, intransitive, irreflexive relation. 'I see marigolds' means: there is an x such that x is marigolds and there is a y such that y is my perception and (ceteris paribus) x causes y. Or, given a number of necessary conditions (light, normal vision on my part and so on), x is a sufficient condition for y. Is x also a necessary condition for y? The causal theory must maintain that it is so. If the marigolds are there, I see them; and if I see them, they are there. x and y then appear to be logically equivalent; nevertheless, they are not identical, and I don't believe any one, however 'vulgar', ever thought they were. But as both Hume and Kant held, the causality of perception is extraneous to its logical, and even its epistemological, analysis. That perception is the effect of some cause external to it seems to be entailed in the character of perception itself; but any detailed account of this causality falls outside the range of philosophical analysis.

What is the use, one may ask, of so empty a theory? If we can know, or formulate, nothing about the 'things in themselves' that affect our senses, what good is it to assert the bare fact that such a causal relation exists? For one thing, it gives us a realist theory, one which interprets perception as a real outcome of our live, ongoing contact with real things in a real world. Thus it places perception where in our ordinary experience it appears to belong, for surely perceptions result from the interaction of sensing organisms with a sensible environment. And as distinct from the so-called 'representative' theory, secondly, the causal theory is comprehensive. My perception of the marigolds may be said to refer to or represent the marigolds; my toothache doesn't 'stand for' or 'refer to' anything, but it is plainly thrust upon me by something other than itself, namely, as I believe on the grounds of experience and authority, by a decaying tooth. Thus even bodily sensations have, or seem to have, causes external to themselves. The tooth is still decaying when it is anaesthetized or when I am asleep; but when the anaesthetic wears off or I wake up, 'it' will hurt again. Indeed, from the point of view of a causal interpretation of perception my body is as much part of the external world as are other bodies. My tooth *is* no more the pain 'in' it than the marigolds *are* my vision of them. I can't always specify, as directly as in these cases, the causes of which my perceptions are the effects. I may even want to assert, with Hume or Kant, that these causal agencies are in fact unknowable. Yet I can still affirm with confidence that, in the most general way, my perception results from my affection by something other than itself.

Even if it doesn't accomplish much philosophically for the analysis of perception, moreover, it is the causal interpretation that governs a good deal of empirical investigation. Take the study of perception through the experimental investigation of illusions. Richard Gregory has constructed, among other ingeniously contrived objects, a hollow face, an inside-out death mask, so to speak. When the experimental subject is presented with this object, he invariably sees it, not as concave – as it in fact is – but as having the ordinary convexity of a 'normal' face.[1] One wants to say, then, that faces usually cause me to see faces, but in this case an inside-out image of a face causes me to see a rightside-out image of a face. One has then to investigate further what makes us react this way, but without a cause-and-effect, interactional framework for our thinking, we couldn't cope with such situations at all.

A causal approach, moreover, enables us to distinguish perception from hallucination. *Does* Macbeth see a dagger? He might be seeing something else, which he took to be a dagger. But if he's hallucinating we would suppose, I think, that he doesn't *see* anything. He only imagines something so vividly that he seems to see it. He will see a dagger all right when he goes to kill Duncan, and feel it in his hand. But pangs of conscience, or events in his nervous system, as you like, trigger an experience that simulates seeing. But it *is* not seeing because its cause is different.

So much, perhaps much more, but so much at least, is to be mined from the involuntariness of perception, its character of being forced on us by something other than itself.

III

In addition to being involuntary, moreover, secondly, perception in its ordinary occurrence is also immediate. True, I happen to know, on reliable authority, that all sorts of complex physiological processes are presupposed in my seeing yellow marigolds. I know, too, that I must have learned to tell the yellow marigolds from the yellow dahlias. Opposite me also is what I believe to be a blue clematis, twined round the post of a wooden bird's nest. But it's so long since I've seen clematis that I'm not sure. Still, there's something I see, and see immediately: this blue flower of this green vine, even though I can't name it with confidence, as I can the marigolds or dahlias. However much has or has not accrued to perception by learning, there seems to be always a core of what is perceptually experienced that is directly and indubitably present.

The immediate character of perception seems somehow to be closely associated with its involuntariness: something is thrust upon me and is quite simply and directly *there*. Yet the philosophical beliefs or arguments associated with the immediacy of perception are very different from the reflections associated with its involuntariness. True, immediacy, like involuntariness, exhibits perception as a passive process: the percipient appears as simply receptive of what flows directly in upon him. Berkeleyan ideas or Humean impressions are pure givens, not products of any (human) agency. But if reflection on the way percepts are thrust on me gives rise to the belief that my perceptions result from my interactions with things 'out there', and hence to causal realism, attention to the immediacy of perception leads to the denial of realism. For, as we have already noticed, it is often difficult to specify just what it is that is 'really' perceived directly and without the mediation of learning. I see marigolds but am not sure about seeing clematis, though both the golden and the blue blossoms are before me at the moment. But both flowers are in the same situation relative to my present perception, so it must be the least common denominator of the two experiences – a colored shape, a yellow shape in the one case and a blue and white in the other – that is 'in fact' the object of my seeing; I judge the one to indicate the presence of a marigold, just as (though with less confidence) I judge the other to be a clematis. This is the slippery slope that leads to phenomenalism, to the view that all I really have are private seemings, on which I have built, by some sort of inference, constructs which I call 'realities', though in fact they are just contrivances that guide me to more such private seemings. I see hats and cloaks under my window, said Descartes, and judge that there are men there. But do I 'see' hats and cloaks? Surely not. I see colored surfaces which, by experience, I have come to associate with other appearances such as the inside of the hats apparent when they are doffed in greeting, or the voices of the wearers when they pass the time of day. Even 'seeing' how far or near things are, Berkeley argued, rests on judgments I learned to make when, as a baby, I crept from a nearby chair-leg to the table-leg over there and put my sensations of touch, inferentially, together with those of vision. And Hume, missing the prop of God's benevolent arrangement which, in different fashions, had come to the aid of Descartes and Berkeley, could only conclude that all our belief in real, permanent entities somehow underlying our perceptions rests on a propensity to feign.

Thus reflection on the immediacy of perception leads to scepticism.

We seem to see and hear and touch things, but all we really have is those seemings themselves. The 'belief' in an external world rests at best on a kind of intellectual game, at worst on blind habit, conditioned reflex, in effect: we expect what we have been programmed to expect, for no good reason. At the same time, however, it is the immediacy of perception that suggests its certainty. What is immediate just is there, incorrigibly. If I feel pain, I feel it; if I see red, I see it; if I hear a noise, I hear it. But if I feel pain *in* my shoulder, I may be mistaken; it may be a referred sensation from some gastric uneasiness. And if I see red tulips, I may be seeing red poppies – or, for all I *certainly* know, hallucinating. (So from this point of view, Macbeth did see a dagger, in the strictest, and incorrigible, sense of seeing.) And if I hear a noise over there, I may be mistaken; it may be over there instead (my auditory localization is bad). All that ties perception (or rather, as this tradition has it, sensation) to reality is guesswork. What is strictly sensed are mere appearances – phantasms, Hobbes called them – but cherish them, insubstantial though they are, for they are all you have *for sure*. We are back with 'Plato's' argument. Sensation by itself, in itself, just is what it is, and is therefore incorrigible. And if we separate knowledge from opinion, and hold that only the indubitable or incorrigible is known, then perception in this limited sense of the term: the presence to me of just what is now present to me, qualifies so far as knowledge. Yet if knowledge is of the real, then this isn't knowledge, for it makes no cognitive claim. It is, indeed, the whole content of experience insofar as our senses provide it. All the rest of what we please to call 'experience' is interpretation, or convention, or mere habit. But it is vacuous content, conveying no information about anything at all. Perceptions (or sensations) so understood are not, as in the causal perspective, one term of a relation to something else, but elements subsisting in and of themselves. They should be, ideally, the 'minimum sensibles' so eagerly sought and devoutly believed in by eighteenth century thinkers, or the designata of 'protocol sentences' vaunted by logical empiricists as late as thirty years ago, or the sensa or sense data of epistemologists like Broad or Price.

Are there any such entities? According to common sense or the arguments of J. L. Austin, we can breathe a sigh of relief and say 'No'. Even the most careful philosophers of this school, such as H. H. Price, have had to rebuild into their theory something called 'perceptual confidence'.

Or, like C. I. Lewis, they have admitted the pure 'given' to be an ideal, an aspect of experience never wholly present in itself, because always already interpreted, like perception at the end of the Kantian Analytic rather than in the more limited analysis of pure receptivity in the Aesthetic.

Still the fact remains that perception is immediate, and as immediate appears incorrigible. Yet when one tries to peel away the possible super-structure of interpretation and get down to *what's* incorrigible one is always driven back to that minimal, self-contained but non-informative, assured but insubstantial, 'feeling to me here now'. Maybe it's not a clematis but a plain morning glory; maybe it's not a back-firing car but a gun. What I'm sure of is something merely subjective; what I build onto this is something purely conjectural. Imagination peoples the world, says Hume. Sensation in itself, the only unquestionable ground of experience, could not by itself support such unjustified conclusions. Even the most general causal theory is undermined by a rigorous phenomenalism. So the reflections suggested by the immediacy of perceptual experience are at loggerheads with those suggested by its involuntariness.

IV

Nor is this the end of the matter. Perception is not only immediate, yet thrust upon us surely by something other than itself; it also informs, or claims to inform, us, about things other than itself. If the epistem-ologist's job is to ask what features in our experience can rightly claim the status of knowledge or claim to contribute to knowledge, he finds in perception not only a subjective certainty, but a claim to objective refer-ence, whose rightness he has also to examine. I see the marigolds: my perception of them refers me to them, it is a sign that they are there. After all, how, except through my senses, do I get information about anything? In his famous argument about a piece of wax melting by the fire, Descartes tried to prove that even what I seem to know by my bodily senses, I really grasp through my mind, not my body. For the shape, size, odor, color and tangibility of the wax all change; I only know it to be something flexible and extended – and those are concepts, not sensible qualities. Yet those very alterations are perceived; only the seen, felt, smelt flow of sensed experience gives me ground and content for my otherwise empty concepts. It informs me, not of my own introspections, but of something out there, melting by the fire.

So considered, however, perception exhibits yet another fundamental character. It appears, again, relational. What I perceive (my percept, as we may call it) appears as one term of an intransitive, asymmetrical, irreflexive relation, this time as the first term of a relation of reference. My perception of the marigolds, taken as a content, as what I perceive when I perceive them, denotes the marigolds; it points to them as its object. In other words, 'I see marigolds' means, there is an x such that x is marigolds and there is a y such that y is my percept and (ceteris paribus) y refers to x. In this case y is, or at least purports to be, a sufficient cause for the existence of x: if I see marigolds, there are marigolds there. But y is not a necessary condition of x. For this aspect of perception it is essential, obviously, that the marigolds may be, and probably are, still there when I'm not seeing them.

Admittedly, I may have got into deep water by introducing 'percept' language here. It sounds suspiciously like 'sense datum' language, which it isn't meant to be. The trouble is that the term 'perception' as we are now considering it means ambiguously either the whole relation, as a transaction of referring to things through the senses, or its first term, as the perception I am now having, the content which signifies its (purported) object. To deal with this equivocation it seems better to call the whole relation perception and its first term a 'percept'. One could also use the same terminology in the context of the causal theory: the object x causes the percept y and the whole relation is perception as a passive process of being affected by things through the senses. But in that context, it seems to me, 'perception' bears primarily the meaning of the product of a causal relation, rather than the relation itself; so one needn't make the distinction there. Here, however, it is unavoidable.

What philosophical theses or arguments does this aspect of perception suggest? It suggests, again, a realistic view of perception, but a different one from the causal; instead of a causal, we have here what is usually called a representative theory. What is central here is not that things cause me to perceive them, but that my percepts inform me, or claim to inform me, about things. Partly, I think, through its insistence on immediacy as the prime character of perception, the empiricist tradition has often confused these two perceptual relations, which it tries (vainly) to assimilate to immediacy. Thus Hume tries to get rid of both causal and representative claims for perception, but both obtrude themselves at a number of junctures in the *Treatise*. Kant, according to Professor Rolf George, did indeed separate these two relations, though not always clearly.[2] When he talks about things in themselves, George argues, Kant is dealing with the causal aspect; when he talks about the transcendental object of experience, on the other hand, he is thinking of the representative function which we associate especially with 'external' perceptions of things and events 'outside us' in space. The pain I feel is caused by a decaying tooth; but the pain doesn't refer to or represent the tooth. Severe pain, indeed, has the property of diminishing to nil the referential character of experience, the coherent structure of the world I usually perceive. There *is* nothing but pain. But my perception of the marigolds refers to the marigolds: it informs me of their presence in the garden. Of course I also believe that my perception of them is caused by their presence, but logically and epistemologically that is quite a different claim. Reference is a very different matter from causality. The one is a logical, and atemporal, relation, the other a temporal one, indeed, one which keeps eluding philosophical analysis. (I have taken it as unanalyzed, but that's not to say its nature is simple or clear.) Besides, as we've seen, perception takes up a different place in the two relations or better, percepts take up a different place: things cause my percepts, my percepts refer to things.

Before we go further, however, we should make clear what sort of 'representation' is involved here. In the empiricist tradition a representative theory often meant a copy theory, like Locke's, where one is given the impression that, ideally, 'representing' means 'picturing'. This was another major error of traditional empiricism which I trust we can discard. Certainly, 'picturing' is not the kind of 'representation' intended here. We are talking about the (alleged) reference of a percept to an object, not about the resemblance of an image (= 'idea') with its object. That would take us into the blind alley of the distinction between primary and secondary qualities: the shape of the marigolds, it is said, belongs to them, but not their color or taste or smell (if they have any). But don't shapes depend for their look on the angle I see them from? So all 'qualities' are only 'in the mind'. We have, once more, only phantasms left. What we're talking about here, however, is representation, not in the sense of a copy-original relation, but in the sense of a sign-signified relation. And a sign need have no more resemblance to what it signifies than a road-sign has to the town it directs us to.

In another sense, Berkeley's theory too could be described as 'representative', since percepts (ideas) are for him signs, not indeed of material objects, but of other ideas to follow. The upshot of this sort of interpretation, however, – where the bulk of our mental possessions are taken to be images, one leading to another, but without any connection with a real external world – the upshot of this view is again to equate perception with the minimal sensed given, percepts being just particular pictures 'in the mind', and to relegate all else to judgment. Since, however, neither the famous 'minima sensibilia' nor any trace of the alleged inferential work of judgment are in fact to be found in our everyday working experience, it seems wiser to attribute the cognitive aspect of our perceptions, their claim that they represent the real, that they refer us to an objective world, to our sense-borne orientation among things, and reckon perception as the whole relation, not just its first term. (Berkeley's argument of course was that there *is* no second term: ideas refer only to other ideas as signs, not to objects at all. That's just why, quite apart from the invocation of Divine contrivance, his view is so counter-commonsensical.)

Eliminating the Lockean or Berkeleyan versions of a representative theory, then, what can we say about representative realism? Its most striking feature, perhaps, is that, as distinct from a causal theory, it lends itself to the interpretation of perception as active rather than as wholly passive. Perception in its function of pointing by means of our senses to things seems to be, at least in part, an activity or the outcome of an activity. (We shall have to consider presently which of the latter formulations to accept.) After all, even though I *have* to see what I see, and hear what I hear, I not only see, but look; I not only hear, but listen; there is an element of attention, and therefore of activity, in much of my perception. From this point of view, perception seems to be a stand I take toward things. In Gregory's terms, perceptions (or percepts?) are hypotheses.[3] Thus when I 'see' the hollow face, I *see* a normal one: I make a wrong hypothesis. True, I am constrained to err; not even the most trained observer can see the object otherwise. But what has constrained me is the general bent of my sensory orientation, rather than a series of judgments over and above sensory experience. When I see an ordinary face, my hypothesis is correct, but again, I do not see something pale and bumpy, and judge it to be a face. I perceive it; that is, my percept, y, refers me to the object, x. I take it to exist, not by a process of reasoning, but through my seeing. Hintikka has reproached Kant for taking as basic to know-ledge perception rather than 'seeking and finding'.[4] And Kant was indeed mistaken in separating as sharply as he tried to do in the first part of the *Critique* the receptivity of the senses from cognitive activity. But perception is, it seems to me, just the most fundamental sort of seeking and finding. If somebody tells me all swans are white, how do I find a counterexample? By looking in New Zealand and finding a black one. True, to see is not yet to say; but since, as Kant noted, apart from percep-tual presentations we would have nothing to say, there seems no reason to cut perception off from the kind of cognitive activity which is made explicit at the linguistic, and logical, level. Perception makes, though tacitly, a cognitive claim.

V

Here, however, we run into grave difficulties. A claim may succeed or fail. Do we call the *claim* perception, whether it succeeds or no? Or is 'percep-tion' a 'success word'? J. J. Gibson in his important book, *The Senses Considered as Perceptual Systems*, defends the view that our senses are systems for information-input from the world.[5] His interpretation seems once more to make the senses wholly passive, and also to connect their informative function too simply and directly with the things they claim to tell us of. These difficulties need not concern us here, however. What matters here is just that Gibson has further interpreted his own theory as a defence of realism. Hintikka, however, objects to this that the claim for 'realism' is merely empirical; the 'logic of perception', he argues, does not demand that the *actual* world contain the realities of which perception informs, or seems to inform us.[6] I cannot go into this argument, since it is based on the articulation of a 'possible-world ontology' which I only partly understand, and, insofar as I do understand it, or think I do, find much too contrived to serve as a philosophical approach to so fun-damental a feature of our experience as perceiving is. My point here is simply that both Gibson's realistic claim (abstracted, if possible, from its questionable features) and Hintikka's counterclaim need to be taken seriously – with paradoxical results. 'Perception' is sometimes, justifiably, a success-word, sometimes, justifiably, not.

On the one hand, when we distinguish between perceiving and imagining, we make the distinction between seeing or hearing what is there and imagining = visualizing or forming an auditory image of what is not now in fact present to our senses. I see the marigolds now when I see them, and imagine them when I recall the scene, or, perhaps, daydreaming or writing a novel, visualize a similar setting. So seeing is what I do only when the marigolds are 'really there'. Again, hallucinating is clearly distinguished from seeing, in these terms, by the failure of a 'real, external' referent in the former case. So we want to say, Macbeth saw a dagger when he picked up his weapon with murderous intent; but when he was frightened by the dagger he saw in his imagination, he was only 'seeing things' in the sense in which one sees pink elephants, that is, not seeing them, since what his vision pointed to did not then exist outside his fevered brain.

There seems to be much to be said for this point of view. Perception is informative of the state of the world, and when it isn't so informative – as in hallucination – it isn't perception. Yet the other side of the case clamors to be heard from too. Indeed, a whole range of counterexamples appears here to overwhelm the happy realist. What about sensory illusions, what about the class of cases where I am not sure what I see or hear, or the class of cases where I turn out to have been mistaken? Let's look briefly at each of these.

For the first class, the hollow face is a good example, or the traditional Müller-Lyer illusion, which one can construct for oneself, will do just as well. One sees a normal face or two unequal lines. By further testing one can establish that the face is hollow or the lines equal. (One handles the actual object or measures the lines.) But one still *sees* the same. This is in sharp contrast to learned perception. The trained radiologist or astronomer sees in a radiogram or through a telescope much that I just don't see, but he really sees it, because it's there. His percepts refer to something genuinely existent which he knows how to find. I don't know how, so I just don't see it. In the case of sensory illusions, however, no amount of knowledge will change what is seen. And yet what I see *is* not what is there. It's all very well to say that I see a concave object as if it were a convex one, or that I see two equal lines as if they were unequal. What I do see is a normal, convex facial configuration, or two unequal lines. I am seeing something, but what I see is not what, on independent evidence, turns out to be the external object to which my perception had referred me. There is a reference, but a mistaken one. Perception seems therefore to consist in purported reference, not in the achievement of correct reference.

But this doesn't seem quite right either, since if we push this line of reasoning far enough, we'll be back with phenomenalism and the undeniable core of perception which is purely subjective, and, indeed, without reference. We'll be unable, again, to distinguish perceiving from hallucinating. Can we perhaps say, then, that we are perceiving when there is *some* external object to which our percept refers, even if it's not the object we perceive it to be?

There is a little nest of problems here, however, that's difficult to disentangle. We can't overlook the fact that perception presents itself as immediate: I just do see marigolds and blue flowers and a normal face and unequal lines. And if it's immediate, surely it's incorrigible. Nobody can persuade me that I'm not seeing what I see, or that I didn't see what I plainly remember that I've seen. Yet perception is also *of* objects. I see marigolds and faces, not colored blobs which I judge to be marigolds or faces. But insofar as perception is of objects, it must be corrigible, since I can discover that I wasn't in fact seeing what ... I did see. But that's absurd. I did see what I did see, I didn't both see it and not see it. Shall we then say that I indubitably saw something, but 'took' it to be something else? That would reintroduce the perception/judgment dichotomy which we wanted, with some reason, to avoid: for then we should be started again on the road to sense-data and subjectivity.

Perhaps we can say that perceptions are hypotheses which refer us with fair reliability to objects or events in the real world, but which we have to verify by looking further into the circumstances of our perception. And here we may reintroduce a causal consideration. A perception is a correct hypothesis when the object to which it claims to point not only exists but is in fact its cause.

But how do we find out whether this is so in any particular case? By further investigation, and so again, ultimately, through perception. And if perception isn't incorrigible in the first place how can our justifying processes rest on it? It seems yet again we'll be driven back to 'mere' phenomena with a purely conventional superstructure to hold them together.

The other kinds of case give rise to similar doubts. We started in the garden with the marigolds and clematis. I see a flower but am not sure if it's a clematis or a morning glory. I see some one coming, George or his brother Gerald? What is it that I see? Some flower or other? One of the Kelly brothers? But surely the object of perception is something *particular*, and what the particulars are are just these shapes and colors, not their classification. There we are again, reducing perception to sensation and separating it from judgment, just what we didn't want to do. And the case where I perceive something, but find I was mistaken – 'it' was something else – is exactly parallel to that of our first class of cases, sensory illusions. We needn't go through the whole gamut again.

The upshot of the matter, however, is clear. Perception, we want to say, makes a cognitive claim. It is our way of interacting, through our sense organs, with things and events in our environment, so as to discover certain features of that environment. There are marigolds in the garden, there is a car backfiring in the street, and so on. The question is whether perception is the successful fulfillment of such a claim, or only the claim itself, whether successful or no. If the former, we cannot explain the many cases of perception which are imprecise or turn out to have been mistaken; if the latter, we may be driven again into the sense-datum corner, where we class hallucinations and perceptions together as 'whatever appears', and that is no account of perception at all.

VI

Where have we got to? We have listed three features of perception: its involuntariness, its immediacy, and its referential claim, and have considered, sketchily, some of the philosophical implications, or associations, of these various features. The first and third suggest a realistic view, in which our percepts appear as one term in a relation between consciousness and things: in the first case, as the effect term of a cause and effect relation, in the third as the first term of a sign-signified relation. The second, on the other hand, suggests phenomenalism, if not scepticism, with respect to our perceptual knowledge. The first and second take perception as purely receptive, the third suggests that in part at least perception is an activity. The third at first sight seems to lead to a 'success' view, the inverse of the realism suggested by the first; but experience and the second feature combine to undermine this notion, and to complicate our specification of causal realism as well. Just what must a perception really refer to to count as a perception? Just what must be there – where? – to count as its cause?

Not that we have noticed by any means all the features of perception. Husserl's account in *Ideas One*, for example, contains much that we have ignored, notably the distinction between perceiving and imagining in terms of the expectation contained in the former that I could see the perceived object from an indefinite, systematically varying range of perspectives (*Abschattungen*). But the three features we have noticed are prominent in, even though not exhaustive of, perceptual experience itself, as well as its traditional philosophical interpretations, and these aspects produce enough puzzles to show us, without the introduction of still further complications, that somehow or other we need to make a fresh start. For we must acknowledge all three features and yet interpret them so as to produce a viable account, not simply paradox. Thus I want to be able to say both that I see the marigolds because they're there and so I can't help seeing them – my percept is the effect of them as cause – and that my perception of them refers me to them; both that my perception happens to me and is something I do; both that the percept 'these yellow marigolds' is just present to me now directly, an undoubted and indubitable subjective presentation, and that as a perception of objects it is nevertheless not (logically) incorrigible; that I am only perceiving, as distinct from dreaming or remembering, when there is something really there to be perceived, yet that the something which my percept allegedly refers to may not be 'exactly' the very thing it seems to refer to – in other words that the alleged referent of the percept and its cause may not in fact coincide. But we want to find a way to say all this at once without getting into the difficulties we have recurrently met so far.

What way can we find? Hume's way, the classical way of the sceptic, is to cherish the negative outcome of 'philosophy', while retaining common sense for the practical concerns of life. But though no one can prevent the sceptic from carrying through his corrosive dialectic if he will, most philosophers would like to achieve some positive illumination of our cognitive claims, including those of perception. Sceptics are irrefutable in their own terms, but aren't there some less vulnerable terms one could employ that would prevent so negative an outcome? One traditional alternative is the way of idealism from Fichte onwards: to admit the Humean denouement, and to find reality constructed by some transcendental ego on the base of purely subjective givens. This is, however, too improbable an escape hatch. I *don't* make the perceived world. The involuntariness of perception is just as plain as its immediacy. Something non-me is there that thrusts itself upon me and to call that non-me a transcendental ego does no good at all. Its issue in idealism, indeed, is, it seems to me, one reason why Husserl's account of perception, for all its excellence of detail, won't save our bacon. It's *some* kind of realism that we need.

Another alternative is to take a given empirical discipline, usually physics, as gospel and pin our account, as well as our faith, on it. That is the way of Hobbes or Russell or Quine, or Carnap and his colleagues in their 'physicalist' period. But this is to abandon philosophy, not, like the sceptic, when it has come to no good, but before one even begins. If the reliability of perception lies at the root of all our knowledge of nature, then we cannot without vicious circularity take one particular branch of such knowledge as it in fact exists or as we hope it will exist for the foundation of our claim to know real things, whether through perception or through some more abstract and intellectual means.

What about so-called 'evolutionary epistemology', which founds our claim to know the real world on an analogy with, or application of, the theory of natural selection? There is some kinship here with Russell's account of what he called 'animal inference'. We simply follow successful hunches and are shocked out of unsuccessful ones. Nature selects the 'hypotheses' that work and eliminates the failures, just as she selects species with a better chance of survival and eliminates the less likely candidates. There are several objections to this now fashionable view. First, it falls before the same objection as its physicalist cousin. One branch of biology has been simply taken as gospel, as in the other case physics had been. So the question of epistemology, by what right we claim to know reality, and in the case of our senses, to know it through perception, remains unasked. We have taken one branch of empirical knowledge as unquestioned and so cannot ask, on principle, on what grounds empirical knowledge in general may be said to rest. Secondly,

natural selection is no guarantee of truth, whether for things or for percepts, since its only criterion is survival, and systematic illusions could
issue in survival just as well as faithful renderings, whether by perception
or by theory, of what there is. Thirdly, this kind of answer, like traditional empiricism, confuses the story of how we come to know with the
question what knowledge is. The former may be interesting, but it fails
to answer, even to ask, the philosophical question: what it is to know, or
to claim to know.

VII

Yet we have to take something for granted to start with. If not subjective
seemings that alone appear indubitable, then what? Why not start where
we are, as live embodied beings trying to orient ourselves in a physical,
biological and cultural environment? But haven't we then begged *all* the
questions, even more thoroughly than do those who take physics or
evolutionary theory as their unquestioned given? No, because we ourselves as questioners are also included in our starting point. Nothing we
have yet learned is taken as exempt from questioning.

But then, it will be objected further, are we not opening the door once
again to the sceptic and his refusal to admit that we know anything, his
denial that perception is more than the subjective presentation of some
airy nothings rather than the outcome of a real interaction with a real
world? Yes and no. There is, we have already admitted, no refutation of
the sceptic on his own terms. And if we start with ourselves as embodied
and encultured creatures in a real environment, with the strange tendency
to critical reflection that we in fact possess, we can see how the move to
scepticism originates. *What* is immediate in perception? Once we start to
wonder, we can end up, logically, with a strict phenomenalism and so a
doubt that we ever perceive real things at all. But if such doubtings indulge
a natural propensity of the human mind, as much as, according to Hume,
our everyday realism results from an equally irresistible propensity, we
can look on the tendency to scepticism with benevolent indulgence – and
forego it if we like. We set the sceptic, not over against the 'vulgar', as
Hume did, but inside the range of common human destiny, where
reflection arises, possibly, but not necessarily, with so corrosive a
result. Moreover, starting in this way, within the medium of our whole
physical, biological and cultural existence, we can also accept empirical

results from the physical, biological, or social sciences as germane to our reflections, but we accept them critically, not as the unquestioned plank, the something firm and permanent, as Descartes put it, on which we hope to build a systematic superstructure.

This approach has affinities with a philosophical anthropology like that of Helmuth Plessner, which I have discussed elsewhere.[7] But it has other philosophical supports as well. To start, as I am suggesting, with what we might call a comprehensive or inclusive realism is in effect to start with Heidegger's being-in-the-world, to insist that we are, in our very nature, *with* things in the world, not over against them. The scandal of philosophy, says Heidegger, is not that the question of the reality of the external world has not been answered, but that it should still be asked. With respect to perception, however, our starting point seems to stand closer to that of Merleau-Ponty, whose magnun opus, *The Phenomenology of Perception*, begins with a theory of the lived body, and shows that such a theory is already a theory of perception and therewith of the perceived world. Akin to Merleau-Ponty's account, finally, but more articulate in its epistemological structure, is Michael Polanyi's theory of tacit knowing, developed in *Personal Knowledge* and in some later essays, notably Chapter One of *The Tacit Dimension*.[8] Let me touch very briefly on this account, in conclusion, to indicate what sort of 'new beginning' it is that I am advocating.

What has led earlier philosophers astray, Polanyi holds, is their attempt, whether in theories of perception or of judgment, and hence of scientific theory as well, to find wholly *explicit* knowledge. All knowledge, beginning with perception, in Polanyi's view, resides in a *from-to* relation, in which I rely on clues already assimilated to my bodily being, in order to attend, through them, or from them, to things out there. Knowledge, including perceptual knowledge, therefore, is never wholly focal, but always relies on *subsidiaries* in order to *focus* on the events or entities to which those subsidiaries point, events or entities of which, indeed, in many cases, they form a part. Take a field naturalist identifying a new species of a familiar genus, an example described by the late C.F.A. Pantin as a case of what he called 'aesthetic recognition'. He spots a worm and exclaims, 'Why, it's Rhyncodemus, but it's not bilineatus, it's an entirely new species.' The features common to the genus are not listed by him one by one, but recognized implicitly in their coherent physiognomy. They are clues he has interiorized, and yet through which he dwells in the world out there: for they point to a member of the class in question as 'that-which-is-now-out-there-before-me'. Polanyi has given a similar example in the case of radiography – an example I have already mentioned in passing – where he contrasts the blur seen by the beginning student with the clear picture seen by the experienced radiologist. Such perception is learned, as perhaps most perception has been, but once learned, it is immediate, just as immediate as the less discriminating perception of the novice. It is still perception, not judgment superadded to sensation, for it has all the hallmarks of perception, not only its immediacy, but its sense-carried involuntariness and its equally sense-mediated referential claim. In support of this view Polanyi points to a case recorded in the work of the Innsbruck school of I. Kohler and their experiments with inverting lenses.[9] When the subject wears a lens that inverts top to bottom or right to left, he is at first confused, and has to be led about by the experimenter, but gradually he adjusts himself, and sees things right again. Now Kohler and most of his co-workers have interpreted this story in the traditional Berkeleyan way. I have, after all, always upside-down pictures on my retina, which I have learned to judge as indication of right-side-up objects. In the experimental situation the work of judgment simply becomes more complicated and therefore more explicit. One of this group, however, named Kottenhoff, furnishes an example that points to a different interpretation. A subject who has learned to see with left-right inverting lenses is asked if he now sees things correctly and answers, 'I wish you hadn't asked me. I was seeing things normally, but now I get inverted images again that disturb my vision'. What he had done, it seems, was to re-orient himself visually in terms of unusual clues. When these were taken as subsidiaries he could focus correctly on things as they were really arranged. When he attended to them, he lost the newly acquired focus. He didn't, normally, look at the images and make inferences from them; he relied on the images to see what was out there. After all, the normal, in fact inverted, flat images on our retinas are what, in ordinary life, we never see and cannot see. We see three-dimensional solid bodies out there in space. We use the subsidiaries assimilated by our senses to find our bearings in a real, three-dimensional world of things and events. Of course in some cases we see 'wrong', just as we can make, at a propositional level, cognitive claims that turn out to have been mistaken. For years people counted – and saw – 48 chromosomes in the nuclei of human cells. Now it turns out there were, and are, only 46. Similarly, Harvard students in the 18th century 'saw' the phlogiston in a retort fill up the vessel and so extinguish the flame. From everyday perception to the most refined theory, there is a continuity: a claim to be in contact with reality which may succeed or fail.

This view takes account of all the features of perception we have noted. Perception is involuntary; things thrust themselves upon us, our percepts are caused by the presence of real things to which, through interiorization of some of their features, we have entered into real relation. There is a real world in the midst of which our senses serve to help us find our way. Perception is immediate: through reliance on the subsidiaries we have interiorized we focus on sights and sounds and feels and tastes and smells which, so far as our attention goes, are just immediately there – or here. And such percepts appear, finally, not as fleeting phantasmagoria, but as signs of things that, our senses tell us, are there around us.

What we have to forego, however, on this account, is the on-principle incorrigibility of the presentations of sense. Something is immediately present to our eyes and ears, palates and noses – but what exactly it is remains intrinsically undecidable. We do decide, of course, but not with logical certainty. For perception itself, being, like all knowledge, a process of orientation in a world outside our subjective awareness of it, is always already interpretation, and so possibly misinterpretation. And at one level of interpretative precision (seeing a clematis) I may turn out to have been mistaken (it was a morning glory) while at a lesser level (blue and white flower) I may have seen aright. Such 'interpretation', however, is not in itself judgmental; it is the sense-carried orientation of ourselves as sensing beings in a real, perceptible environment.

One could, as I've suggested, call such a realism with respect to perception a comprehensive or inclusive realism, or even, if one wants to be modish, an ecological realism. Sense-data theorists themselves, we have noticed, if they don't want to fall into extreme scepticism, have to add a realistic thrust of sensation somehow to come out where we are: as for instance in Price's notion of perceptual confidence. Or again, an empiricist like C.I. Lewis, who analyzes the ingredients of experience into the given (which isn't literally given) and the *a priori* through which we organize it, also arrives finally, though lamely and without much argument, at a reference to 'reality' to bolster up the construction he has offered. Here, on the contrary, instead of ending up apologetically with a patched up reference to reality, we start with realism, not, indeed, as a postulate, that says, blindly, let's take it that there is an external world, but with the acknowledgement of ourselves as *in a world*, a world in which through our senses, as much as through our judgment, we are trying to find our way.

Such a realism, moreover, is compatible with the core of Gibson's theory of the senses as conveyors of information and also – and more fundamentally – with Gregory's interpretation. Perceptions *are* 'hypotheses'. What we have in perception is not judgments added to sensations, but efforts to orient ourselves through our senses in a real physical, biological and cultural world. Like all hypotheses, perceptions lack infallibility. But like scientific hypotheses that make it into the permanent lore of scientific knowledge they are generally reliable and even if their alleged reference has to be modified in the light of further critical experience, we are confident that *some*thing solid will remain. If it is not clematis I'm looking at, it *is* a blue and white flower. Because my attempt at reference doesn't always hit the bull's eye, I needn't therefore give up all claim to real orientation in a real world of objects and events, and fall back on 'colored surfaces', uninterpreted noises, or 'mere feels' as my only objects. After all, every kind of activity that aims at an achievement can go wrong, sometimes altogether (as in hallucination), sometimes partly (as in mistaken perception or illusion). And such partial errors can sometimes be avoided, sometimes not, depending on the constancy or flexibility of the orienting mechanism in the particular kind of case.

So we are back almost where we started, dropping the demand of incorrigibility, admitting that all knowledge is subject to correction, and that perception can claim cognitive status as well as any other information-seeking (and finding) activity can do. Perception informs us about things and events in the sensible world, as theories do about the laws that govern the behavior of those things and events. Perception on this view is thus an analogue of scientific knowledge, or in general of knowledge in the more fully intellectual sense; and it is also the coping stone of such knowledge, insofar as, embodied beings that we are, we have to find our way among real perceptible things in a real, perceptible as well as intelligible world. Thus, finally, the tension among the features of perception is not, indeed, abolished, but is sufficiently alleviated to take the sting out of scepticism and allow us to live with fallibility, as in any case we have to do.[10]

NOTES

1 R. L. Gregory, 'Perceptions as Hypotheses' (preprint). Cf. R. L. Gregory, *The Intelligent Eye*, Weidenfeld & Nicolson, London, 1970, pp. 126–129.

[2] Rolf George, 'The Thing in Itself and Kant's Transcendental Object', paper presented at the 1974 Kant Congress. I am very grateful to Professor George for letting me read his MS; in fact I owe to his clear formulation there the awareness of the distinction between the 'causal' and referential aspects of perception and of the philosophical import of this distinction (or of its neglect).

3 R. L. Gregory, *op. cit.* In *The Intelligent Eye*, Gregory speaks of 'object hypotheses', e.g. on pp. 64, 69–72, 150.

[4] Jaakko Hintikka, 'Quantifiers, Language Games and Transcendental Arguments', in *Logic, Language Games and Information*, Clarendon Press, Oxford, 1973, pp. 98–122.

[5] J. J. Gibson, *The Senses Considered as Perceptual Systems*, Houghton Mifflin, Boston, 1966.

[6] Jaakko Hintikka, 'Information, Causality and the Logic of Perception', preprint.

[7] Cf. 'People and Other Animals', below.

[8] Michael Polanyi, *Personal Knowledge*, University of Chicago Press, Chicago, 1958; *The Tacit Dimension*, Doubleday, New York, 1966.

[9] Discussed in M. Polanyi, 'Sense-Giving and Sense-Reading', in *Knowing and Being*, University of Chicago Press, Chicago, 1969, pp. 181–207, especially 198–199, ref. on p. 207. Much further empirical support for this general approach to perception can also be found in the works of Hans Wallach and his collaborators. Cf. e.g. Hans Wallach and David Huntington, 'Counteradaptation after Exposure to Displaced Visual Direction, *Perception and Psychophysics* 13 (1973) 519–524, as well as Gregory's work.

[10] Along with the reading of Professor George's manuscript mentioned above, this paper owes its inception to a discussion with Tom Kuhn, who is seeking (if I understand him) a theory of perception that will provide both realism and incorrigibility. I don't suppose my argument will convince him, but I am grateful to him too for making me try. Professor Lorenz Krueger read the manuscript and made some very apt suggestions, and it was Professor Michael Frede, who also read the manuscript, who warned me against uncritical acceptance of the Cornford interpretation of the *Theaetetus*. Since I was using it only as a framework for my own argument, however, I have allowed my report of it to stand with only slight modifications; classical scholars may take my Plato for a fictional character if they like. I am grateful to both Professors Krueger and Frede also for their careful reading and criticism.

CHAPTER III

BIOLOGY AND THE PROBLEM OF LEVELS
OF REALITY

This paper is intended to sort out some problems in the philosophical foundations of biology. For the reader's convenience, the questions referred to in the text are listed here:

(1) Is a one-level ontology adequate to account for the major areas of human experience, both in and out of science?

(2) If not, how can we formulate adequately a many-levelled ontology?

(3) Is biology reducible to physics and chemistry?

(4) Is biology a molecular science?

(5) Are all biological explanations mechanical or are some irreducibly teleological?

(6) Did life in its present form originate from non-life and by what means?

(7) Is *all* biology molecular science? i.e., is *every* biological discipline in principle molecular?

(8) Is *some* biology molecular? Or is every biological discipline in principle non-molecular?

(9) Meta-question: Are questions seven and eight philosophical or empirical?

(10) Are physics and chemistry molecular sciences?

(11) Is the distinction between the living and the non-living primarily *morphological* or *functional*?

No reader of this essay, I trust, will want to deny that there is *some* positive relation between metaphysics and science. Metaphysical arguments – like Descartes' – have been used to support scientific advance, and revolutions in science – as in twentieth century physics – have demanded the rethinking of metaphysical as well as epistemological issues. What concerns me here is the philosophical problem – or problems – inherent in contemporary biological theory, particularly in molecular biology. The situation here differs from the crisis surrounding quantum mechanics in that only a minority of the scientists involved in the recent advances in molecular biology consider that there *is* a crisis.

An articulate minority do believe, however, that a crisis exists. It is, moreover, in the last analysis, the *same* crisis that obtains (although, again, only a minority admit its existence) in behavioral science, and the same crisis which constitutes, in my view, the central problem of contemporary metaphysics. I can put this problem in the form of two questions: first, is a one-level ontology adequate to account for the major areas of human experience, both in and out of science, and secondly, if not, how can we formulate a many-level ontology in a fashion consonant with our general *Weltanschauung* – with the set of fundamental beliefs derived in part from science itself as it has filtered down to and transformed common sense, as well as from the heightened influence of historical thinking on our basic attitudes? (Questions 1 and 2.)

Of those philosophers who would answer 'no' to my first question, a number would assert that the second *has* also been answered: we need only adopt Whitehead's cosmology and the results of recent biochemical research are smoothly assimilated, as specifications of the necessary, though not sufficient, conditions for the emergence of novel nexus of actual entities – of those societies which we call living. This *may*, in the end, be the best answer – but I have two objections to accepting it at this stage. First, there are formidable difficulties in the Whiteheadian scheme – notably, in my view, the doctrine of eternal objects. Secondly, as current discussions bear on the fundamental issue – the issue of what it means to accept a hierarchical ontology, and where the essential cuts are to be made in it – they raise precisely the question whether or not a Whiteheadian metaphysics, which assimilates all reality, non-organic *and* organic, to a hierarchical scheme, is adequate to interpret the facts that confront us. I prefer, therefore, to look at the current dispute without an *a priori* commitment to one metaphysical solution.

The dispute in question concerns the age-old dream of reducing all knowledge to physics, and all there is to particulate physical entities. The latest impulse to claim such a reduction comes, of course, from the recent dramatic advances in biochemistry, and in particular from the Watson-Crick theory, which has, as the press put it, 'cracked the genetic code' and shown that DNA holds 'the secret of life'. (And at least as important, though less widely popularized, is the work of Monod and Jacob on allosteric enzymes.) In this situation some biologists, as well as philosophers, feel impelled to ask once more: is biology vanishing as, for example, astrology or alchemy have already done? Is there 'really' only one science, which is, in principle, mathematical physics? The question itself, however, needs sorting out – as it stands several issues get themselves entangled in it. So I propose to start with an alternative question put by one of the biological dissenters, Barry Commoner. He asks: *not* (at least, not directly) is biology reducible to physics and chemistry (which we shall table for now as question *three*), but: *is biology a molecular science*?[1] Taking this question (our fourth) as our starting point, I hope we can sort out some of the problems in which it involves us.

To begin with, I want to set aside two further questions with which our original problem is sometimes confused.

First, the question whether biology is a molecular science is not to be identified with the question whether biology demands teleological as well as mechanistic explanation, or – its metaphysical equivalent – whether there are irreducibly telic phenomena (let us call this question *five*). The reinstatement of some concept of final causality is doubtless an important part of the conceptual reform which the proper assessment of biological knowledge would demand – especially, for example, in embryology or psychology – but the problem is distorted, I believe, if one sees this part of it as the whole. For both sides in the mechanism-teleology controversy have generally failed to see that what is needed for the adequate philo-sophical foundation of biological thought is neither to get rid of teleology, nor to rely on it as the self-sufficient partner of causality, but to supple-ment both cause-and-effect thinking and means-end thinking by reference to the still more basic concept of *standards* or *norms*.

To put it in Aristotelian terms – which are not strictly adequate in this context, but will perhaps indicate roughly what is involved – biology needs to rely not only on material and efficient causes, but also on final causes, and not only on final causes but on formal causes as well, and on formal causes most fundamentally. Formal cause for Aristotle is not, of course, or not primarily, shape or contour: it is the defining principle, the operat-ing principle that makes a given kind of thing the kind of thing it is. In the *Metaphysics*, Aristotle draws together the concept of form – which is the equivalent of formal cause – and his peculiar concept of each kind of thing's *being what it is*: that is, the type or norm or standard that makes it *this* kind of thing and not another. It is form in this sense, meaningful or significant form, form as generic standard, that is fundamental to the study of living things. It is true, of course, that Aristotelian eternal species have vanished from our world; living things for us are products of evolution, concretions out of flux. But, short of the problem of the origin of life (about which a little more in a moment), we do have to deal with a multiplicity of present kinds of plants and animals, and, did we not, in ordinary life and in biological research, rely on our recognition of such kinds, of the differing styles of living that separate them from one another and that unite the phases of individual development, and separate the individual from its environment, we could neither begin nor continue to practise biology at all. Of course, we want to find out with all possible precision what organisms are made of – Aristotelian material cause – and the sequence of processes – efficient causes – which subtend the ends of life, its final causes: as, for example, the processes of metabolism subtend the ends of nutrition and growth, or nervous processes the ends of perception, action, and thought. Yet all these have their ultimate importance within a different context: with reference to the intrinsic significance of each variety of life itself, the standard by which we judge each kind of organism to be or to fail to be the kind of thing it is. The basic question, therefore, is: without reference to such norms, is there and could there be any biology at all?

Nor, secondly, is our question to be identified with the problem of the origin of life or its evolution, i.e., again to use Aristotelian terminology, with the problem of efficient causes. We are *not* asking what we may list as question *six*: did the present forms of life originate from non-life and by what means? *If* we can articulate properly our understanding of what organisms are, we may then be able to formulate with philosophical consistency and adequacy a theory of their history. But, as Kant argued against Locke, the question, what is necessarily presupposed in our understanding of the objects of experience, takes philosophical priority over the question how we came to see them thus, or for that matter – to introduce a non-Kantian addendum – over the question how they came to be. We can, indeed, take as proven the core of contemporary evolutionary theory, which Medawar recently stated as consisting in the two propositions: (1) that the terrestrial populations existing at a given future time will differ statistically in some degree from those existing today; and (2) that the genetic constitution of those populations will have *some* connection with their changed phenotypic characters.[2] These are perfectly harmless statements which no one, philosopher or otherwise, would want to challenge. But they tell us nothing at all about the epistemic relation between discourse about gene pools or populations of gene pools and discourse about cells or organ systems or organisms, nor about the ontological import of such discourse, let alone telling us how from a time when there were no living cells or organ systems or organisms there came to be such entities. The muddle at the center of neo-biological thought needs to be disentangled before an adequate theory of evolution, emergent or otherwise, can be formulated.

I have tried to analyze this peculiar muddle a number of times elsewhere; so, of course, have numerous other people, but perhaps I had better clear the ground by attempting it briefly once more.

Behind Medawar's harmless statement, the principal presupposition of modern evolutionary theory is the thesis formulated in 1932 by R. A. Fisher: 'Evolution *is* progressive adaptation, and consists in nothing else'.[3] Yet adaptation on its own, or progressive adaptation on its own, as neo-Darwinism takes it, is by no means self-explanatory, or indeed explicable. For one thing, the concept of evolution as identified with progressive adaptation is basically ambiguous. On the one hand, such adaptations are supposed to be mechanically self-generating – through mutation, natural selection, recombination and isolation – and so to entail no teleological reference. Yet on the other hand, adaptation is itself a teleological concept: it is adjustment of something to something for some end. To interpret organisms as adaptation machines, as neo-Darwinism does, therefore, is to interpret them as complexes of means for ends, and therefore teleologically. Secondly, if the means-end reference is admitted, one must ask further: means to what end? But the 'end', for Darwinism, dare not be some 'higher' form of life, the next 'level' to which evolution aspires: that would be to reintroduce a forbidden version of 'unscientific' teleology. It must then, and is usually said to, be *survival* that adaptation is 'for'. But in that case the whole 'theory' becomes a tautology, a complicated way of saying simply that what survives survives. (And Medawar's two propositions, which were stated in response to the accusation of tautology, simply open out the tautology a little way, only to let it form again when we try to interpret those propositions in their full theoretical context.) Finally, and fundamentally, the trouble is, really, that the concept of adaptation is essentially a relative one – relative to the existence of two things: the *subject* to be adapted and the *environment* to which it is to be adapted. Helmuth Plessner makes a similar point, in an early work, *Die Einheit der Sinne*,[4] in the context of an analysis of sense perception. Referring to the physiological investigations of perception, he points out that perception cannot be explained wholly in terms of adaptation, since there must be *something there already to become adapted. Adaptability*, which is a potential relation between organ, medium, and object of perception, must precede adaptation. Thus the adaptable entity, the organism which *can* achieve adaptation, must be assessed in its own right, by its appropriate norm, before the detailed conditions of its adaptation can be specified. The same is true in the context of evolutionary theory, or indeed wherever the attempt is made to rely on adaptation as a principle of ultimate explanation. Adaptation is a cryptoteleological concept, but teleology even when explicit is itself dependent on the prior evaluation of the ends evoked. And in the case of evolutionary biology, the end is not simply survival but the survival of – a type, a mode of living adjudged as significant in itself. That is the only judgment that can fill in the tautology of survival-for-survival-for-survival.

Our question concerns, then, in the first instance, neither the *telos* nor the genesis, but the *form* of organic being. Still in Aristotelian terminology, it is the question whether material causes are ever adequate, without reference to formal causes, to explain the nature and existence of organized beings.

Let us return, then, to Commoner's question: is biology a molecular science? First, we had better agree on what is meant here by 'molecular'; a routine usage for biologists but somewhat puzzling to outsiders. On principle, the term appears to mean 'such that the whole is determined by its parts'. Biochemists work with macromolecules; they are not primarily concerned with atomic or subatomic events; so they need not distinguish for their purposes between 'molecular' and 'atomic' or 'sub-atomic'. But what is meant by the thesis that biology is a molecular science is in effect the old atomist's thesis that the study of the parts is equivalent to the study of the whole, that things are particulate, are 'no more than' aggregates of least bits. The alternative to 'molecular' we may call, as some psychologists do, 'molar'; that is, a property is molar if it is not accountable for solely in terms of the aggregation of its least parts.

Secondly, the question itself needs to be divided into two alternatives, which are often confused. It may mean (this we will call question *seven*): is *all* biology molecular science; that is, is every biological discipline on principle molecular?

Or it may mean (question *eight*) is any biological discipline molecular? That is, is it false that all biology is non-molecular? For the defender of the affirmative – i.e., of the identification of biology as molecular – this is of course a much weaker claim.

Before I proceed to consider either of these questions, however, let me interject a meta-question (which we may label question *nine*): are questions seven and eight empirical or philosophical? I hope to show that they are philosophical questions: that the question of the uniqueness of biology – i.e., in our present formulation, of its non-reducibility to the chemistry of macromolecules – is to be answered in terms of a difference in kinds of knowledge, and in particular of the logical levels entailed in various orders of knowledge; and in terms of a metaphysical theory of the kinds of entities – the levels of reality – to the comprehension of which these several orders of knowledge are directed.

But I must add two warnings here. First, in calling our pair of questions 'philosophical', I am not supporting a hard and fast distinction between empirical and conceptual questions – a distinction of the kind which either an old-fashioned Hegelian or a new-fashioned ordinary language philosopher would draw in order to exempt himself from any need to know anything about the progress of scientific knowledge. Empirical advances *do* affect philosophical positions and hence philosophical arguments. Aristotle could argue philosophically that the world is unique, finite, and eternal; we cannot. Yet, on the other hand, there are genuinely philosophical questions which arise precisely out of and in conjunction with advances in scientific knowledge, and this seems to me to be one of them.

Nor, of course, secondly, am I suggesting that if we succeed in stating a philosophical refutation of a one-level ontology, this will mean legislating a place at which biochemical analysis of organic processes must stop, as Drieschian vitalism proposed that analysis must do. The necessary conditions for the existence of organic phenomena can of course be analyzed *ad infinitum* and with continuing increase in biochemical and biophysical knowledge. It is a question of the epistemological and ontological relation between different branches of knowledge, not of asking one to stop in order to allow the other to begin.

What, then, of question seven: is *all* biology molecular? There is a very simple answer here, as straightforward as it is venerable. Commoner starts by asking whether the explanation of events in the living *cell* can be reduced to molecular considerations only, but he suggests that the answer to this inquiry is relevant to a series of further questions, and ultimately to the question: what is involved in observing a human being, and I would add, understanding and in some sense explaining the behavior of a human being? Can *this* kind of observation, understanding, and explanation be rendered in molecular terms? No, it cannot. For, short of the articulation of a complete ontology, which would enable us to fit artifacts, other organisms, and ourselves into their proper places in a cosmological scheme – and apart from the particular question about biochemistry suggested by recent advances in that field – the old epistemological argument (as old at least as the *Theaetetus*) is valid here in relation to *any* 'molecular science', or to the claim of any science that it is molecular because all science is so, and that all science is so because molecules – or their constituent particles – are all there is to know. As Erwin Straus puts it, 'physics refutes physicalism'.[5] If there is any knowledge, including, if that were possible, the 'knowledge' that there is nothing but material particles in motion, then there must be something other than material particles in motion, namely something – I don't mean some 'stuff', but some process, some real existent who can make a competent, if not a veridical claim that this is so. But molecules can make no claim to truth, any more than they can err. So if there is any knowledge, even 'molecular science', there is *something* more than the subject matter of molecular science. There are at least molecular scientists. In other words, either there is no knowledge (including the knowledge of philosophical atomism), or there is at least the knowlege that philosophical atomism is false.

Now obviously this is not an answer to Commoner's particular question in its immediate context: i.e., it is not an answer to question eight. Commoner is talking about the biology of the cell as distinct from its biochemistry, and the question whether this particular reduction is possible is not answered by the epistemological argument alone. What that argument does show, however, is that there is a sound philosophical objection to a universal reductivism (to an affirmative answer to question seven) and that, therefore, whatever empirical results research on DNA or on enzyme specificity or the like may produce, we shall need some philosophical interpretation of them which permits the existence of more than one level of reality if such results are to be consistently assessed as knowledge. In other words, the thesis: that *all* biology, including the biology of organisms sufficiently complex to be accounted capable of knowing, is molecular science is demonstrably unsound, and the question, is this or that branch of biology a molecular science, needs to be treated with this comprehensive denial in view.

It may, of course, be objected – as in effect Nagel objects in the *Structure of Science* – that we have no right to say what physics will *never* do, that we are advancing *a priori* dogmas where we should wait and see – that we have appropriated to philosophy a question proper to the empirical sphere.[6] To this I would reply that – quite apart from the unsoundness of a theory of knowledge which leaves open a proof of the impossibility of knowledge – those who consider our question empirical are just as plainly arguing *a priori*: for they are saying, it is on philosophical grounds perfectly feasible to reduce biology – or in the case of the *Theaetetus*-Straus argument – philosophical psychology – to physics, but it just hasn't yet been done. Nobody, except perhaps Francis Crick or E. B. Skinner, would claim it has been done. But the question, can it or can't it be done, is a question of principle.

It is important to stress this point even at the cost of reverting for a moment to my earlier meta-question. Many biologists nowadays hold, some joyfully, some regretfully, that it is only a matter of time before their science ceases to exist, yet at the same time they continue to practise it by methods which entail a recognition of its uniqueness: that is, of its resistance on conceptual grounds to such extinction. (More of this later.) Others, like Commoner, take both positions at once: if the DNA dogma were really successful, he says, biology would be a molecular science and so would cease to exist as biology. But it is not successful, so biology is still on the scene. At the same time, he pursues this line of argument – and conducts his own research accordingly – just because he is convinced that living things, from the cell to human beings, *cannot* be understood purely in molecular terms. In that case it is not empirical results, but the epistemology of biology, and the metaphysics inherent in it, which have to guide us in interpreting whatever results empirical research may offer. A similar confusion, in my opinion, mars the argument of Charles Taylor about the explanation of behavior.[7] It is a purely empirical question, he believes, whether traditional causal methods can succeed in psychology. At the same time he admits that his basic position is dependent on that of Merleau-Ponty. But in *The Structure of Behavior* Merleau-Ponty has already made a philosophical distinction of three levels of order inherent in the analysis or understanding of behavior. And while these distinctions are rather asserted than argued for there, if we put the earlier book into the context of its sequel, *The Phenomenology of Perception*, we see it assimilated to a total philosophical position out of which the causal reduction of behavior appears comprehensively impossible. So does it, of course, in the light of Chomsky's refutation of Skinner. This is not my theme here, however; I only want to indicate some of the reasons why I consider it important to emphasize the philosophical nature of the argument here.

Let us move on then to question eight: is *some* biology a molecular science, or, on the contrary, is all biology molar? Or better, what about, jointly, questions eight and two: is all biology non-molecular, and if so, in the light of what kind of many-levelled ontology ought we to interpret it? My answer to this question is based on metaphysical, empirical and ontological grounds.

First, we must answer no to question seven, and this means also no to question one. A one-level ontology contradicts itself. So we need, in some form, a many-levelled metaphysic. But there might, for all we have seen so far, be cognitive events and molecular events and nothing in between. This was, of course, Descartes' position. The Cartesian solution, however, is inadequate by the criterion of metaphysical coherence as well as on massive empirical and methodological grounds. The incoherencies of Cartesian dualism – both mind-body and god-world – I need not stop to enumerate. Let me just announce baldly that the metaphysics we need is many-levelled but not Cartesian. That is my first point on question eight. Secondly, our experience of other animals, both practical and scientific, sets a massive obstacle in the way of a theory that embraces the 'bête-machine'. I shall not attempt to elaborate a practical example, but take it that there are plenty. (If the reader is in any doubt, let him consult any standard work such as Hediger's *Wild Animals in Captivity*).[8] As one scientific example, again to take the place of many, let me just mention the behavior of dolphins in learning experiments. Moreover, whatever our interpretation of the mechanism of evolution, we know that we *are* animals, descended from other, ultimately from unicellular, animals, and so our kind of inwardness has arisen from primordial beginnings which we have no right to cut off radically from its intensification, or transmutation, in our case. In short, the concept, not necessarily of consciousness, but of some sort of 'inwardness', of what Adolf Portmann calls 'Weltbeziehung durch Innerlichkeit' (and which I have rendered as 'centricity') demands extrapolation to the whole animal kingdom.[9] Whether we can meaningfully extend this generalization to the plant kingdom as well is another question. Portmann declares that we must do so, and one can perhaps see some logical sense in it: 'plant' and 'animal' are not hard and fast dividing concepts; compare for example *Euglena* and *Paramecium*. But the notion of a plant's centricity is, so far, wholly beyond my own comprehension. Centricity seems to me an emergent of animal life. Here Plessner's distinction between open and closed positionality seems apposite.[10]

If we look, however, at *form* and its development rather than behavior, experience provides a panorama in which *all* organic phenomena resist reduction to one-level explanation. For with respect to Portmann's second basic criterion of living things, *Selbstdarstellung* (which I render as 'display', in a broader than its usual zoological sense) the plant kingdom rivals, if it does not outshine, ourselves and our fellow animals. Orthodox evolutionary theory takes *form* as a dependent variable, secondary to survival value. Of course this is often the case. Thus plant color and pattern attract insects, insects carry pollen, and so on. The instances are legion, and undeniable. But, as Portmann argues, the vast range and elaborate detail which are constantly maintained in being by both plants and animals, by means of immensely complex and diversified genetic mechanisms, have by no means been shown to be dependent in all their details upon such a function. *Some* striking color, yes, but why such constancy and minuteness of pattern and variety of genetic mechanism in maintaining it? There *seems* to be something more here than means to perpetuation of the species. Only if one accepts to begin with the *a priori* principle that all forms *must* be a function of adaptation, must one argue otherwise. One might adduce here also the argument of Agnes Arber in her *Natural Philosophy of Plant Form*,[11] or for that matter the experience of J. C. Willis, who accompanied Francis Darwin to India to study river plants.[12] Finding the bewildering variety of form and color in one environment too multifarious to reduce dogmatically to the presumably one-level adaptational terms of Darwinian theory, Willis abandoned Darwinian explanation for a statistical theory of his own, the theory of the hollow curve: a theory which no one, so far as I know, has accepted. Indeed, it seems a mad theory, but it did arise from crediting organic form with an inherent significance which its reduction to a function of adaptation refuses to allow it. And I think that Willis – like Semmelweiss – deserves to count as one of the martyrs of science, driven in his case to what amounted to professional suicide by his refusal to accept a reductivist dogma with respect to the colors and shapes of plants.

Thirdly, and finally, *methodological* evidence lends further, and decisive, weight to a non-reductivist view. For if we examine cases in which an alleged 'molecular' explanation of 'life' is said to have succeeded, we find in fact that a second level has been surreptitiously introduced, and there seems good reason why this should be so. Take, for example, the paper of Crick and others on the form of the genetic code, which appeared in *Nature* in December, 1961. The problem that Crick was working on at that time was the question: If the information necessary to development is conveyed by the nucleic acid in the nuclei of the germ-cells, how do the four bases, CGAT, which are spread out along these giant molecules give instructions for the formation of the twenty amino-acids which make up the protein of the developed cell? Crick and his team concluded that the code is built of triplets strung together without 'commas' and 'read from a fixed starting point'. They say:

any long sequence of bases can be read correctly in one way, but incorrectly (by starting at the wrong point) in two different ways, depending on whether the 'reading frame' is shifted one place to the right, or one place to the left.[13]

The whole business, then, consists in 'cracking a code'. But a code is a message which can be deciphered only if it is a totality with a meaning, as distinct from a jumble of particulars, a mere noise. Without this reference to a meaningful pattern, to a form, the investigation would be impossible. *Any* arrangement of the four bases is compatible with the laws of physics and chemistry, and must be so, else it could not exist at all. And if the particular arrangement is destroyed – as in a lethal mutation, for example – the laws of physics and chemistry continue to operate exactly as they did before. Thus the arrangement that constitutes a 'code' is physico-chemically indifferent. But in terms of biological information it is just the particular order that makes all the difference. Of course, most biochemists would probably deny the application of such reflections to their subject: they are confident that they need only work out a sequence of spatial arrangements of nucleic acid molecules to explain development. But it is just such spatial arrangements that are physico-chemically indeterminate, and which do nevertheless determine the shape and functioning of the organism in question. They entail geometrical or *morphological* as distinct from merely physico-chemical distinctions.

There is more to it than this, moreover. Not only must the investigator recognize the existence of a code, but his language suggests that in some sense the organism must 'recognize' it too. Why all the talk about 'reading' the code? About a message sometimes becoming 'nonsense' and so on? Not of course because biochemists think that bacteria phage can read. Why then? In reading a code we use the sounds or shapes presented to us as clues to something beyond them which we call their *meaning*. This is both a two-level structure and a two-level process which presents an achievement: it can succeed or fail. Thus the use of words like 'reading' strongly suggests that, whatever they claim, biochemists are in fact interpreting development, not only on the level of micromorphology – with reference to a lower level of physico-chemical conditions and an upper level of geometrical order constituting a code – but with reference to a third level as well. For they are interpreting the code as a sequence which, as a lower level in its turn, subserves the emergence of a new pattern, the mature organism. It is form again that constitutes the raison-d'être of the process – this time a more complex and comprehensive form, the form of the organism, with its many cells, within each cell a nucleus, and within each nucleus the code which has been successfully read off in the achievement of its own maturity. So even in the simplest organism there are at least three structural levels of complexity: the physico-chemical particuculars of the genetic material, the codes in which it is arranged, and the whole organism of which these nuclear compositions again form part. And there are at least two functional levels: there is the message to be read, and the development which is the result of successful reading. Without these references to forms, both micro- and macro-morphological, the whole thing would be gibberish; there would be neither scientists to know, nor organisms to be known.

On all these grounds, then, metaphysical, empirical, and methodological, I conclude that our answer to question eight: is *any* biological science molecular, ought also to be: no. One-level concepts are adequate, if at all, only for the interpretation of inanimate systems. Life is radically non-interpretable, on this view, except in terms of matter *and* form: physical-chemical conditions *and* biotic principles. In R.O. Kapp's terms, living things, unlike non-living, are *doubly determinate*.[14] Commoner – contradicting his own empirical approach to the question – adduces here Bohr's own suggestion that a principle of complementarity applies to analytical as against organismic explanations in biology. The embryologist C. P. Raven takes a similar view.[15] The term 'complementarity', however, does not adequately define the conceptual situation they are describing. I think we should speak instead of *ordinal* complementarity. The relation, say, between cellular function, and the chemical laws which specify necessary conditions for its operations, is *not* a complementarity like that of wave and particle. It is a hierarchical complementarity, in which the lower level leaves open boundary conditions to be specified by the laws of the higher level. The higher level depends for its existence on the lower, but the laws of the lower level, though presupposed by, cannot explain the existence of the higher – although they may suffice to explain its failures.[16]

To sum up, then: a one-level ontology is untenable, and it is false that all biology is molecular. Further, at least a two-level ontology is needed for the interpretation of all living things; in answer to question eight, no biological science is wholly molecular, and an organicist rather than a Cartesian solution is therefore needed to provide a satisfying answer to our question two.

This would seem to be the close of my argument. But in fact it only brings me to the two questions for whose sake I wanted to present this article – because I hope the discussion may shed some light on them. The first problem I have in mind brings us back to question three: is biology reducible to physics and chemistry? As against the layman's and biologist's simple picture of classical physics as equivalent to the 'new mechanical philosophy' of Boyle and Newton, biological explanation exhibits a hierarchical structure which strongly suggests the existence of doubly determinate entities, and, indeed, of intricate hierarchies of such entities.

But where does the complexity begin? As early as one hundred and fifty years ago the creed of philosophical atomism, in its Democritean and Newtonian sense, was by no means universally accepted even in the physico-chemical sciences themselves – see, for example, Pearce Williams' account of Boscovichean atomism in his *Michael Faraday*.[17] And after the recent revolutions in physics it may seem to have been superseded altogether, so that a one-level corpuscular ontology becomes inappropriate for physics itself. That is the lesson, for example – in harmony with a Whiteheadian metaphysic – that Sewall Wright derives from Čapek's work.[18] It is also a theme of E. E. Harris' recent book, *The Foundations of Metaphysics in Science*.[19] In this belief, also, Commoner adduces the example of superconductivity to show that wholly particulate explanation is sometimes inadequate even in physics. Wholes, even here, he argues, sometimes determine parts rather than parts wholes – or, form regulates matter rather than matter form, and this before (both logically and ontologically) we consider the still more complex systems of the organic world. And Wright, allegedly following Čapek, argues similarly: we already find in the inorganic world hierarchical relations of which biological hierarchies may be the continuation and intensification.

Against this view, however, others – Kapp, Bertalanffy, Elsasser, Sinnott, Jonas, Polanyi, for example – insist that biological explanation is unique. This appears to entail the view that all physico-chemical explanation, as distinct from biological theory, is one-levelled. That does not mean, of course, 'one-levelled' in the sense of a classical atomism, but, presumably in the sense that all non-organic phenomena can be explained as functions of conditions of minimal energy, whereas *no* organic phenomena are wholly explicable in these terms. The DNA molecule may again be taken as a case in point. The genetic code is a function of the *particular arrangement* of the four bases CGAT; but, as we have seen, many of the possible arrangements would be equally compatible with the laws of physics and chemistry. So the existence of this particular arrangement rather than that is not determined by the physico-chemical conditions on which admittedly its existence depends. DNA is, in other words, a *morphological* structure, and not simply a chemical molecule. The principles explanatory of superconductivity, on the other hand, appear as necessary results of certain physical conditions; there is, so to speak, no 'choice'.

The alternative before us is simple. Biology, we have concluded, is not a molecular science; we answer 'no' to both questions seven and eight. But, Commoner and Wright argue, neither are physics and chemistry. Therefore biology may be reducible to physics and chemistry – we may answer 'yes' to question three – but still biology is not reducible to a one-level physics and chemistry. On the other side, the position is, as Polanyi argues,[20] for example, in reply to Commoner, that the hierarchical nature of biology entails at its very foundation the specification of a new kind of order which is alien to the basic concepts of any merely physical science, however non-classical. In the case of superconductivity, for example, very low temperatures *produce* the condition in question. In the case of DNA, on the other hand, as we have seen, physico-chemical laws *permit* a great number of alternatives, only one of which – only one *morphe* – is in fact realized, and thereby sets the further conditions for the development of a particular species: this bacterium, this amphibian, this mammal, and so on. This is, then, really a case of double determinateness, while the other is only a case of a modified heuristics, involving, however, the assertion of no new ontological principles. (The alleged difference is even clearer if we consider modern genetic theory, which acknowledges that any one phenotypic character may be maintained in being by a vast number of alternative genetic determinants: as a regiment remains 'the same regiment' as one hundred years ago although not a single soldier in it remains the same.) No levels of reality, therefore, no nexus, no societies or conceptual aims *below* the level of life. As a matter of fact, Wright's analysis at one point also suggests such a solution, since he remarks that the hierarchy of biological forms should be seen 'at one side' from the cosmological hierarchy of the physical world. And so do Commoner's remarks about the uniqueness of biology, which, as I have already pointed out, contradict his empirical approach to the problem of reductivism. Thus we have a tenth question: in effect, 'are physics and chemistry molecular?' which will affect materially our answer to question three.

It is a matter of the greatest metaphysical – and epistemological – moment, I believe, to decide this issue. And although I find myself convinced, on the whole, by Polanyi's argument, I must leave the question for philosophers professionally cognizant of contemporary physics to clinch if they can, simply adding two remarks. First, Polanyi's objection to

Commoner's reading of the superconductivity story is confirmed by Professor Kenneth Greider of the Davis physics department, who is working in the same field.[21] The paper on which Commoner chiefly relies, he points out, while admitting the heuristic usefulness of the 'macroscopic' theory, confidently predicts that such a theory will ultimately be derivable in 'microscopic' terms.[22] But this is precisely the attitude of reductivist microbiologists and thus offers, logically, no case for their opponents. Whether a revolution in physics of the kind Commoner envisages may be ahead in some other area is another, and open, question.[23] Secondly, of course, we *may* be able to answer this question both ways at once – or, if you prefer, to achieve a synthesis of the two contrary positions: to find the universe hierarchical in a Whiteheadian way, yet differently so in the non-biological and biological realms.

(Not, again, that this means a discontinuity in the history of the universe – that life did develop from the non-living I am, once more, not for a moment denying. Nor am I making any suggestion about how it happened – only that life as we now have it presents a hierarchically ordered reality more complex than the order of the non-living.)

Finally, at the eleventh hour, an eleventh question: if we place an ontological cut between the living and the non-living, is our distinction a *morphological* or a *functional* one? The example of DNA suggests that this is a question of morphology; but since living systems are historical, either innovating novel types of order or maintaining in being their already existent forms, morphology may be taken as itself a process and thus already functional. Thus in Ledyard Stebbins' recent suggestion that the basic characteristic of life is what he calls relational order (for example, colinearity between DNA and RNA), we have a thesis which combines both structural and functional aspects. He defines relational order as '... the kind of order which exists among the parts of any compound structure, and which is significant only or chiefly in relation to similar orders in other comparable structures of the same rank, permitting the structures to cooperate in performing the more specific functions'.[24] Now plainly what is involved here is a structure which functions to perpetuate itself and to initiate new structures. Again, as in the previous question, we may be able to have it both ways. But again, this is a question I wish to leave open for further development. For it is just in this area that the most pressing task of the contemporary metaphysician lies. I have been trying in this article to clear away some of the neighboring under-growth, not because I believe this is all philosophers are entitled to do, but because I wanted, if I could, to bring into clearer focus the space where the still debateable and, I hope, promising questions lie.

NOTES

[1] MS of a paper presented to the Study Group on Foundations of Cultural Unity at Bowdoin College, Brunswick, Maine, August, 1966. Cf. B. Commoner, *Science and Survival*, New York, 1966, Ch. 3.

[2] *Mathematical Challenges to the Neo-Darwinian Interpretation of Evolution* (ed. by P. S. Moorhead and M. M. Kaplan), Wistar Institute, Philadelphia, 1967, p. 12.

[3] In a discussion of evolution at the Royal Society. *Proc. Roy. Soc.* **B 121** (1936–7), 58.

[4] H. Plessner, *Die Einheit der Sinne*, reprint of 1922 edition, Bonn, 1965.

[5] E. Straus, *The Primary World of Senses*, New York, 1963, pp. 298–304.

[6] E. Nagel, *The Structure of Science*, New York, 1961, Ch. 12.

[7] C. Taylor, *The Explanation of Behavior*, London, 1964.

[8] H. Hediger, *Wild Animals in Captivity*, New York, 1964.

[9] See A. Portmann, *Neue Wege der Biologie*, Munich, 1961.

[10] See M. Grene, this volume, chapter XVIII.

[11] A. Arber, *The Natural Philosophy of Plant Form*, Cambridge, 1950.

[12] See J. C. Willis, *The Course of Evolution*, New York, 1940.

[13] *Nature* **192** (1961), 1229.

[14] R. O. Kapp, *Science versus Materialism*, London 1940, Section 2.

[15] C. P. Raven, 'The Formalisation of Finality', *Folia Biotheoretica* **B 5** (1961), 127.

[16] This argument is developed by M. Polanyi in Part IV of *Personal Knowledge*. Chicago, 1958 and in his more recent writings. (See *The Tacit Dimension*, New York, 1966.)

[17] C. P. Williams, *Michael Faraday*, New York, 1966.

[18] S. Wright, 'Biology and the Philosophy of Science', *The Monist* **43** (1964), 265–90. Cf. M. Čapek, *The Philosophical Impact of Contemporary Physics*, New York, 1961.

[19] E. E. Harris, *The Foundations of Metaphysics in Science*, London, 1965.

[20] MS reply to Commoner's paper.

[21] Oral communication.

[22] J. Bardeen, 'Developments of Concepts in Superconductivity', *Physics Today* (January, 1963), 19–28.

[23] G. F. Chew, 'Crisis for the Elementary-Particle Concept', submitted to *Science and Humanity*, Year Book (Moscow), University of California Radiation Laboratory Preprint 17137.

[24] G. L. Stebbins, *The Basis of Progressive Evolution*, University of N. Carolina Press, Chapel Hill, 1969, p. 4ff.

CHAPTER IV

REDUCIBILITY: ANOTHER SIDE ISSUE?

Charles Taylor, in a paper entitled 'Mind-Body Identity: A Side Issue'?,[1] argues that those who oppose mind-body identity can grant to its defendants all the concessions they desire and still maintain the theses *they* wish to defend: that, in other words, there is no substantive disagreement between the contestants in the case, and so philosophers of mind would do better to ignore this alleged dispute and turn to still open and more interesting questions. A similar situation obtains, I propose to argue, with respect to the much debated question of the reduction of theories, in particular, the question of the reducibility or the irreducibility of biology to physics.

My exposition has three parts: first, an analysis of the present state of the controversy, in which we appear to find a fifth antinomy (not, however, quite antinomic in its structure); second, a suggestion of an alternative approach which, by treating the problem as a side-issue, manages to evade it; and third, a brief case-study in confirmation of the thesis of part II. No part claims any philosophical originality; the first is an attempt to sort out some of the current arguments in both sides of the issue; the second, an experiment in applying Rom Harré's theory of scientific inquiry to this particular problem; the third, an unphilosophical postscript. But I hope I can clear away some cobwebs so that we can look at more specific problems in the philosophy of biology and psychology without this impediment to our vision.

I

A. First, then, about the current state of the controversy. And first, here, a preliminary adjuration which *ought* to be unnecessary; we are not dealing with the old 'mechanism'-'vitalism' quarrel. This adjuration *is* necessary, unfortunately, in view of the crude reductivism of such statements, for example, as Crick's *Of Molecules and Men*:[2] the alleged 'vitalists' he is attacking – people who hold that there is some mysterious something about living things exempt from physico-chemical laws – are scarce on the ground these days. There is of course the strange evolution-ism of Teilhard de Chardin, but its import is, to say the least, ambiguous. It looks to many like a perversely mystical 'theory' of emergence; but as Huxley's enthusiasm for it indicates, it can also be taken as supporting, or at any rate as not contradicting, more orthodox biological opinion. An English version of Schubert-Soldern's *Philosophie des Lebendigen*, published in 1962 under the title *Mechanism and Vitalism*, does indeed represent a more traditional vitalist view;[3] but its quaint Aristotelianism is scarcely representative of the best anti-reductivist arguments by cither philosophers or biologists. And on the other side, the case made by most contemporary reductivists – notably by Nagel in *The Structure of Science*[4] or by Oppenheim and Putnam in their 'Unity of Science' paper[5] – is much more temperate than Crick's. In short, the issue is not quite what it was between Loeb and Driesch in the early years of the century. Both sides nowadays agree that living systems are made of the same kind of matter as non-living, and obey physical laws; the question is only: do those laws state the sufficient as well as necessary conditions essential to the descrip-tion and explanation of biological phenomena?

Leaving antiquated and extreme positions out of it, therefore, we may ask: how does the case stand nowadays for and against the reducibility of biology to physics? We can ignore chemistry, since those who hold the affirmative view consider chemistry *a fortiori* reducible – there must be *one* lowest-level set of laws into which all others are translatable and from which they can be derived.

As the arguments fly back and forth we seem to have at first sight something like a fifth Kantian antinomy. The thesis: biology *is* reducible to physics. The antithesis: biology is *not* reducible to physics.[6] Let us take, in Kantian fashion, first thesis and then antithesis as *dogma* and see what happens. The thesis in its dogmatic sense is vulnerable on two fronts, one epistemological and the other cybernetic.

The epistemological attack is an ancient one which can be stated in many forms. Let me remind you of two of them. The dogmatic reductivist asserts: Biology is reducible to physics – call this R. He will also assent to the formulation: it is true that biology is reducible to physics (it is true that R), or to the formulation: I know that biology is reducible to physics (I know that R).

What does R mean? It means that all scientific laws, including those of biology, can be translated into and are derivable from basic, universal laws of matter in motion. In effect, then, whatever one truly asserts about the world expresses, in the last analysis, some change in the configurations of matter described by physics. Usually such thinking is also particulate; that is, it is the laws of the fundamental particles of which 'matter' is composed from which the laws of macroscopic behavior are to be derived. (More of that later.) It makes no difference, however, to this argument whether or not one thinks of the fundamental level as atomistic: the point is that there is just one level of existence from the laws of which all the processes of all 'higher' levels or, if you like to avoid 'level' language, of all larger systems can be derived. But the reductivist's assertion is that R is itself an event, whether of emission of sound, or making of marks on paper. Therefore, it follows, like all other processes, from the laws of the fundamental level. The alternative formulations, likewise, 'it is true that R', or 'R is true', and 'I know that R' are equally necessitated by the laws of physics. But how can a truth claim or a knowledge claim be necessitated? That it cannot be so is plain from the fact that the contradictory 'claim', non-R, R is false, I know that non-R, enunciated by the antireductivist, is equally necessitated. In other words, to make a claim to knowledge is to perform an action, and, as has been argued *ad nauseam*, 'actions' demand reasons, rather than – or as well as – causes. It is the *reasons* for the act of claiming that R that constitute *evidence* for R as vs. non-R or non-R as vs. R, while both statement *and* denial are equally *caused*. Thus the reductivist's thesis, since it admits its contradictory as well as itself, contradicts itself; it can be uttered as a noise but not reasonably asserted as a claim to knowledge.

If the reductivist answers, moreover, that he must be right because, were we all mistaken in our scientific theories, natural selection would have long since exterminated us, the antireductivist has only to reply: here I am too! Either of us may be right, or neither, but we are equally successful. Nor is 'success' equivalent to truth: either or both of us may be guided by a useful illusion. Neither physical necessity nor practical success can give evidence for the truth or R or non-R.

This argument examines the status of the reductivist thesis in its own terms and reduces it to absurdity, much as Socrates does with the relativistic thesis of Protagoras in the *Theaetetus*. One can also show, more generally, that *any* statement presupposes the existence of more than matter in motion: that, in other words, if there is language, including physics, there must be systems (ourselves) capable of rule-governed behavior as distinct from aggregates of matter determined by physical laws (and again it makes no difference here whether the laws in question are deterministic or statistical). On this question the arguments of Chomsky and Chomskyan linguists seem to me definitive. In short, were reductivism true, knowledge would be impossible, including the knowledge that reductivism is true. And were reductivism true, language would be impossible, including the formulation of the reductivist's thesis.

The second argument depends on the claims of contemporary reductivists themselves. Since, they declare, the processes of growth and heredity have been shown to be determined by a sequence of DNA molecules, biology has already been reduced, on principle, to biochemistry; the completion of the job is routine. Granted, however, that DNA has precisely the power they claim for it, its operation demonstrates, on the contrary, that biology is *not* reducible to biochemistry (and ultimately, to physics). What makes DNA do its work is not its chemistry but the order of the bases along the DNA chain. It is this order which functions as a code to be read out by the developing organism. The laws of physics and chemistry hold, as reductivists rightly insist, universally; they are entirely unaffected by the particular linear sequence that characterizes the triplet code. Any order is possible physico-chemically; therefore physics and chemistry cannot specify *which* order will in fact succeed in functioning as a code. This argument, which appears incontrovertible, was stated by Michael Polanyi in *Science* in 1968.[7] The orderly structure of a chemical molecule, Polanyi points out, 'is due to a maximum of stability, corresponding to a minimum of potential energy'.[8] Such order, wholly determined by the laws of physics and chemistry and the boundary conditions of the system, is incapable of functioning as a code. The particular sequence of the bases on the DNA spiral, however, is not so determined. There is simply no question of energy here at all. One can say that, statistically, all sequences are equiprobable; any one, accordingly, is highly improbable, and the measure of this improbability is precisely the measure of the information it provides. Polanyi writes:

As the arrangement of a printed page is extraneous to the chemistry of the printed page, so is the base sequence in a DNA molecule extraneous to the chemical forces at work in the DNA molecule. It is this physical indeterminacy of the sequence that produces the improbability of occurrence of any particular sequence and thereby enables it to have a meaning – a meaning that has a mathematically determinate information content equal to the numerical improbability of the arrangement.[9]

We conclude, then, that the thesis contradicts itself whether as a claim to knowledge or simply as a piece of rational discourse, and the empirical evidence said most dramatically to confirm it proves in fact incompatible with it.

What about the antithesis: Biology is not reducible to physics? In other words, biology has its unique subject-matter and its unique methods which cannot on principle be translated into or derived from the laws of physics. To this statement three counter-arguments may be presented: First, there is an argument from the universality of physical laws. It runs: All material systems are governed by the laws of physics. All living systems are material. Therefore, all living systems are governed by the laws of physics. This too appears incontrovertible.

Second, moreover, there is the empirical evidence that analytical biology, which studies living systems in terms of their macromolecular parts, has in fact produced unquestionable advances of immense range and value. But the anti-reductivist, it is argued, holding as he does that it is 'wholes' not 'parts' that merit the biologist's study, would discourage such techniques and so would put a stop to research just where it is proving most fruitful.

Again historically, thirdly, science has advanced in virtue of the belief in a single scientific method and the ideal, at any rate, of a single unified body of scientific knowledge. It is the more wide-ranging hypothesis or the more powerful, i.e. more comprehensive, deductive theory that wins out in the end. Indeed, science originated from this demand for unity and cannot exist without it. As the gas laws have been reduced to the kinetic theory, so must all laws in the end be reduced to one fundamental plane. Such is the march of science! But the anti-reductivist's assertion that there are many sciences with many methods would cut off this essential *Leitmotif*. By telling him, so far and no farther, it would stifle the infant *Newton des Grashalms* in his cradle and prevent those future triumphs which might have continued to flow from the all-important drive toward unity.

B. Here, then, is our antinomy. But is it an antinomy? In one sense, yes:

both thesis and antithesis are valuable as regulative maxims. On the one hand, there is no end, on principle, to the knowledge to be gained by studying the parts of living systems by biochemical and physical methods, and on the other, there is no reason to cease and every reason to continue studying complex systems, including living systems, at their macrolevels and in their own terms. This point is made, for example, by Robert Rosen in a paper in *Journal of Theoretical Biology*.[10] Many living systems, he points out – e.g. the CNS – appear 'intractable', i.e. irreducible to least parts from which, and from the laws of which, those systems could be reconstructed. But it would be folly to wait for someone to show that such systems are, in his term, 'tractable'. They can be fruitfully studied by exact methods in terms of systems theory. Why forbid such investigations in the name of an abstract and unattainable 'unity of science'?

At the same time, however, there is an uncomfortable asymmetry about the two sides of our 'antinomy'. The reductivist thesis is refutable, yet to many irresistibly convincing. The arguments *against* reductivism are valid; yet the reductivist position fits so smoothly into what seems, and has been for countless scientists and philosophers, 'the world view of science' that the counter-arguments, for all their logical power, are in fact powerless against its authority. The arguments against the antithesis, i.e. against anti-reductivism, on the contrary, fail logically to refute it, and yet it is persistently difficult to maintain.

Let us look briefly at these alleged arguments. The first argument I mentioned depends on the fallacy of equivocation: 'All material systems are governed by the laws of physics. All living systems are material. Therefore all living systems are governed by the laws of physics'. If we substitute 'obey' for 'are governed by' we obtain a valid syllogism, which however, has no bearing on the problem of reducibility. All living systems do indeed obey the laws of physics, but without countervening the laws of physics they may well obey other laws as well. To say that 'all living systems are *governed* by the laws of physics', however, while it says at least that living systems obey physical laws, seems also to say that all living systems are exhaustively *explained* by them, i.e. that their regularities are functions of the laws of physics and of no other laws. That does not follow unless we know in addition that the laws of physics are the only laws there are.

The second and third 'arguments' against the antithesis, moreover, from the success of analytical biology and the power of the unity of science ideal, are not so much arguments as historical analogies. They may well serve to support the thesis as a methodological maxim, but they fail to contradict its denial as the epistemological and cybernetical arguments contradict the denial of the antithesis.

Yet the anti-reducibility position is, for many people, impossible to accept; it is too uncomfortable. Why? Because it breaks through the defenses of a simple, one-level physicalism without providing an alternative metaphysic to take its place. To think anti-reductively demands thinking in terms of hierarchical systems, of levels of reality and the like; but we don't know any longer how to think in that way – and to be told, even to *know*, that the contrary position is absurd does not in itself allow us to embrace wholeheartedly what ought to be the more reasonable alternative. For anti-reductivism is 'reasonable' only in the perverse sense that its negation is self-contradictory, not in the more substantive sense of fitting smoothly into a *Weltanschauung* in which, as people educated in the ideal of a 'scientific' world view, we can feel at home.

II

A.   How can we escape from this uneasy situation? There are two alternatives. One is to develop a comprehensive metaphysic consonant with the anti-reductivist view. There is one such system available, so to speak, in the twentieth century, and that is Whitehead's 'philosophy of organism', a system, however, with some grave weaknesses (notably the doctrine of 'eternal objects'). Another attempt at such a comprehensive ontology is that of David Bohm. Bohm's cosmology seems to be a kind of modern Spinozism, as rigorous, but also as obscure, as the original; even were it systematically stated, as it has not yet been, it would be unlikely to convince most science-oriented philosophers.

There is another possibility, however, and that is to undercut the conception of science from which the apparent antinomy flows. This too needs a fundamental conceptual reform, but at least it evades the leap into metaphysics and works instead by taking a new look at the methods of science itself. I am referring here to Rom Harré's self-named 'Copernican revolution' in the philosophy of science, which, in what follows, I shall try to apply to the problem in hand.

B. I can give only a very schematic account of Harré's position, partly because of the limits of space, but also because I know it only from a brief oral statement and from the manuscript of one chapter of its forthcoming book-length exposition.[11] But I hope I can say enough to show how it bears on our problem.

It has always been admitted in discussions of scientific explanation, Harré points out, that one should distinguish between the role of deductive *theories* and of theoretical *models*. Generally, however, the former have been held to be central, and the latter have been viewed as a sort of *ad hoc*, inferior addendum to them. This emphasis, he argues, is mistaken and misleading. It is chiefly through the search for and elaboration of models that science exists; compared with these activities, the so-called hypothetico-deductive method and the elaboration of formal systems of axioms or postulates and theorems deduced from them are relatively peripheral. His theory – or model – of model-building is careful and elaborate; I shall try to sketch in very rough outline its major features.

A scientific explanation, as Harré sees it, may be metaphorically described as a 'statement-picture complex' consisting of three major parts. First, there are sentences describing the puzzling phenomena needing to be explained. Second, there are sentences describing a model which might explain these phenomena, and thirdly, there are sentences belonging to other, non-problematic disciplines or sub-disciplines, or even areas of common experience, from which the model is drawn. Thus we must distinguish between the *subject* of the model and its *source*. These may be the same, but in most interesting cases are distinct. To explain the multiplication and variety of species, for example, Darwin introduced the model of Natural Selection, whose sources were the selection of favorable varieties by plant and animal breeders, the Malthusian doctrine and Paley's watch (though without its maker). In this case, as in many, the model was presented as a hypothetical mechanism which was said to stand to its subject as cause to effect. It is Natural Selection, says the Darwinian theory, which has *produced* the multiplicity of species. And if the model fits well – if, also, the sources it has drawn on are authoritative: indeed, a great many different factors may be operative here – but if, let us say, to put it broadly, the hypothetical mechanism is successful, it ceases to be hypothetical; it becomes an accepted fact. Descartes thought of the heart as a furnace; Harvey, as a pump. Descartes was wrong; it *is* a pump: it is the pumping action of the heart that *makes* the blood circulate. Such existential choices form, in Harré's view, the very heart of science: for scientists want to find out, in particular contexts, for particular ranges of phenomena, how nature really works – and sometimes, indeed, often, they succeed.

C.   The superiority of this model as an explanation of explanation leaps to the eye – but that is not my point here. The question is: what bearing has Harré's proposal on the problem of reducibility? Let us look at our problem in the light of the two concepts of scientific inquiry, the orthodox, hypothetico-deductive model, on the one hand, and, on the other, Harré's heterodoxy.

Hume's account of causal inference, he argued, in his chapter on reason in animals, was superior to its rivals because more phenomena were derivable from it: he had one theory for men *and* animals, not one for each. This demand for economy or 'simplicity' is a familiar one. Thus Reichenbach explained the superiority of general relativity, e.g. by pointing out that though we *could* keep adding additional forces to Newtonian gravity, it was much more efficient to think in more comprehensive terms, letting the inverse square law become a limiting case of a broader theory.[12] And of course that is how everybody explains the original power of the Newtonian synthesis itself. On the analogy of these examples, the ideal, therefore, is clearly *one* theory that will comprise *all* of science. This impulse to unification, we have already seen, is one of the primary objections to the antireductivist claim.

And, plainly, if there is to be *one* science, physics, which is the only universal science, is the one candidate. For since everything is made of matter, the laws of quantum mechanics, i.e. the fundamental laws of physics, plus initial conditions and boundary conditions, ought to give us the laws of all systems. On the other hand, the laws of special systems cannot as such be universalized. The laws of the nervous system, for example, or of the migration of peoples, cannot as such give us universal laws unless we can first break them down into, precisely, initial conditions, boundary conditions and the laws of physics – and then, those laws, the laws of nerve action or of migration, would have disappeared. So if science is to be unified, it is only through the reduction to physics that it can be unified. Such unification, the ideal of one great comprehensive theory drives us to attempt.

There is more to it than this, however: there are, and have been since Leucippus' time, a paradoxical pair of principles at work in the support of the reductivist's ideal. On the one hand, the orthodox view of scientific inquiry is presented as phenomenalism: science is shorthand for observations; theoretical constructs are conventions, enabling us to get from one stand in observation to another. And on the other hand, it is the smallest, invisible parts of things that alone are allowed to count as 'real'. Only an atomistic metaphysic, it seems, is *no* metaphysic, but a defense of phenomena *against* metaphysics. Thus when I tell my students, there are quail in my garden in the morning and robins in the afternoon, they say, nonsense, there is one gene pool in your garden in the morning and another in the afternoon. That's what you're *really* observing. In the name of scientific observation I must admit that my eyes deceive me: the real phenomena are the invisible ones demanded by the most unifying and most economical theory, the phenomena I *see* are only apparent and must be explained away.

Now I cannot indeed claim to understand this strange combination of principles; but I think we must admit that it has long been, and continues to be, with us as a powerful intellectual force, from Lucretius, through Hobbes, to Carnap or Nagel or Oppenheim and Putnam. On the one hand, science is not allowed to tell us about the real world, only about aggregates of phenomena, and, on the other, it is only allowed to tell us about the 'primary qualities' characterizing the least parts of which all 'things' are composed: all 'reduction' of one theory to another is and must be 'microreduction'. In psychology, indeed, in associationism and C-R theory (which are still surprisingly influential, e.g. in ethology) the two motives appear to coalesce in forming the unit of the psychological, or behavioral, atom; but, again, only a metaphysical bias could so persistently misread the phenomena as such theories have to do.

Now, admittedly, the best liberation from a monolithic metaphysic would be a better metaphysic – more 'adequate', to use an old-fashioned term, to experience, both everyday and scientific. But, for one thing, as I have already confessed, I have no such to offer; and for another – and this is more important here – it is the authority of science as (mis)understood by scientists and philosophers alike that forces on us the straitjacket of a universal physics; a different – and sounder – approach to science may liberate us from this cramping restraint.

D.   What becomes of the demand for reducibility if we think of science in Harré's terms? It becomes thin air into which it vanishes.

First, Harré's approach brings back into play a maxim contrary to the unity of science ideal, a maxim which might be christened the plurality of science. This is a maxim which Kant recognized – indeed, it is the principle of the antinomies: no knowledge can be established, or even sought experimentally, about the *whole* of nature. Given our finite powers, every investigation is and must be partial and perspectival. 'Objective' knowledge works through abstraction; it must, to be manageable, neglect some features of experience in order to establish others sufficiently exact for the scientist to manipulate. The data of science, moreover, vary indefinitely and are unbounded in their scope: only some segment cut out of them can be subjected to the scrutiny of any investigator or team of investigators at a given time and place. Thus on grounds both of method and content, science is partial and plural. This maxim, stressed also by Max Weber, is at least as essential to the practice of science as is the hope of unity. And it is primarily on this maxim that Harré's model of science relies. The subject of a theoretical model – the puzzling subject matter to be investigated – is always some special set of phenomena which, at some time in the development of some particular discipline, have aroused the interest of some particular investigators. Not that the ideal of unity is neglected either: for the *source* of a model may come from anywhere – as, e.g. cognitive psychology may rely on computer science; or Freudian theory on the physical concept of energy; or evolutionary theory on the work of the great cattle breeders – and so on. But this is an *open* unity, confined at the same time by the demands of a given problem. In this spirit, e.g. psychologists can use cybernetical models without thereby asserting a metaphysical dogma of 'mechanical man'. Thus Ulric Neisser writes:

There is an important place for eventual neurological interpretations of cognitive processes ... *but we should strive to establish a mechanism and discover its properties first.*[13]

In other words, let us look for a mechanism which might underlie the phenomena we hope to understand, seeking wherever we may relevant sources from which to derive, first, an analogue of a possible mechanism, and then, if we are shrewd and lucky and experience bears us out, maybe a description of the mechanism itself. There are no strict rules for this procedure, as Harré emphasizes; we have to rely on what he calls 'plausibility control'. The enterprise is personal and irreducibly pluralistic, but also open-ended. Anything *may* be relevant to anything: in *this* sense, science is one, but also many.

Secondly, in so far as the demand to reduce all theories to one level springs from a phenomenalistic program, Harré's revolution cuts it off at its root. The discovery of stable mechanisms in nature, not the summary of one flat level of pure phenomena, is what science is after. The inverse square law, or the principles of evolutionary theory, or the psychology of association, or the laws of good closure, or the kinetic theory of gases, may each embrace a wide range of phenomena in its explanatory scope. And from this scope in each case, the law or theory in part derives its explanatory power. A model that modelled just *one* particular phenomenon would be admittedly of little use. But on the one hand, no such law or theory – and no law or theory – comprehends all the phenomena of every kind that any scientist wants to, or might want to, explain. And on the other hand, the explanation works in each case not just by bringing together many observations into one otherwise meaningless and conventional formula. It works by leading us to see, in the case of a particular set of puzzling phenomena, *how in fact those phenomena are produced.* I remember hearing a speaker explain, at a symposium of biology teachers ten or twelve years ago, in the early days of DNA-RNA research, that what biochemists were doing was to ask 'what is the genetic material that in fact produces the effects which we see'? This is a case not of looking for a sentence or sentences from which the phenomena in question can be deduced, but of seeking the then unknown processes in nature which determine – in the sense of having the power to produce – certain perceptible effects.

Indeed, it is strange that the orthodox theory of explanation, which is phenomenalistic in intent, supposes itself also to be causal. To predict one darned thing after another is not in any sense to say *why* one follows the other. By themselves, a sequence of phenomena can give at most constant conjunction, not necessary connection. That is just what Hume discovered. What Humean 'necessary connection' adds to constant conjunction is simply the force of habit, or as Russell calls it, 'animal inference'. If *B* has always followed *A*, I may come, on *A*'s occurrence, to expect *B*, but such a sequence, on its own, is not explanatory: only the fitting of both *A* and *B* into a rational context enables one to say, in the light of *C*, when *A*, then *B*. Indeed, it has never been possible, in all honesty, to give a reasonable account of scientific method in pure phenomenalist terms. Even Hempel and Oppenheim, in their famous paper on explanation, admit that in addition to the observables in the case, we must have a law-like statement, from which they 'follow', and even (apologetically) one which is 'true'.[14] And Hempel in 'The Theoretician's Dilemma' is still wrestling with this concession.[15]

Admit, on the contrary, that scientists are trying, not to tie phenomena together into convenient bundles, but to look for the hidden mechanisms that produce them, and you are rid of the anomalies of the phenomenalist position – and at the same time of one of the prime motives for demanding universal reduction. For again, the central task of science in Harréan terms is the imaginative construction of theoretical models which suggest ways in which particular sets of phenomena may be *produced*. Such models are apposite for, and succeed in explaining, in the first instance, the phenomenon they were intended to explain – and perhaps, if they are powerful enough, others as well: but not *everything*. To find that the heart pumps blood explains a lot about the circulation but not, e.g. the action of the liver (nor even for that matter of the lungs – which was still a mystery to Harvey). To find that the anopheles mosquito carries *plasmodium malariae* explains how malaria is contracted, but not the cause of cancer. When we look at science from this perspective, the overabstract ideal of one unified system drops away, and with it the demand that all sciences be reduced to physics – and *a fortiori*, the need to combat that demand.

By the same token, finally, the dogma of microreduction lets go its stranglehold. On the model-model of scientific method, as we may call it, we can see both why microreduction is so powerful a tool of scientific advance, and why it has not been and need not be the single and exclusive technique for relating one science to another. Every model, we have seen, models its subject by drawing on some source, usually distinct from that subject. Such a model *may* just give us a new way of thinking about the subject (as light waves do for color), but often the model is proposed as a hypothetical mechanism which, if it exists, bears a causal relation to the subject as its effect. Dalton's atomic theory itself was such a hypothetical mechanism which has come, in a form severely modified by comparison with his own version, to be accepted as real. However intricate the modern 'atom', on the one hand, and however justified, on the other, is the movement away from particulate thinking in contemporary physics, the fact remains that for ordinary purposes common salt is correctly described as NaCl or water as H_2O (not HO as Dalton thought). There was substantial resistance to Dalton's theory as late as sixty years ago – that's another fascinating story which should be told in Harréan terms; the history of atomism is far from the triumphal progress Reichenbach supposed it to be. But let that slipshold remark suffice here; the point is simply that the decision has been made, there *are* atoms, not indeed as ultimate indivisibles, but as units of chemical reaction, and many properties of familiar compounds are successfully explained in these terms. Given this fact, plus the natural tendency of many minds (whatever the reason) to think of explanation as analysis, plus the association of corpuscular thinking with such names as Newton and Boyle, it becomes habitual to draw one's models for *other* sciences, more or less directly, from the successful principles of atomism. See Nagel's favorite case, the kinetic theory of gases.[16] See, in reliance on a more remote analogy, the cell theory, or the theory of the gene. One may even be tempted to assert, with Gillispie, that all advances in science are advances to, and in, particulate thinking, and all moves away from such thinking retrogressions.[17]

But that would be a serious error. Not only is it being impressively argued in some quarters that particulate thinking has partly been, and needs to be wholly, replaced in physics.[18] The reliance on atomism as a source of theoretical models has never been as exclusive or as complete as its defendants assume. Oppenheim and Putnam, for example, suggest in their 'Unity of Science' paper six levels of microreduction, of which the first three are: communities to living organisms to cells.[19] This is an oversimplification extreme to the point of absurdity. True, communities are made up of individuals; there is no existent superindividual such as Teilhard envisaged with his 'noosphere'. Yet ecology as the study of animal and plant communities is not a summation of the facts of individual behavior. Take a work like Fraenkel and Gunn's *Orientation of*

Animals, which is explicitly 'reductivist' in its import.[20] It is precisely the emphasis on individuals – who might be suspected, heaven forbid, of 'purposive behavior' – that these authors wish to avoid. A traffic jam, they point out, is not explained through any given drivers, or any group of drivers, trying to get to work; it is the overall laws of the collective that explain the situation, despite, not because of, the behavior of each individual. In general, I think it may be safely said, ecology draws its models from many sources: from thermodynamics, from statistics, from chemical engineering, from computer science, not by any means wholly or even primarily from the laws governing the community's parts. In particular, the theory of natural selection, although strongly influenced in our time by the particulate theory of the gene, constantly and necessarily escapes reductivist thinking. Darwin's orginal model, to begin with, by showing how predator-prey relations and more generally organism-environment relations generate increased adaptation to new environments, typically exhibits *hierarchical, not reductive* thinking. A hierarchy (I am following here Howard Pattee's description) is first a collection of elements such that the elements are subject to the laws peculiar to their level; while, secondly, the interactions between them produce patterns imposing restraints upon those very elements; restraints, thirdly, which must operate with a certain reliability if the hierarchy is to survive.[21] Thus even though there are no elements in the collection other than individual organisms and their physical as well as organic environments – there are only, for example, bees, clover, cats, mice and spinsters in Darwin's famous example – the interaction between these constrains the operation of the elements themselves in a way which from the study of the elements at their own level we could neither have predicted nor understood. Or to put it in the fashion beloved of contemporary Darwinians, only thinking in terms of populations, *not* individuals, allows us to formulate evolutionary theory. And if evolutionary theory is the proper framework of biology, as its proponents maintain, then biology is typically macro – not micro – reductive, or better antireductive in its chief import. Indeed, so powerful, because of its central role in this central theory, is the concept of 'population' that it can itself serve in turn as a source for other models: as in the study, through computer techniques, of 'populations' of muscle cells. Such thinking, like population-genetical theorizing in terms of gene pools, has of course the advantage of *appearing* also to be particulate and so acquiring some of the prestige of atomism: for it works in terms of collections of bits. Yet it is at the same time, and fundamentally, hierarchical thinking in so far as it is the laws of the *collection*, not of its elements, which are being investigated.

With respect to the next alleged reductive level, organisms and cells, the oversimplicity of Oppenheim and Putnam's hypothesis is even more apparent. As formulated, it has a quaintly nineteenth-century ring; but suppose we put it instead in Crickian terms, in terms of the reduction of biology to a molecular science, substituting the DNA molecule for the cell. Every molecule can then be studied in terms of physics and chemistry. True; there is no special 'organic matter'. But to function in such a way as to produce biological structures, DNA molecules, or any other molecules, must be programmed. And a programmed system, again, is hierarchical. It can be made, predicted or understood only in terms of engineering principles as well as of physics and chemistry. It is, as Polanyi calls it, a system with dual control. This has been argued conclusively not only in the paper I quoted earlier, but also, for example (if in a different spirit), by Richard Gregory in a paper on 'The Brain as an Engineering Problem'[22] or by Longuet-Higgins in Waddington's Theoretical Biology series.[23] Thanks to the prestige of reductivism, it is a difficult argument to grasp, and I shall not attempt to restate or reinforce it here. Let my report of the code argument, which is a corollary of it, serve as its representation. I only want again to point out that in Harré's terms this argument, otherwise so hard to swallow, presents no difficulty to our intellectual digestions. We are entirely at liberty to draw our models whence we please: why not from cybernetics or general systems theory if from such sources we acquire plausible suggestions for a mechanism which, if correctly guessed at, would in fact 'produce the effects that we see'? To define living systems, with Longuet-Higgins, as machines capable of improving their own programs may sound to 'organismic' biologists like one more reductivist slogan. Yet it says in effect: look to engineering, to blueprints and operational principles – *not* to chemistry and physics – for the sources of your theoretical models in biology, much as Darwin drew on the work of sheep breeders and pigeon fanciers as a source for Natural Selection. The concept of genetical selection works of course (more or *less* successfully), by combining atomic and selectional sources, and much biochemical research works, admittedly, by relying on particulate thinking – and hence on the success first of atomism and then of quantum mechanics – as its principal inspiration. But there is no reason why any discipline should draw all its models from one source. 'Contemporary biology', Robert Rosen remarks in the paper already mentioned, is still 'dominated by the viewpoint that a total understanding of biological activity is to be found in the systematic physico-chemical fractionation of organisms, and the intensive study of the resulting fractions by means of the standard techniques of physics and chemistry'.[24] But, he argues, such techniques of fragmentation have limits as well as strengths. Though reduction is never impossible *a priori* in any given case, what it produces is often artifact, so that the properties of the fractions of a system studied 'give no information at all concerning ... the overall system'[25] on which it was thought to bear. In such cases, in other words, to put it in Harréan terms, analytical techniques produce models, drawn, indeed, from an impeccable *source* but with limited if any bearing on their subject, models, moreover, which may well fail to give us information about their subject, precisely because they are of our own making; as natural mechanisms, they do not exist. True, that is no reason to abandon microreductive models, but it *is* good reason to supplement them with models drawn from other sources. Indeed, Rosen argues, such alternatives, by suggesting new questions, may shed important light on physics itself. In short, once one looks at scientific inquiry pragmatically and pluralistically, as Harré suggests, as an effort to solve particular problems by seeking wherever one plausibly can for hypothetical mechanisms that might explain them, there is no compulsion to look always to one source – and that a source which often supplies over-abstract and artificial analogies. It is true, of course, as Rosen concludes, that the analytical approach 'has already told us much that we wish to know about the intimate details of biological processes, and promises to tell us much more in the future'.[26] At the same time, the very achievements of this method tend to generate over-confidence in its power. 'Many important aspects of biological activity', Rosen argues, 'are certain to be refractory to reductionist techniques, and must be treated holistically and relationally'.[27] Moreover, these aspects, he believes, 'raise general methodological and system theoretic questions of the greatest interest.'[28] Thus it is important to the development of 'a true theoretical biology' to understand the limitations of analytical methods as well as their strengths.[29] Reductivist or analytical *techniques*, in other words, have been and will be of great heuristic value; the same is true, however, of systemic or relational techniques as well. Only an outworn and over-abstract theory of theories compels us wholly to exclude the second in favor of the first.

III

So much for my general argument. Let me conclude, then, by mentioning an example which strikingly supports my thesis: the study of perception, and in particular J. J. Gibson's *The Senses Considered as Perceptual Systems*.[30]

I have often wondered why eighteenth-century philosophers were so convinced that there must be minimal sensibles. Partly of course through the influence of the 'new mechanical philosophy', which in turn had been influenced by the empiricist transmutation of the Cartesian clear and distinct idea. But the good Bishop Berkeley, for instance, who was no great admirer of the new science, filled his commonplace book with the record of his search for the minimum visible, the minimum audible. Only through describing the aggregation of such minima, it was held, could one construct a true account of perception; only through an aggregation of sensations do we perceive. I need scarcely reiterate here the commonplace that the distinction between elementary recountable sensations and perceptions somehow put together out of them, whether by 'association' or 'unconscious inference', dominated psychology at least until the advent of the Gestalt school, and in many quarters even longer.

Now along comes Gibson and tells us that such elementary sensations don't exist – well, hardly ever. The senses, in their more interesting and 'normal' role, are not purveyors of sensations, but detection systems through which organisms sometimes passively receive, but more frequently and significantly act to obtain information about their environment.

How do the old and the new theories look in Harréan terms? The subject-matter, the puzzling phenomenon to be investigated, is the question: how we use our senses to find out about the world. The hypothetical mechanism traditionally suggested was based on a building brick model; its sources were presumably the prestige of atomic thinking in physics, the successes of technology and the like. And so convincing was this hypothetical mechanism in its day that the bricks it needed were taken as facts. There *must* be least sensations (in philosophical parlance: sense data), so there *are*. Or in terms of physicalism or classical behaviorism, there must be least units of the CNS, reflexes, out of which behavior is constructed, so there are. In the case of visual perception, moreover, there was one given, or apparent given: the visible image at the back of the ox's eye. Here it was; Descartes himself had seen it. Out of such items, therefore, visual perception must be built. And if the image exists in this case, so must it in others. As a source for interpreting the retinal image, further, there is physical optics: a branch of the one sacrosanct universal science. The statement-picture-complex was authoritative; and the existential choices it demanded, compelling.

Such choices, however, are not necessary. Other experimental work (Gibson cites Köhler, Michotte and Johannsen, for example) and the principles of other disciplines (evolutionary biology and ecology, for instance; perhaps also general systems theory and information theory) may suggest a very different hypothetical mechanism or even a set of hypothetical mechanisms, which, if they prove acceptable, entail very different existential choices – and suggest also the development of other disciplines which may in their turn serve as the source for further models.

I cannot fairly paraphrase Gibson's argument here, and I hope I need not. Let us take the hypothetical mechanism he proposes for vision as summarized in his chart of the perceptual systems.[31] The essential opposition to traditional theories appears most plainly under the column 'Stimuli available'. These are, in his view, 'the varieties of structure in ambient light'. These, he holds, are acted on by the 'ocular mechanism', which includes 'eyes, with intrinsic and extrinsic eye muscles, as related to the vestibular organs, the head and the whole body' to provide information about everything that can be specified by the variables of optical structure (information about objects, animals, motions, events and places)'.[32] The sources for this hypothetical mechanism, I have suggested, are evolutionary theory and ecology. It needs to be substantiated, its author admits, by the development of what the calls 'ecological optics', and this in turn has as its sources 'parts of physical optics, illumination engineering (again) ecology and perspective geometry',[33] all 'respectable' fields to which an experimental innovator may reasonably look for support. And of course the parallel hypothetical mechanisms for the other perceptual systems will also confirm and be confirmed by the acceptance of the visual system. Finally, the theory entails predictions[34] capable of experimental testing, which will strengthen or weaken the case for its acceptance and the acceptance of the existential choices which it carries with it. These include, for example, the denial of sensations as universally existent, the acceptance of immediate contact with the external world, the acceptance of the structures of ambient light as real aspects of the natural world.

These are all revolutionary changes; but why not? In terms of Harré's theory of scientific method, there is no one impeccable and universal source from which the scientist must draw his models. It is high time that our obeisance to evolutionary theory and to the importance of organism-environment relations should really direct our thinking and allow us to make more sense of the senses than the search for sensory atoms permitted. Here as elsewhere, reductivist techniques have been and may yet (in their proper place) prove fruitful, but here as elsewhere they often produce artifact.

NOTES

[1] C. Taylor, 'Mind-Body Identity: A Side Issue?', in *Philosophical Review* **76** (1967), 201–213.

[2] F. Crick, *Of Molecules and Men*, University of Washington Press, Seattle, 1966.

[3] R. Schubert-Soldern, *Mechanism and Vitalism* (transl. by C. E. Robin), University of Notre Dame, South Bend, 1962.

[4] E. Nagel, *The Structure of Science*, Harcourt, Brace & World, New York, 1961.

[5] P. Oppenheim and H. Putman, 'Unity of Science as a Working Hypothesis', in *Minnesota Studies in Philosophy of Science, II, Concepts, Theories, and the Mind-Body Problem* (ed. by H. Feigl, M. Scriven, and G. Maxwell), University of Minnesota Press, Minneapolis, 1963, pp. 3–36.

[6] Kant of course held that there would not be a Newton of a blade of grass; the question for him was not arguable. Had he thought it arguable, however, he would, one supposes, have reversed the two theses, taking reducibility, with its 'materialist' and anti-theistic implications, as antithesis and the more conservative alternative as thesis. But not only can I say what I want to more easily if I take it the other way around – I would also suggest that if in our situation any 'religious' bias comes into play, it is not so much the opposition of religion to science as the religion *of* science that is involved. What used to be the 'new corpuscular philosophy' has become the faith of the orthodox, which a more pluralistic conception of reality appears to challenge.

[7] M. Polanyi, 'Life's Irreducible Structure', *Science* **160** (1968), 1308–12. Essentially the same argument has been stated, for example, by David Hawkins in *The Language of Nature*, W. H. Freeman, London and San Francisco, 1964; by H. Pattee in *Towards a*

Theoretical Biology, I and II (ed. by C. H. Waddington), Edinburgh University Press, 1968 and 1969, respectively; a similar argument is stated by C. P. Raven in 'The Formalization of Finality', *Folia Biotheoretica B* **5** (1960), 168.

[8] Polanyi, *op. cit.*, p. 1309.

[9] *Loc. cit.*

[10] R. Rosen, 'Some Comments on the Physico-Chemical Description of Biological Activity', in *Journal of Theoretical Biology* **18** (1968), 380–6.

[11] R. Harré, *Principles of Scientific Thinking*, Oxford University Press, Oxford, 1970.

[12] Cf. H. Reichenbach, *The Philosophy of Space and Time*, Dover, New York, 1958.

[13] U. Neisser, *Cognitive Psychology*, Appleton-Century-Crofts, New York, 1966, p. 20.

[14] C. Hempel and P. Oppenheim, 'Studies in the Logic of Explanation', *Philosophy of Science* **15** (1948), 135–75, p. 137; this also appeared in *The Structure of Scientific Thought* (ed. by E. H. Madden), Houghton, Mifflin, Boston, 1960, pp. 19–29.

[15] C. Hempel, 'The Theoretician's Dilemma', in *Minnesota Studies in Philosophy of Science, II, Concepts, Theories and the Mind-Body Problem* (ed. by H. Feigl, M. Scriven and G. Maxwell), University of Minnesota Press, Minneapolis, 1963, pp. 37–98.

[16] Nagel, *op. cit.*, pp. 338–45.

[17] C. C. Gillispie, *The Edge of Objectivity*, Princeton University Press, Princeton, 1960.

[18] Cf., e.g., J. M. Burgers, *Experience and Conceptual Activity*, MIT Press, Cambridge, Mass., 1965 and M. Čapek, *Philosophical Impact of Contemporary Physics*, Van Nostrand Press, Princeton, 1961.

[19] Oppenheim and Putman, *op. cit.*, p. 9.

[20] G. S. Fraenkel and D. L. Gunn, *Orientation of Animals*, Clarendon Press, Oxford, 1940.

[21] Cf. H. Pattee, 'Physical Problems of Heredity and Evolution', in *Towards a Theoretical Biology*, II (ed. by C. H. Waddington), Edinburgh University Press, 1969, pp. 227–32.

[22] R. H. Gregory, 'The Brain as an Engineering Problem', in *Current Problems in Animal Behaviour* (ed. by W. H. Thorpe and O. L. Zangwill), Cambridge University Press, Cambridge, 1961, pp. 307–30.

[23] C. Longuet-Higgins, 'What Biology Is About', in *Towards a Theoretical Biology*, II (ed. by C. H. Waddington), Edinburgh University Press, 1969, pp. 227–32.

[24] Rosen, *op. cit.*, p. 380.

[25] *Ibid.*, p. 386.

[26] *Loc. cit.*

[27] *Loc. cit.*

[28] *Loc. cit.*

[29] *Loc. cit.*

[30] J. J. Gibson, *The Senses Considered as Perceptual Systems*, Houghton Mifflin, Boston 1966.

[31] *Ibid.*, Table 1, p. 50.

[32] All of the above are taken from Gibson's Table 1, *loc. cit.*

[33] *Ibid.*, pp. 221–2.

[34] *Ibid.*, see for example p. 215 on perception of the blackness of a surface with varying illumination, depth, information and so on.

CHAPTER V

ARISTOTLE AND MODERN BIOLOGY

I

Science had its origin, if not in opposition to Aristotle, at least in opposition to Aristotelianism. But science in its most authoritative form was what came to be called physics, not biology. Faced with the recent crisis in biology, in which the life sciences have been threatened with reduction to microversions of themselves and ultimately to chemistry and physics, one wonders if the besieged biologists, or at any rate their philosophical defenders, might not after all learn something to their advantage by reflecting on the one great philosopher who was also a great biologist. And we *can* learn from Aristotle; not, however, in a simple or straightforward fashion. There is no use just contrasting, as some have been tempted to do, Democritean with Aristotelian science and putting physicists in the former class, biologists in the latter. Even if we reject Simpson's alleged reaffirmation of Roger Bacon and stoutly deny that 'the study of Aristotle increases ignorance',[1] we must nevertheless admit that in some important respects biology, like all modern science, really is, and must be, un-Aristotelian. This thesis could be defended in a number of ways; let me select four.[2]

First, the role of abstraction and the relation of scientific reasoning to everyday experience differ deeply in Aristotelian and in modern science. Only mathematics, Aristotle insists, abstracts *from* most of the ordinary perceptible properties (qualities, relations, states) of the things around us. Natural science, as he understands it, remains *within* the framework of everyday perception and makes more precise, within that framework, our formulation and understanding of the essential natures of quite ordinarily accessible entities. Modern science began and continues, on the contrary, precisely by closing its eyes to all but a few highly selected features of the world around us, abstracting from all but those variables which give promise of susceptibility to some sophisticated, usually quantitative, manipulation by the experimenter or the theorist.

Secondly, because he sticks so closely to the concrete, limited, and limiting physiognomy of things encountered in the everyday world, the Aristotelian scientist never confronts the teasing problem of induction which the modern scientist, or better, philosopher of science, necessarily has to face. The possible data of scientific calculations are infinite, the calculations themselves and their results are finite. However ingenious the arguments that have been used to comfort us before this gap, the gap remains. For Aristotle it does not exist. Admittedly even the most ingenious experimenter must rely, in Aristotelian fashion, on the stability of his surroundings: on his materials, his apparatus, as being reliably not simply this-here, but this-such.[3] Every time you pick up a handful of $CuSO_4$ crystals, Norman Campbell argued, you are in effect acknowledging a *law* of nature: you confidently expect these blue crystals to have the same chemical properties as they did yesterday and will tomorrow.[4] Confidence in such stabilities does indeed depend, if you like, on Aristotelian induction, where we move from a rough and ready perception of the character of a thing to a more precise knowledge of just what kind of thing it is. But – *pace* Hume – it is only after this everyday induction that the problem of induction in science first begins: that is, the problem evoked by the necessary disproportion between data and hypothesis, between evidence and conclusion. Just because it remains within the horizon of everyday things-to-hand, Aristotelian science evades altogether the problem of induction in this its modern form.

A third way to emphasize the same contrast is to stress the role of productive imagination in scientific discovery. Aristotelian *phantasia* is powerless to excel what Kant called reproductive imagination. The productive power of that faculty, not *a priori* and once-for-all, as in the Schematism of the *Critique of Pure Reason*, but advancing hazardously to and beyond ever new frontiers: that is the moving force behind the scientific adventure, a force wholly beyond the ken of the much cosier *Prinzipienforschung* of Aristotle, cradled as it is within the comforting embrace of the familiar everyday world.

All this concerns Aristotelian methodology. Cosmologically, too, and especially for biology, there are features of Aristotelian science which the modern mind radically rejects. Aristotle's world is finite, unique, eternal, consisting of a finite number of eternally existent species, 'endeavoring' in their re-production to simulate the eternal circling of the celestial

spheres (so he argues at the close of the *De Generatione et Corruptione*[5]).
For modern biology, this eternal frame is shattered. All things flow. For
many modern biologists, indeed, the theory of evolution is comprehensive
for their science; all biological research, they feel, somehow derives from
and contributes to it. That claim may be exaggerated. But even research
seemingly unconnected with evolution is nevertheless related to it in some
degree, as figure to ground. Some of Aristotle's detailed biological work
too was sound and still retains its validity despite the incorrectness of
his cosmological theory, e.g., his description of chick development or his
account of the life history of *Parasilurus aristotelis* or of *Mustelus laevus*.
Much modern research, similarly, may reach correct conclusions independently of the general conception of life's development from the non-
living or of the transmutation of species. Yet the overall thrust of modern
biology has been radically altered by the idea of evolution, much as the
Aristotelian science of nature was guided by the contrary view of a static
and finite cosmos. In their overall implications, the two are incompatible.

Despite these contrarieties, however, there is much to be learned from
Aristotle in relation to the philosophical problems of biology. I want to
discuss in this context three Aristotelian concepts: τέλος or the οὗ ἕνεκα,
that for the sake of which; εἶδος in contrast both to ὕλη and to γένος,
that is *eidos* as form and *eidos* as species, and finally that most puzzling
of Aristotelian phrases, τὸ τί ἦν εἶναι, the 'being-what-it-is' of each kind
of thing.

II

First, *telos*. Again, it was Aristotelian teleology that seventeenth-century
innovators were most emphatically determined to abolish from the study
of nature. And again, as Aristotelian teleology had come to be understood, or misunderstood, this was a necessary move. Yet some sort of
teleology, or teleonomy, as some modern biologists prefer to call it, keeps
creeping back into biological language and thought. Let us consider,
then, how and where something like Aristotle's *telos* occurs in modern
biology and how modern usage compares with his.

Two misconceptions must first be set aside. (1) *Telos* is not in the first
instance – and in the study of nature is not at all – 'purpose or plan'. In
nature, 'that for the sake of which' a series of events takes place is the

intrinsic endpoint in which, if nothing fatally interferes, that series normally culminates. What usually happens to a fertilized robin's egg, for instance? A baby robin hatches out of it; that is its *telos*. There is absolutely no question of any kind of 'purpose' here, either man's or God's. To suppose otherwise is to introduce a Judaeo-Christian confusion of which Aristotle must be entirely acquitted. (2) Nor is Aristotle interested primarily in one over-all cosmic *telos*. Despite the passage from *De Generatione et Corruptione* already referred to, and despite the 'teleological' causality of the unmoved mover, the kind of 'ends' that usually interest Aristotle are the determinate end-points of particular processes within the natural world. True, the stability of the universe is, for him, a necessary condition of the orderly processes of its components; but this is by no means the target of his primary interest. On the contrary, it was, again, the Judaeo-Christian God who (with the help of neo-Platonism) imposed the dominance of a cosmic teleology upon Aristotelian nature. Such sweeping purpose is the very contrary of Aristotelian.[6]

The concept of *telos* in exactly Aristotle's sense, however, does occur in exactly the area where he himself invokes it: in the study of ontogenesis. Here, indeed, the contrast of Aristotelian and Democritean science still appears valid at least to a first approximation. The embryologist must put questions to the living embryo, in terms to which, and in which, it can respond.[7] And this is impossible on principle in terms of a thoroughgoing atomism, since in those terms there *is* nothing alive. What might happen to a *really* Democritean scientist faced with a biological problem was suggested by Frank Baker in a paper I have quoted elsewhere. Imagine, he says, an observer looking in the field of a microscope at the filaments of a fungus. 'He witnesses', says Baker,

that at the tips of the filaments are disposed a number of radiating branches more frequently segmented than those of the stalk to which they are attached; which adjoined elliptical segments are easily set free by pressure in the surrounding medium. But, supposing that he decides to investigate these segments, what kind of ideas are going to control his choice of further observations; how will be proceed, loosely speaking, to discover 'the nature' of these structures; and, in brief, in such a context, what does this notion of 'investigation' already imply?[8]

An old-fashioned chemist, Baker suggests, might throw these segments (which, unknown to him, we may call spores) into concentrated H_2SO_4. He would learn something; but would this 'lay an effective basis to the study of mycology'? And why are these not as good facts to start with as any other? But suppose our investigator places the segments on jam (where he first found them) or sugar, and in the warmth, and watches what happens. He may then discover 'their relation to the life cycle of the fungus'. Actually, Baker remarks, it would take a whole series of investigators 'animated by a single scientific impulse or tradition' to lay the foundation which would so much as show him where to start. When he gets this far he can then, but not before, undertake his chemical analysis. Only, in other words, when the concept of germination is understood and its designatum assumed to exist, do the details 'fall into order and acquire a significance', such that detailed analysis of some parts of the process of germination can be undertaken. The orderly development of the organism under investigation must have been assumed, Baker concludes, before the right 'facts' could be selected for further investigation and analysis.[9] An orderly development toward a normal end, therefore, is necessarily presupposed by the biological investigator before he can set out to make his investigation. A concept of 'that for the sake of which' the development is occurring, of its natural *telos*, is contained in the very question asked.

Such considerations place teleological (or teleonomic) thinking in the position of at least a *regulative* idea (in the Kantian sense) at the beginning of biological research. From this point of view *telos* is a signpost to the study of nature: a 'reflective concept' (*ein Reflexionsbegriff*) as Wieland argues.[10] Looking at the endpoint of the series helps us to start looking for its necessary antecedents; there is nothing 'unscientific' about this, not even anything very un-Democritean. But is that all there is to it? A Kantian regulative idea – say, the infinite divisibility of matter, or indeed natural teleology as Kant conceived it – is a pure *as-if*. And many modern thinkers would be content with this, with 'the appearance of end', as Waddington calls it.[11] It's a makeshift, they say, a crutch to lean on until we have mastered the necessary and sufficient conditions, or until we have constructed a machine to simulate an animal, or until we have synthesized life. Then we can throw away the crutch and walk alone.[12] Or perhaps, as Piaget argues, the very idea of a final cause, even in this seemingly harmless form, is based on a logical error and we don't need it at all. It rests, he maintains, on a confusion of three different notions: physical or physiological causality (cause *a* produces effect *b*), logical implication (the use of A implies the consequence B), and instrumentality (to get B 'we must' use A).[13]

Both these views are mistaken. To assert that a robin's egg hatches out a robin and not an oak tree is to state not a regulative idea but a fact of nature. Nor, *a fortiori*, does such an assertion represent any logical howler. At the least, it locates in the real world an orderly process, the details of which the biologist may undertake to study. It selects a certain segment of orderly temporal process *in* its orderliness as the locus of an inquiry. To this extent at least it locates real, not apparent, ends and suggests really, not seemingly or misguidedly, teleological questions.

But is that all? If there are real processes with natural endpoints, real τέλη in nature, are there not also teleological *answers* to the questions we put to nature? This seems to me a much more difficult problem. Professor C. P. Raven of Utrecht has written of the application of information theory to biology as 'the formalization of finality', and I had formerly taken him at his word.[14] Yet now I wonder. What is 'finalistic' about information theory? Admittedly, both teleological explanation and cybernetical explanation are complementary to classical causal explanation; in this sense, both resist a one-level, Democritean approach. But that does not make them equivalent. I prefer to leave this question open here, and postpone a consideration of multilevelled explanation in general until I come to the concept of *eidos*. For the moment we may take it that in the study of individual development, the concept of a normal endpoint of development helps methodologically to locate the place of the inquiry and to locate it really in nature, whether or not the concept of *telos* is also embedded in the solution to the embryologist's problem.

III

The most vexing problem with respect to Aristotelian *telos* and modern biology, however, concerns not ontogeny, but phylogeny. Granted, there is a definite, if perhaps limited, role for teleology (or teleonomy) in the study of ontogenesis; can one transfer the concept of end to the study of evolution, or can one at least discover in evolutionary explanation an analogue of teleonomic thinking? It has repeatedly been claimed both that Darwinian evolutionary theory rejects any cosmic *telos* and that it retains the concept of *telos* in some more acceptably 'scientific' sense. Indeed, it has even been argued that it is precisely by virtue of its teleological structure that evolutionary theory, and only evolutionary theory, rescues biology from reduction to physics and makes it 'an autonomous science'.[15] These claims may perhaps be clarified if we compare the 'teleonomic' thinking of modern evolutionists with Aristotelian teleology.

The fixed endpoint of a natural process, for Aristotle, is the mature form of the adult individual (strictly, of the adult male!) of the species in question. 'Nature is like a runner', he says, 'running her course from non-being to being and back again'.[16] The being in the case is the developed adult of such and such a kind. In modern evolutionary biology, however, there is no such fixed form; the *eidos* itself, which is the *telos* of individual development, is transitory. What remains? For Darwinism, the *telos* that remains when the eternal species is removed is simply survival. Survival of what? The individual perishes; what survives? In Darwinian terms: the descendants of the slightly more 'successful' members of a species. In neo-Darwinian terms: the alleles which made possible the development of phenotypes carrying the slightly more 'successful' characters, or rather, statistically, a higher ratio of those alleles in the gene pool of the next generation. A robin's egg is not, it seems, the way to make a robin, but a robin is a robin's egg's way to make more, and more probably surviving, robin's eggs. The locus of the goal of biological process is not, as it appeared to Aristotle, in the mature individual, who is, as such, mortal and of no concern to evolution, but in the future gene pool of the population of individuals of a potentially interbreeding population.

Take, e.g., Kettlewell's classic study of industrial melanism.[17] If tree trunks are blackened by industry, birds take more peppered moths, and hence more genes for peppered wings, than they do the *carbonaria* mutant of the species. Hence, fewer mutant genes are eaten and proportionally more survive. The *telos* of this process is the greater ratio of *carbonaria* genes in the next generation. Evolutionarily speaking, that is what the whole business of being a moth is *for*; not, indeed, just 'for' the survival of *this* gene, but if we had a complete count of all the genes in the population at time t_0 and time t_1 the differential ratio would give us the 'end' of the story: the differential survival of some genes rather than others.

But which genes? Whichever ones survive, of course. If we clean up industry and the tree trunks bleach again, more peppered and fewer *carbonaria* genes survive, and the endpoint goes the other way. Similarly in peacetime healthy human males are, other things being equal, better adapted – and that means of course in evolutionary terms more likely to leave descendants – than sickly ones, but in war time the contrary holds: the halt, the lame, and the blind are better adapted than the healthy. Biological process is first and last and always evolution; evolution is first and last and always a chronicle of survival, the survival of whatever survives.

A strange *telos*: we are told simply, what survives survives. But this, it has repeatedly been objected, looks like a mere tautology. And at first sight, at least, it has also been repeatedly objected, a tautology seems to have no explanatory power, let alone the explanatory power that would be characteristic of a teleological account. For a teleological account distinguishes, and sets out as aimed at, a goal to which it can then relate the antecedent steps. Here, however, goal and steps are collapsed into an empty identity.

Yet that identity, we are told, presides over a rich and precise elaboration of 'evolutionary mechanisms', and hence of teleonomic patterns of structure or behavior. The case for this view is argued in a thought-provoking book by George Williams, *Adaptation and Natural Selection*. Although he accepts, and celebrates, the tautological character of the principle of natural selection, as the survival of the fitter in the sense of the more probable survival of what will more probably survive, he insists nevertheless that this principle, correctly used, can preside over a vast range of teleonomic investigations:

A frequently helpful, but not infallible rule is to recognize adaptation in organic systems that show a clear analogy with human implements. There are convincing analogies between bird wings and airship wings, between bridge suspensions and skeletal suspensions, between the vascularization of a leaf and the water supply of a city. In all such examples, conscious human goals have an analogy *in the biological goal of survival*, and similar problems are often resolved by similar mechanisms. Such analogies may forcefully occur to a physiologist at the beginning of an investigation of a structure or process and provide a continuing source of fruitful hypotheses. At other times the purpose of a mechanism may not be apparent initially, and the search for the goal becomes a motivation for further study. Adaptation is assumed in such cases, not on the basis of a demonstrable appropriateness of the means to the end, but on the direct evidence of complexity and constancy.[18]

The study of the lateral line of fishes, he suggests, is a good example of this kind of reasoning:

> The lateral line is a good illustration. This organ is a conspicuous morphological feature of the great majority of fishes. It shows a structural constancy within taxa and a high degree of histological complexity. In all these features it is analogous to clearly adaptive and demonstrably important structures. The only missing feature, to those who first concerned themselves with this organ, was a convincing story as to how it might make an efficient contribution to survival. Eventually painstaking morphological and physiological studies by many workers demonstrated that the lateral line is a sense organ related in basic mechanism to audition (Dijkgraaf, 1952, 1963). The fact that man does not have this sense organ himself, and had not perfected artificial receptors in any way analogous, was a handicap in the attempt to understand the organ. Its constancy and complexity, however, and the consequent conviction that it must be useful in some way, were incentives and guides in the studies that eventually elucidated the workings of an important sensory mechanism.[19]

How does this kind of teleonomic thinking compare with the use of *telos* in Aristotle? Aristotle presents his concept of 'that for the sake of which' as a guide to the study of nature in opposition to the thinking of Empedocles, who would elicit the phenomena of the living world, without ordered ends, out of a combination of chance and necessity. At one stage in cosmic history, Empedocles imagines, there were heads and trunks and limbs rolling about the world. Those that happened to come together in a viable combination survived; the others perished. This was a very crude theory of natural selection, to be sure, but a theory of natural selection, nevertheless. Aristotle as a practising biologist objected: ox-headed man progeny and vine-bearing olives, such as Empedocles envisages in his transitory world, are an absurdity. What we *always* have in nature is the ordered passage to a definite endpoint: man to man, cattle to cattle, grape to grape, olive to olive. Only where there are such functioning, ordered series does the study of life *begin*. Williams would agree. Where we can use only the concepts of chance and necessity, he insists, we should. Thus the descent of flying fishes can be explained in terms of physics alone; their flight, however, which is 'contrived', in analogy to human contrivance, needs, he argues, another and teleonomic principle of explanation, as any piece of machinery does.

So, as we saw with the case of ontogenesis, we need, it seems, a teleological approach to locate a biological problem. But is the explanation, in the case of selection theory, teleological as well? Have we even found, as in individual development, a directed process to describe – however we may eventually explain it? I think not; for explanation in terms of orthodox evolutionary theory collapses pretty quickly into pure Empedoclean chance-times-necessity. Had fishes not had the sensory mechanism of the lateral line they would not have 'heard' their predators coming and would not have survived. Or better: those whose 'hearing' was slightly more acute left descendants in the gene pool, those not so gifted left fewer and finally none. Chance mutations necessarily sorted out by the compulsions of environmental circumstances: that is a pure Empedoclean, anti-teleological process. The peppered moth case is a striking example. We have here, we are told, 'evolution at work': now we see the whole process in little. Extrapolate this 'mechanism' to the whole story of life and you have the vast panorama before you: no other principles are needed. But what is this story? The environment, for extraneous reasons (in this case the industrial revolution) changes; the gene pool is always changing; the changed environment necessitates changed predation (birds can't as easily see black moths on black trees as peppered ones); changed predation necessitates differential survival of some genes rather than their alleles. So we *necessarily* get more black moths than peppered ones. Extrapolate this process to the whole of evolution and you see a vast sequence of necessities. True, the sequence is triggered, and kept going, by a set of curious chances. These, however, are 'chances' only in the sense of being at a tangent to the 'normal' sequence of development. They are to be explicated, on principle, in terms of natural, that is, physico-chemical laws. Thus, given the nature of bituminous coal, the tree trunks had to be blackened. Given the chemical nature of DNA, one supposes, the 'errors' which occur in its replication will ultimately be explained also in physico-chemical terms. They happen by 'chance' – just as in Aristotelian chance – only in the sense that they are outside the usual sequence of events to which one has been attending, in this case the 'normal' development of the peppered moth. But they have, or will have, their physico-chemical explanation, which must ultimately exhibit their necessary occurrence. Again, extrapolate this reduction to the whole history of nature. Where is the teleology? It has served as a heuristic maxim to start us off on our inquiry, but in the sequence of survivals *it* does not survive even as a factor in the phenomena described, let alone as an explanatory principle.

Is there any other way to introduce teleology into evolution? Many philosophers and some biologists have tried to do this in terms of a theory of 'emergence'.[20] But these theories, compared either with the precise and limited teleology of Aristotle or with the vanishing teleology of Darwinism, are vague, and empty of explanatory power. If one says, for example, with Vandel (following Bergson) that life has moved toward an increase of 'le psychisme', with two high points, the insects (generic inventiveness) and man (individual inventiveness), two objections at once arise.[21] First, some of the diverse branches of evolution have gone that way, but not by any means all. What of parasitism, what of long stable forms like Lingula, what of the vast variety of birds or 'lower' mammals, what of the evolution of plants, etc., etc.? Second, even if we can see, very generally, some such tendency in the history of life on earth, *how* did it happen? What does such an assertion of the 'emergence' of psychic powers explain, and how? This appears an even stranger extrapolation than the Darwinian. For one *can* imagine the melanism story stretched back to the beginning of time. I suspect (indeed, I have argued elsewhere) that this extrapolation entails untenable pseudo-reductions of richer to poorer concepts; but still one can see how it's done. The 'emergence' extrapolation, however, I for one simply cannot follow at all. There is a goal, mind or thought or inventiveness, we are told, *for* which evolution happened. It is the achievement of this goal that we are studying when we look at evolution's course. But *whose* goal? *Whose* achievement? The giraffe, we know *contra* Lamarck, didn't get a longer neck by *trying*; and are we to believe that the brachiopods tried to achieve thought and left it to us to succeed, or tried to achieve social rituals and left it to the ants to carry them through? Achievements must be some *one*'s achievements. A goal, even if it is an Aristotelian *telos*, not a conscious purpose, must be the endpoint of some *entity's* becoming. Whose achievement is evolution? Whose goal, on an evolutionary scale, is thought? In any terms available to this writer at least the very question is nonsense. The concept of *telos* is intelligible and useful, I submit, only with reference to something already in existence. In the study of evolution, on the contrary, where we have no fixed individuals and therefore no fixed endpoint of process, we have, whether in the Darwinian view or in the efforts of 'emergence' theorists to revise it, only the appearance of teleology, not its flesh and blood.

IV

Still, that appearance keeps reappearing. Why? In emergence theory it is a case of metaphysical aspirations as yet unfulfilled. All honor to them; it may well be that this controversy will only come to rest once one has accepted a cosmology of some Whiteheadian kind. This, to most of us, has not yet happened, certainly not in such a fashion as to affect the practice or the thinking of biologists. But why does neo-Darwinism, as distinct from those broader and vaguer views, recurrently lay claim to being teleonomic in its structure?

The answer is not far to seek. To give it will permit one more comparison with the teleology of Aristotelian science.

Darwinian evolutionary theory *appears* teleological because it is first and last a theory of adaptation. Deriving from Paley, it views all organisms as adaptation machines, aggregates of devices for the adjustment of the organism to its environment. On this view it is, as Williams insists, thinking in terms of adaptation, and this alone, that distinguishes biology as a science from physics and chemistry. Yet adaptation in evolutionary terms is for survival and survival only. Everything non-trivial in specifically biological processes reduces to this one phenomenon. But explanation in these terms, as we have already seen, either collapses into tautology or is reduced to necessity, and so in either case fails to retain its alleged teleonomic structure.

How does this situation compare with that of Aristotelian teleology? First, in Aristotle we find for each kind of thing a given normal endpoint of development, and relate to it a set of what Aristotle calls 'hypothetical necessities'. Given, for instance, that a creature needs to hear – or be somehow sensitive to environmental vibrations – it will develop *some* kind of auditory organ, whether a vertebrate ear or a piscine lateral line. In the modern version, however, such hypothetical necessities become simple necessities. Since the endpoint to which one might refer them as means is not fixed in advance, it becomes simply the ineluctable issue of the preceding steps. Instead, therefore, of the necessary conditions being relative to the end, the end is the automatic product of its necessary (and sufficient) conditions. In Aristotle, secondly, the end being given, its achievement happens 'always or for the most part' – but it may fail. All along the way, there is room for abnormality, for chance. In modern theory, however, there is no such leeway. Even though, in terms of our present knowledge, most mutations may be 'chance' occurrences, that is so only in a sense analogous to Aristotelian *tyche*; that is, they are caused by some cause and effect sequence outside the usual pattern of development in the given case. But they are not – or cannot survive – as *to automaton* in Aristotle's sense, as sheer random happenings. Did they not prove to be adaptive as the environment changes, natural selection would eliminate them. And if they do prove adaptive, on the other hand, they have to survive. Depending on how you look at it, in other words, everything is teleological (adaptive) or everything is necessary. There is no middle ground for the merely contingent in the interstices of an otherwise orderly sequence.

It is just this two-edged comprehensiveness, finally, combined with the authority of an algebraic formulation, that lends to modern natural selection theory its great explanatory power. First, there is a formal, mathematical instrument, the algebra of Fisher or Haldane, which may be used to measure natural selection and which lends weight and precision to experimental results in this field. Such a formula, intrinsically tautological, is used to measure changing adaptive relations and therefore serves the appearance of teleology. But these relations in turn, when viewed as a series of organism-environment interactions, appear thoroughly necessitated, not teleological at all. When, however, such interactions are summed up, for long periods, in algebraic formulation, the results, neatly ordered, present apparent trends and thus once more give the appearance of teleology, an appearance once more reduced to necessity when we visualize the whole sequence of action/reaction from which they have eventuated. Thus if one accuses Darwinians of being 'mechanistic' they point to the 'trends' in evolution as they see it; if one accuses them of being 'teleological' in their thinking, they point to the necessity of the whole show in terms of environmental pressure and the consequent changes in gene ratios. And if we try to bring this to-and-fro to rest, what have we? Once more, tautology: well, after all, what survives survives. If you look at the process *a tergo* it appears teleological; if *a fronte*, necessitated. And *sub specie aeternitatis*, when the theory is summed up in a formula for measuring differential gene ratios, you have a theorem universally applicable because empty, totally comprehensive because it expresses a simple identity.[22]

Why do we keep going round this merry-go-round? 'Adaptation' is a means-end concept. Yet if all adaptations are for no specifiable end except survival, one keeps falling back into a universal necessity which is in turn reducible to the same old tautology. Stretch it out: it's teleology. Collapse it one level: it's necessity. Collapse it still further: it's tautology. What is lacking to stabilize this endless vacillation? This brings me at last to my second major Aristotelian concept: *eidos*. For what is lacking in the modern concept of adaptation is precisely the definite *telos*, which in Aristotle is the mature form of the species, of the type, the τοιοῦτον of the τόδε τι.

Nor, of course, do I mean by this some cosmic goal. Again, there are no such goals within Aristotelian biology. True, there are such in Aristotelian cosmology, as e.g. in respect to the proof of the unmoved mover. In the framework of biological investigation, however, we need not, indeed we may not, invoke such dialectical arguments. Modern biology, however, lacks even the more limited and concrete end-points of Aristotelian science. In short, when Darwinism evicted the watchmaker of Paley's famous watch, it threw out as well the *telos* of the watch itself. But without a *terminus ad quem* of development, without a *terminus ad quem* for our understanding of the organization of a living system, of an organism, of an organ, of an organelle, one has no univocal concept of adaptation, of the adjustment of these means to that end. True, the organism, the organ, the organelle is continuously adapting itself to its environment, both internal and external; but what for? To what end? In ontogenesis, to the end of maturation and self-maintenance of the organism, the organ or the organelle, that is, to the end of the origination and conservation of some *form*. The processes of adaptation, as distinct from their result and adaptedness, are thus related as dynamic processes to their goal, that is, to the actuality of the organized system which comes into being or maintains itself in being through those processes. And the result, adaptedness, as differentiated being-such-and-such of the parts of the organism, the organ or the organelle, is also subordinated, therefore, to the development or maintenance of the form, the *eidos*, of the whole. It is, then, precisely the Aristotelian concept of form, or some modern analogue thereof, which is lacking in the modern concept of adaptation, or better, of the organism as a pure aggregate of adaptive mechanisms.[23]

V

Let us look a bit more closely, then, at the relation between modern biology and Aristotle's concept of *eidos*.

Perceptions of form – the shape of an oak leaf, the walk of a cat, the metamorphosis of a butterfly – the grasp of such configurations and changes of configuration, are among the basic insights by which the subject-matter of biology is singled out. A certain freedom of form within form, Buytendijk has shown, is the criterion by which we see a shape as 'alive'.[24] Biological knowledge, the knowledge of men like Ray or Hooker, was a refinement of such elementary perceptions, a refinement to the point of genius, but not different in kind from its everyday counterpart. Modern biologists, however, at least the more theoretical, and more articulate, among them, sharply reject such old-fashioned connoisseurship. Form, and the recognition of form, are not only not (according to their own professions of faith) their central concern; they exhibit a positive dread of form. In the polemics that characterize contemporary taxonomy, for example, and in evolutionary controversy also, the favorite epithet of the combatants is 'typology'. To call a man a typologist is the worst insult you can bestow.[25] It is hard sometimes to tell quite what is meant by the term; it is not necessarily Platonic realism, perhaps something like Aristotelian realism. In any case, it's deadly. Perhaps, indeed, it is most of all this *eidophobia*, if one may so christen it, that makes biologists shudder at the very name of Aristotle. Yet the Aristotelian concept of *eidos* could teach reflective biologists much about the foundations of their discipline.

Eidos in Aristotle is used differently in different contexts, but when used technically it seems to represent a single concept, although it is already rendered in Latin (as in modern European languages) by two separate terms: *forma* and *species*. I have not found, however, any indication that Aristotle took the term *eidos* to be in any formal sense equivocal. Thus he includes *genos*, for example, in the philosophical dictionary (Delta 28), but uses *eidos* in that discussion as if there were no problem about its meaning.[26] Since he deals so carefully with the several meanings of equivocal terms, I can only conclude, therefore, that he thought of *eidos* as one comprehensive concept, with different applications in respect to different problems, and perhaps with two major applications which correspond, for us, to 'form' and 'species'. Or, if even this partial separation is incorrect, we may separate, for ourselves, the two aspects of his one concept when we try to see what it can tell us in the context of modern biological methods.[27]

First, then, form in contrast to matter. *Eidos* in this context functions in a number of striking respects in the same way as the concept of organization (or information) in modern biology.

A.   For one thing, form and matter in Aristotle constitute a pair of concepts used relatively to one another and relatively to the problem at hand in a great variety of different contexts. The *eidos* of an entity or process is its organizing principle, the way it works to organize some substrate capable of such control. Though it is sometimes equivalent to *morphe* or shape, that is by no means always the case. Nor is form in nature a separate, self-subsistent 'absolute'; on the contrary, it must once more be emphatically affirmed, it exists in, and only in, that which it informs. In the context of the entity or process in question it exists *as* the organizing principle of that process, just as its matter exists as the potentiality of such (or other) organization. Thus in noses (one of Aristotle's favorite examples) snub is the form of the matter flesh and bone. In bone, however, boniness is the form of earth and fire or whatever elements compose it. Biological systems lend themselves *par excellence* to this dual – but not dualistic – analysis. It depends on the particular system one is studying what will be form and what matter in a given case; but the two-level analysis is always apposite. On the one hand, this matter as the matter of this form is by no means to be ignored, since natural form exists only as actualized in an appropriate matter. As the matter of this form, indeed, it can be studied with profit for itself – as modern biology studies, in much greater exactitude than Aristotle could dream of, the physico-chemical substrate of living systems. But the form too can be studied for its own sake, as it is by modern systems-theorists, even though it exists only as enmattered and depends for its existence and continuance on the laws governing the matter of which it is the form. And again, be it noted, if Platonists and scholastics generalized the Aristotelian concepts to form a cosmic hierarchy from the abstraction of prime matter to the Divine Mind, this was not Aristotle's primary intent. *Eidos* and *hyle* were for him a pair of analytical tools, to be applied in the study of

nature relatively to one another and relatively to the particular inquiry.

Despite the simplicity of his examples and the crudity of his 'chemistry',
Aristotle's methodological thesis is an important one. *Eidos* in the sense
of organizing principle is indeed a definitive concept for biological
method. True, in view of the advance of scientific knowledge since
Aristotle's time, its modern counterpart is couched in very different terms.
Thus, for example, G. L. Stebbins in his *Basis of Progressive Evolution*
describes the principle characterizing living systems as that of relational
order. 'In living organisms', he writes, 'the ordered arrangement of the
basic parts or units of any compound structure is related to similar orders
in other comparable structures of the same rank in the hierarchy, permitting the structures to co-operate in performing one or more specific functions'.[28] This is a much more precise statement, indeed, but it plays the
same role relative to the chemistry and physics of living things as does
Aristotle's concept of form in relation to matter. The colinearity of the
DNA chain is a relatively simple example of such order. The concept of
information may also play a similar part. In fact, it is even closer to
Aristotelian form. For information can be found in any sort of system
(and any system, in Aristotle's view, can, and should, be studied in form-
matter terms), yet in living things the quantity of information is vastly
greater – and more interesting – than in non-living systems.[29] From this
point of view, indeed, Raven's 'Formalization of Finality' might better
be called 'The Formalization of Form'. Just as, from the Aristotelian
scientist's point of view, matter *is* the possibility of taking on one form or
another, so the elements of an information-bearing system can assume
any one of many equiprobable states, one of which, by virtue of its very
improbability and in proportion to that improbability, becomes a bearer
of information. In short, the relation between entropy and negentropy in
biological processes expresses a quantitative equivalent of Aristotle's
qualitative distinction between the material and formal aspects of a given
system or subsystem, of an organism, organ system, tissue, etc.

B.   As either of the instances just cited indicates, moreover, (that is,
Stebbins on relational order or Raven on the information-theoretical
aspect of biology), the role in biology of a concept of form, organization
or information should demonstrate once for all the irreducibility of

biology to physics and chemistry (at least in their classical, reduced, and one-level form). This was Aristotle's thesis also, against Democritus. Organized systems cannot be understood in terms of their least parts alone, but only in terms of those parts *as organized* in such systems. Organized systems are *doubly determinate*; they exist on at least two levels at once. True, the form-matter pair of concepts do not of themselves generate a stratified *cosmos*; but they do show us how to resist reduction to *one* single cosmic level, whether of Democritean atoms or of the fundamental particles of modern physics. There may or may not be a cosmic hierarchy; the fact remains that whenever we study living systems we are studying particular, limited systems that are hierarchically organized, organized on at least two levels. We are studying systems composed of elements obeying their own laws, but constrained at the same time by arrangements of those very elements which constitute, as such, laws of a higher level. The higher level – form, organizing principle, code, fixed action pattern or what you will – exists only *in* its elements, and depends on them for its continuance, yet the laws of the elements in themselves, corresponding to Aristotelian *hyle*, permitting any number of informing arrangements, do not as such account for the principle which in this case happens to constrain them.

If, moreover, the question of teleological explanation left us puzzled, here, it seems to me, the case for non-'mechanistic' explanation becomes much clearer. The concept of *eidos*, like that of *telos*, is indispensable to help locate a biological problem: it has *heuristic* value. If used with good judgment it locates, as *telos* does in limited areas, a real phenomenon, a structure or process, in nature; it has *descriptive* value. But much more clearly than *telos*, it also has *explanatory* power. To discover the working principle of an organized system, as in the specification of the DNA code, or in the functional explanation of the lateral line (apart from the question of its origin), is to *explain* the system just as truly as one explains it by analyzing its physico-chemical parts. To learn how an organized system operates is just as conducive to the understanding of it – the *scientific* understanding of it – as is the analysis of the same system into its elementary components. Indeed, the teleonomic study of biological systems is probably reducible, I would suggest, to the diachronic rather than the synchronic study of their form. Teleonomy sits uneasily on evolutionary theory, one would then suspect, because, since the sequence of living systems that have inhabited this planet does not itself constitute a living system, it has no *eidos*, and therefore no *telos*, to which the study of its necessary conditions could be referred.

C.   Biological explanation, then, works in terms of form *or* matter, systems-theoretic study of wholes or part-analysis (ultimately physico-chemical) of their constituents, with the two kinds of explanation comple-menting one another. That complementarity, thirdly, however, is not symmetrical. Again, there is a striking resemblance here to the Aristote-lian form/matter pair. All natural things, organized parts of such things and processes exhibited by them, are inherently informed matter and can be studied on both levels; but form is prior. In non-Aristotelian language: although the upper level of a doubly determinate system depends on the lower level, and the laws of the lower level, for its existence, and is inse-parable from it, it is the upper level that makes the system the kind of system it is. We have to refer to the upper level, we have seen, to generate a problem, to describe the system we are studying, and in certain cases at least to explain the operation of the system as such. Thus in the code case or in physiological explanation, the problem-location, the description *and* the explanation all refer to form: in*form*ation. In some cases perhaps only the first and second obtain as explanatory principles. Research on the the lower level, physics in relation to molecular biology, or physiology in relation to behavior, or chemistry in relation to metabolism, may indeed go on indefinitely without explicit reference to the higher level, the struc-ture of an enzyme, the typical course of a nesting behavior, the normal growth process of an embryo; but it will not be study of *this system* with-out some implicit reference at least to the organizing principle concerned. The higher level, though dependent on the lower, is both epistemologically and ontologically prior to it. For Aristotle, of course, it is also prior in time: man begets man eternally. That is what evolution has altered: for us potency is, in nature as a whole, prior to actuality. But given the existence of an organized being or process, form is then prior to matter, not only cognitively but ontologically as well. Prior cognitively: to know the system is to identify, describe, and understand it in terms of its operating princi-ples, of the way it uniquely constrains its components to make this system of *this kind*. Prior ontologically: since the principles of 'matter' on their own could logically, and in terms of the laws of probability, take on any 'form', it is the existence of *this* form that makes the system what it is.

Once more, of course, we are talking about *relative* levels of a system or subsystem e.g., organic bases vs. their arrangement in a code, or the reactions entered into by chemical trace elements vs. the structure of the metabolic pathways in which they serve. Whatever system or subsystem we happen to be attending to in a given inquiry, however, this relative priority of the higher level obtains. One might call this relation, as I have suggested elsewhere, a principle of ordinal complementarity.[30] Aristotle understood it well.

D. Finally, there is a special area where the Aristotelian concept of form has proved strikingly parallel to some aspects of modern thought: that is, in his doctrine of soul, the form of organized bodies. We may leave the vexed question of active reason aside and note briefly two theses in Aristotle's 'psychology' to which modern reflection about biology seems to be slowly and painfully returning.

First, Aristotle's concept of psyche as such is functional. This is a concept of mind suggested by thinkers as different as Putnam, Ryle, and Polanyi.[31] Polanyi puts this thesis in his essay, 'Logic and Psychology', in terms of his distinction between focal and subsidiary awareness. We can formulate the distinction between mind and body, he says,

as the disparity between the experience of a subject observing an external object like a cat, and a neurophysiologist observing the bodily mechanisms by which the subject sees the cat. The experience of the two is very different. The subject sees the cat, but does not see the mechanism he uses in seeing the cat, while, on the other hand, the neurophysiologist sees the mechanism used by the subject, but does not share the subject's sight of the cat ... to see a cat differs sharply from a knowledge of the mechanism of seeing a cat. They are a knowledge of quite different things. The perception of an external thing is a from-to knowledge. It is a subsidiary awareness of bodily responses evoked by external stimuli, seen with a bearing on their meaning situated at the focus of our attention. The neurophysiologist has no experience of this integration, he has an at-knowledge of the body with its bodily responses at the focus of his attention. These two experiences have a sharply different content, which represents the viable core of the traditional mind-body dualism. 'Dualism' thus becomes merely an instance of the change of subject matter due to shifting one's attention from the direction on which the subsidiaries bear and focusing instead on the subsidiaries themselves.[32]

Thus, mind is not a separate something but is what Ryle calls 'minding'. It is the higher-level, operating principle of a complex system:

Some principles, – for example, those of physics – apply in a variety of circumstances. These circumstances are not determined by the principles in question; they are its boundary conditions, and no principle can determine its own boundary conditions. When there is a principle controlling the boundary conditions of another principle, the two operate jointly. In this relation the first can be called the higher, the second the lower principle.

Mental principles and the principles of physiology form a pair of jointly operating principles. The mind relies for its working on the continued operation of physiological principles, but it controls the boundary conditions left undetermined by physiology.[33]

This approach is generalizable, moreover, to living things as such. In general the 'soul' of any living thing is its style of operating on and in its environment, no more, but also no less.

Secondly, the 'kinds' of 'soul' distinguished by Aristotle appear to correspond in general to the major divisions which, however we see the problem of 'higher' and 'lower' among organisms, do in fact seem to obtain. If we look, independently of any special theory of the *how* of evolution, at its general course, we find, I think, three and only three really 'surprising' 'advances'. First, there is the origin of life. Here we get what Aristotle calls 'nutritive soul'. Living things grow and reproduce. These are the minimal functions of all life. Secondly, living things acquire the principle of sentience and self-locomotion. This is 'sensitive soul'. We get organisms capable of behavior, centers of irritability, appetite, and self-motion. And thirdly we have the origin of man, of culture, of the human social world, of what Aristotle calls 'passive reason'. Surely these steps, distinguished by Aristotle, are just those that may well puzzle the evolutionary theorist.[34] These are very crude distinctions if you like, but it may be worth noting that with painful deviousness we are coming back to the simple divisions in the levels of organization in the world around us which Aristotle had recognized long ago, divisions which he made, as we are trying to do, not dualistically, like Plato, but in terms of function, of the inherent organizing and operating principles that mark off kinds of complex systems as unique.

VI

So much for *eidos* in contrast to *hyle*. What about *eidos* in contrast to *genos*? It is here that the opposition to Aristotle is centered. For if modern biologists in fact use concepts like organization and information in ways that resemble Aristotle's use of 'form' in the form-matter pair, they strongly object to the Aristotelian species concept, even though in Aristotle this appears to be, if not the very same concept, at least the same concept under another aspect. I want to make two points in this connection: first, to indicate how the use of the species concept in modern biology does still resemble the Aristotelian, and secondly, to explain, or at least to locate clearly, the modern resistance to Aristotelian thinking on this score. Again, however, let me reiterate briefly what I said at the outset about the different cosmologies associated with the ancient and the modern view. Aristotelian species are certainly eternal, modern species certainly are not. This meta-scientific contrast should not be underrated. A modern biologist can no more be a complete Aristotelian than he can be a complete Cartesian. Yet in the routine use of the species concept there is nevertheless a residual, though not a merely vestigial, similarity, and at the same time, in the epistemological foundation of that use, a very deep-seated contrast.

First, the similarity. *Eidos* and *hyle* form, we have seen, throughout the range of Aristotelian sciences, an analytical pair to be used relatively to one another and to the subject matter in question in a particular investigation. When it comes to the *eidos/genos* contrast, however, *eidos* assumes a different and less relational aspect. (This holds, I think, even if we admit that *genos means hyle*, as it sometimes does; see note 27.) Only individuals are real for Aristotle – the modern biologist would agree – but they are individuals of *such and such a kind*. The *infima species*, like any form, exists only in, and as form of, the individuals who exemplify it. The species is the sum total of its specimens, past, present, and future, and they are the individuals they are in virtue of their membership in that species. But there is nothing relative about this. *Eidos* interpreted as species takes on an absolute character which, in the natural world at least, eludes *eidos* as paired with *hyle*. This of course, in its explicit enunciation at least, is just what modern taxonomists, whether evolutionists or pheneticists, so stoutly object to. And yet practising taxonomists of whatever school do in fact continue to treat 'species' as having a special role, a role in some way less conventional and closer to the real ways of nature than the concepts designating 'higher categories'. True, as Simpson points out, all categories are 'objective' in that all taxa are collections of real organisms. And they are all 'subjective' insofar as they are all concepts in the minds of taxonomists. Nevertheless, he admits, 'species' is more clearly 'non-arbitrary' than other categories.[35]

Admittedly, some biologists, from Darwin himself to Ehrlich and Holm, have predicted that the species concept would wither away and we should be left with a classless aggregate of biological particulars.[36] Yet biologists still classify and argue about the foundations of such activities; and in the view of most of them the species concept has been refined, indeed, even transformed out of recognition, but not abolished. Perhaps this is correct. Certainly, the 'biological species concept', defined in terms of potentially interbreeding Mendelian populations, or the 'multi-dimensional species concept', tailored for the inclusion of non-sexually reproducing as well as of Mendelian populations, looks at first sight very unlike the traditional, originally Aristotelian, concept.[37] I want to point out only that it is not as *wholly* unlike as it is usually painted.

Aristotle is usually accused of tagging species by means of one single character selected *a priori* and abstractly.[38] This is unfair. He is certainly no *a priorist*. Again, modern science had to reject him at its outset because he was not 'a prioristic' enough: he stayed too close to the concrete pronouncements of everyday experience; he was too good an empiricist.[39] And even though he writes in the *Topics* of, and is perpetuated in the tradition as insisting on, definition *per genus et differentiam*, he himself suggests both in the *Post. Anal.* and in *De Part. Anim.* that the 'substance' of a thing (*ousia*) or its nature cannot be captured by specifying any one differentia alone.[40] Indeed, his opposition to Platonic 'division' is based at least in part on the insistence that we divide up natural things as nature demands, not by one character and its contrary, but by the cluster of characters which helps us to single out a natural kind within a larger group.[41]

In this piecemeal and empirical approach to classification he is not so different from modern taxonomists as it is now fashionable to consider him. And again, like modern biologists, he is driven, despite his insistence that only individuals are real, to grant to the species concept some kind of uniqueness, as the least and most 'real' of universals. In short, it is in *eidos* as species that the relativity of form is somehow or other anchored in reality.

Just how this anchoring comes about is hard to say. Just how is the relational concept which we render 'form' to be identified with the non-relational concept called 'species' in the inclusive but univocal concept *eidos*? I can give no satisfactory answer, but only suggest the location of an answer in the even more puzzling conception of the τί ἦν εἶναι, which I shall discuss briefly in the concluding sections of this paper. Even more difficult would be the task of specifying the lesson to be derived for modern biology from the plurality-in-unity of Aristotelian *eidos*. I can only register tentatively the suggestion that one might profitably reflect on the link, whatever it turns out to be, between organization or information on the one hand and species on the other.

Biologists study throughout the widely varying phenomena of living nature the organization of systems or subsystems at any number of levels. Living things, as information-bearing systems, have arisen gradually from non-living systems much poorer in information content, and, once evolved, they continued to vary continuously, to throw up, in correlation with their changing environments, myriad new patterns in every conceivable direction of novelty. At the same time, there are cuts in this continuum, not only the infinite number which we might make anywhere, but a few (in relation to the infinite possibilities) which present themselves as preeminently 'natural' or, in Simpson's term, 'non-arbitrary'. These we designate as cuts between 'species' that is, between carriers, for a time, of distinctive patterns of information. One can study the organization of any organized system or subsystem, of chloroplasts, cell membranes, muscle cells, populations of genes, populations of whole organisms, communities of populations of many species, etc., etc., but there are also some points at which the transfer of a stable pattern of organization (or information) from one living individual to another stops – stops 'really', not because we decide to stop analyzing just there, but because there is a gap, a real discontinuity. To populations confined by these plain discontinuities (plain at least in sexually reproducing organisms with a relatively long generation span),[42] we give the same name that Aristotle gave them; *eidos*, *species*, the very name he gave to organization, *eidos*, *forma*. But here we are not, as in multi-levelled analysis of forms and their matter, singling out such patterns as we discover by applying 'form' and 'matter' as shifting and relational tools for our own study of nature, tools that locate form-in-matter here, there, and everywhere. We are finding certain forms singled out for discontinuity within the continuity of the phenomena, as it were, by nature itself. It may of course be objected, as Simpson remarks about higher categories, that in a sense all forms discovered everywhere are equally objective. Every cell is really bounded by a real membrane: the cytologist studies in the parts of one kind of organism the structures of cells as such. Geneticists study drosophila not because the species of that genus are themselves of greater intrinsic interest than elephants or antelopes, but because they are good experimental subjects and so from them much can be learned about the organizing principles of all heredity. Yet there *is* something unique – even uniquely obtrusive – about species. The developmental biologist – as distinct from the old-fashioned zoologist or botanist – studies organized processes that range much farther than any given species, or even phylum or kingdom. Yet even he has to select, and learn to know, *some* species in order to study *in* it the universal life pattern that interests him.[43] Much as he would like to, he cannot evade these fundamental gaps. In practice he is still an Aristotelian in spite of himself.

VII

Yet in their *attitude* to taxonomy, ancients and moderns are very different. If forms are really pinned down into discontinuous species in a fashion not so very unlike that recognized by Aristotle, why should modern biologists so emphatically deny that any shadow of Aristotelian thinking lingers in their own methods? Partly because they take a crude and truncated 'Aristotelianism' as identical with the thought of Aristotle himself. But there is another and more deep-seated reason, and that concerns, in Aristotle, the relation of knowledge, especially the knowledge of species, to perception. Modern science, let us recall once more, began by rejecting the Aristotelian approach to nature, in part at least because it was too directly tied to everyday perception of natural entities and processes and so prevented the flights of abstractive thought and creative imagination on which, as we can now see, the development of science largely depends.[44] The ideal of 'scientific method' for many philosophers and scientists has become the correlation, not of perceptions, usually directed as they are to complex, concrete individuals, but of sheer particulars, of 'hard data', with abstract laws, whether universal or statistical: correlations peculiarly susceptible, it seems, to quantitative manipulation and experimental control. For Aristotle, on the contrary, knowledge, however theoretical, is rooted in the full, concrete, perceptual world; it analyzes that concrete world and gains new insight into it, but never leaves it as ultimate, as well as initial, dwelling place. Now such perceptual insight is indeed essential to certain kinds of biological practice: from the macroscopic recognition of specimens in the field to the recognition of structures in electronmicrographs. The late C. F. A. Pantin called such biological connoisseurship 'aesthetic recognition'.[45] The double meaning of 'aesthetic': informal or connoisseurlike on the one hand, and having to do with 'aesthesis', perception, on the other, should be kept in mind in considering what this means. Pantin recalls, for example, seeing a worm in the field and saying, 'Why, that's a *Rhynchodemus*, but it's not *bilineatus*, it's an entirely new species'. This, he points out, is not the yes-no procedure of the museum taxonomist, nor does it resemble at first sight the generalizing procedures of the exact scientist.[46] It is, precisely as for Aristotle, a case of seeing a *this-here* as a *such-and-such* – or in this case, not quite a *such-and-such* but a *somewhat-different*. But just such perceptual recognition of real kinds is what modern theorists profess to abhor. They seek to produce scientific knowledge in an abstracter and more completely specifiable way. Scientific knowledge has its pedigree, they claim, by mathematical thinking out of bare particulars, rather like Love in Diotima's story, who was begotten by Resource on Poverty.

This distrust of anything but bare particulars on the one hand and high flights of theory on the other comes clearly into view in the contemporary taxonomic controversy between the *phenetic* and *phylogenetic* schools.[47] The pheneticists profess to take all and any particulars, without prior weighting derived from taxonomic skill and experience, feed them into computers (those praiseworthy inorganic animals) and come out with classifications better (for what purpose?) than those derived from less restricted starting points and less quantitative manipulation. If they sometimes admit sadly to producing by this method something justifiably called 'types', they are at any rate, they claim, '*empirical* typologists', with no initial predilection for one cluster of characters rather than another. Phylogeneticists, on the other hand, hasten to cover their undoubted, but theoretically suspect, taxonomic insights, derived from the aesthetic recognitions of field experience, under the convenient bushel of evolutionary descent. Darwinian-Mendelian theory, in other words, serves them as an abstract and therefore scientifically respectable cover for their delicate perceptual discriminations: for the heritage of Ray, Hooker, and even, on the side of biological practice as against theory, Darwin himself. Particulars tied together by computer techniques, says the one side; particulars tied together, says the other, by lines of descent inferred though necessarily unobserved. Some writers, notably Gilmour, try to go between the horns of the phenetic-phylogenetic dilemma by espousing a pleasant pragmatism.[48] We classify as we need to for our uses, says he; as the use shifts, so does the classification. This easy way out, however, while correct in a way, since there are of course many possible classifications of anything for many possible uses, fails to still the controversy. For it neglects the fact that some classifications do seem to be, quite apart from our wants and uses, less arbitrary than others. To give due weight to the role of aesthetic recognition in taxonomy, I submit, and to acknowledge its rootedness in the perception of real *this-suches*, would permit taxonomists to make more sense than they have recently done of the real nature of their calling.

It should be duly noted, in passing, that the grounding of scientific knowledge, and especially of scientific discovery, in perception (rather than in sensation or the bare observation of bare particulars) is beginning at last to be acknowledged by philosophers of science. In a general way, perception as the paradigm of discovery was the *Leitmotif* of Hanson's writing.[49] The 'primacy of perception' as our chief path of access to reality was the central theme of Merleau-Ponty's work.[50] A similar theme dominates Straus's phenomenology.[51] And in Polanyi's *Personal Knowledge, Tacit Dimension*, and other essays, both earlier and more recent, one has, as distinct from those more general intuitions, a carefully articulated epistemology which explicitly makes of perception, understood in a Gestalt-cum-transactional fashion (not unlike Aristotelian aesthesis), the primordial and paradigm case of knowing, and explicitly makes the achievement of perception the primordial and paradigm case of discovery.[25] These lessons are beginning to have some impact on philosophers of science, especially on those who base their philosophy largely on physics. But biology, from which, through Aristotle's biological practice, the acknowledgement of the primacy of perception took its start, still (with a few honorable exceptions like that of Pantin) stubbornly resists this fundamental insight.

I am not, of course, alleging that perception in its newly-discovered role plays the *same* part as it did in Aristotelian science. It *was* the basis, and the home, of discovery and of knowledge; it *is* the primordial case of discovery and of knowledge, and all discovery and all knowledge are structured as it is. In Polanyi's terms: in all knowledge, as in perception, we rely on subsidiary clues within our bodies to attend focally to something in the real world outside. However 'abstract' that something be, both the bodily base and the from-to structure characteristic of sense perception persist. In the present context: there is no disgrace, therefore, in acknowledging the perceptual skill of field naturalists and taxonomists as part of science. It is not a 'primitive' survival, but a visible analogue of the achievement of knowledge, the paradigm case of our way of gaining contact with reality.

VIII

Let me return now to the question raised earlier: how does *eidos* as the correlate of *genos* escape the relativity of the *eidos/hyle* pair? *What* is unique about species that makes it the paradigm case of natural form? Here we come to that most idiosyncratic of Aristotelian terms: the τί ἦν εἶναι, what it is for a such-and-such to be a such-and-such. 'Essence', with its age-old accretions of misleading connotations, is a poor translation of Aristotle's phrase; perhaps 'being-what-it-is' is the best one can do.[53] When one speaks of 'form', this is what one's discourse is aiming at; the form of a given kind of thing is just what it is for that thing to be the kind of thing it is.

According to the *Topics*, the 'being-what-it-is' of a kind of thing is what its definition designates. Indeed, a definition is there defined as 'a phrase indicating the being-what-it-is'. To some interpreters, therefore, τί ἦν εἶναι appears to be primarily a logical term. Aristotelian science starts from first principles, including real definitions. These specify, among the properties of a thing, certain characters which are 'essential' to it, and that means characters which can be deduced from the initial predicating statement. Definitions, in other words, are premises of scientific demonstrations, and the τί ἦν εἶναι is of interest as the reservoir, so to speak, from which the predicate of such a premise can be drawn.[54] Such a rendering seems to bring the Aristotelian approach close, in form at least, to the so-called 'hypothetico-deductive method'. It must of course be noted, however, that Aristotelian demonstration is anchored, through *nous*, in the direct, 'intuitive' knowledge of first principles. It is not, like dialectical syllogism, merely hypothetical. Yet there is a parallel. For in both cases it is the deduction of some properties from others that is chiefly of interest, and the τί ἦν εἶναι is thought of in this context as ancillary to this primarily logical game. So far as the importance of deductive method goes, Aristotle himself, at least in his logical writing, certainly reinforces this impression.

Yet if one searches the corpus for scientific demonstration, one finds relatively little of it. Most of the arguments of most of the treatises do not look like assertions of defining phrases followed by deductions from these. Most of them appear to be not strictly demonstrative, but inductive, dialectical, or aporetic. They move from common experience or common opinions, weighing the views of others, analyzing difficulties, in the hope, it seems, of arriving at (not starting from) an insight into some specific nature. This *searching* nature of much of Aristotle's writing leads W. Wieland to a different interpretation. For him the τί ἦν εἶναι is not so much a guide to definition, and therefore to demonstration, as it is a heuristic tool, a *topos* or path along which the thinker may seek insight into some special problem.[55]

For form and matter this heuristic or methodological interpretation is indeed fruitful, as I have emphasized. Yet *via* form as species we have seen that there is also a resting place for form-matter analysis, a place at which form becomes uniquely non-relational. And it is here that form exhibits its *ontological* foundation in 'the being-what-it-is' of each kind of thing, the very foundation which, according to Aristotle, his predecessors lacked, the ignorance of which prevented them from discovering the right method for the investigation of nature. Neither the logical nor the methodological approach to Aristotelian science makes sense without this frankly ontological foundation. The τί ἦν εἶναι is expressly what definition is about, both its target and its presupposition. Indeed, without the τί ἦν εἶναι as the referent of definition, the real being-what-it-is of this kind of thing which the defining phrase designates, the demonstrations that follow on definition would be simply 'hypothetico-deductive' in the sense of positivism or phenomenalism. They would not be rooted, as Aristotelian science was certainly meant to be, in the natures of the things themselves, and in our understanding of these natures. Without the τί ἦν εἶναι as endpoint and foundation of inquiry, moreover, the inquirer would be confined to an endless and directionless groping; but that is not the case. On the contrary, the real being-what-it-is of the kind of entity under investigation is constantly guiding his search, and directs it to its successful issue. Despite the apparent formal correctness of the logical interpretation, therefore, and despite the importance of the methodological aspect, the traditional ontological interpretation of the τί ἦν εἶναι is still fundamentally sound.[56]

What does it teach us? We would not link the being-what-it-is of things, as Aristotle did, to eternal kinds, nor would we restrict these 'kinds' to kinds of substance. Everything becomes, including species; and *what* becomes may as well be events or processes as more literally 'things'. But despite these deep differences, there are, I believe, two important lessons to be learned from the Aristotelian concept of being-what-it-is as the designatum of definition and the target of inquiry. Aristotle understood, as most modern philosophers of science until recently have not, that the investigation of nature arises out of puzzlement about some particular problem in some limited area: nobody investigates, or can investigate, everything at once. And in such an investigation, further, it is the real nature of the real entity or process that the investigator seeks, and sometimes finds. Science is *pluralistic* and *realistic*, not uniform and phenomenalist, as modern orthodoxy has supposed.

IX

If we put these brief remarks together with our earlier reflections about form and matter, we find, in conclusion, three important methodological lessons to be derived from the study of Aristotle. Through the concept of form as an analytical tool, correlative with matter, Aristotle can remind us of the many-levelled structure both of inquiries into complex systems and of the systems themselves, and thus of the inadequacy of a one-levelled atomism for the understanding of such systems. In conjunction with the grounding of form in the τί ἦν εἶναι of each kind of thing, further, he can remind us of the falsity of two other modern misconceptions: the unity of science concept on the one hand, the claim that the subject-matter and method of science are everywhere the same, and, on the other, the insistence that science must renounce any claim to seeking contact with reality: that theories float, as pure constructs, on the surface of the phenomena, with no mooring in the real nature of the real events or things.

The first of these reminders is plainly related to the concept of organization or information and hence (as I have already argued) to the subject-matter of biology, and to the question of its reducibility or irreducibility to chemistry and physics. The second reminder, of the plurality of science, a reminder of the good Kantian principle[57] that we can have no systematic knowledge of the whole of nature, should help also to liberate biology, or thinking about biology, from the overabstract and reductive demands imposed by taking one science, classical physics, as the ideal of all. And lastly, the acknowledgement of scientific realism should release the biologist to admit the insights into the concrete manifold of his subject-matter, from which his work originates and in which, however abstract and sophisticated it may become, it still is anchored. Nor, finally, as I emphasized at the outset, is this a plea for a return to Aristotelianism. It is a plea for us to listen, despite our fundamental differences of metaphysic and of method, to some of the tenets that Aristotle, as a biologist-philosopher, advocated long ago, and to try to interpret them in ways that could be useful to us as we attempt to articulate and revise our conception of what the investigation and the knowledge of living nature are.

NOTES

[1] G. G. Simpson, *Principles of Animal Taxonomy*, New York and London, 1961, 36n: 'I tend to agree with Roger Bacon that the study of Aristotle increases ignorance.' In fact Bacon was objecting to the current translations of Aristotle, not to Aristotle's teachings themselves. His statement reads as follows: 'Si enim haberem potestatem super libros Aristotelis ego facerem omnes cremari qui non est nisi temporis amissio studere in illis et causa erroris et multiplicatio ignorantiae ultra id quod valeat explicari. Et quoniam labores Aristotelis sunt fundamenta totius sapientiae, ideo nemo potest aestimare quantum dispendium accidit Latinis quia malas translationes receperunt philosophi.' ('Compendium Studii Philosophiae', *Bacon Opera Inedita*, Rolls Series number 15, 469). I am grateful to my colleague, Professor John Malcolm, for finding this passage.

[2] See my *Portrait of Aristotle*, Chicago and London, 1964, esp. Ch. VII.

[3] Wolfgang Wieland (*Die Aristotelische Physik*, Göttingen, 1962, 95n.) argues that the fashion in which modern physicists take a given experiment as general is still Aristotelian: 'Auch die neuzeitliche Physik liest an einem speziellen experimentellen Fall eine allgemeine Regel ab und prüft dann, ob es sich um die wahre Allgemeinheit handelt. Erst dann nämlich kann sie sehen, ob sie von einem speziellen Fall ausgegangen ist oder nicht; d. h. ob sie diesen Fall von seinen besonderen oder von seinen allgemeinen Merkmalen her gedeutet hat.' It is, in his view, modern *theories* of induction that mislead us here. But those theories *have* recognized a logical gap which Aristotle failed, and for his purposes did not need, to recognize.

⁴ Norman Campbell, *What is Science?*, New York and London, 1921, Ch. II.

⁵ *De generatione et corruptione*, II, 11, 338b7ff.

⁶ See the exposition of Wieland in the work referred to, note 3, above. For a close study of Aristotle's use of *telos* in the explanation of generation, as well as, and in relation to, *eidos* and *hyle*, see the excellent paper by Anthony Preus, 'Science and Philosophy in Aristotle's *Generation of Animals*', *J. Hist. Biol.* 3 (1970), 1–52; see also his *Science and Philosophy in Aristotle's Biology*, Darmstadt, in press.

⁷ Cf. M. Grene, *The Knower and the Known*, New York and London 1966, p. 237.

⁸ Quoted *ibid.*, pp. 235–236, from A. F. Baker, 'Purpose and Natural Selection: A Defense of Teleology', *Scientific Journal of the Royal Coll. of Science* 4 (1934), 106–19, 107–108.

⁹ Baker, *op. cit.*, 108–10.

¹⁰ W. Wieland, *op. cit.*, 254–77.

¹¹ C. H. Waddington, *The Strategy of the Genes*, London 1957, p. 190.

¹² See the (by now classic) interpretation of organicism in Ernest Nagel, *The Structure of Science*, New York 1961, pp. 428–46.

¹³ Jean Piaget, *Biologie et connaissance*, Paris, 1967, pp. 225–226. Piaget refers here to an argument by J.-B. Grize: 'J.-B. Grize, qui a étudié ces trois relations au point de vue du calcul logistique, montre de même que la relation de 'cause finale' est mal formée logiquement, du fait qu'elle mêle les relations réelles de la 'langue' (instrumentalité et causalité) avec les relations d'isomorphisme appartenant à la 'méta-langue' et utilisées pour mettre en correspondance cette causalité $a \rightarrow b$ et l'instrumentalité $B \rightarrow A$.' (*ibid.*, 226n.)

¹⁴ *The Knower and the Known*, Ch. IX, 238.

¹⁵ F. Ayala, 'Biology as an Autonomous Science', *Amer. Scientist* 56 (1968), 207–21.

¹⁶ *De Gen. Anim.* 741b 21ff.

¹⁷ H. B. D. Kettlewell, 'Selection experiments on Industrial Melanism in the Lepidoptera', *Heredity* 9 (1955), 323ff. 'A résumé of investigations on the evolution of Melanism in the Lepidoptera', *Proc. Roy. Soc. Lond. B.* 145 (1956), 297ff.

¹⁸ George Williams, *Adaptation and Natural Selection*, Princeton, 1966, p. 10.

¹⁹ *Ibid.*, 10–11.

²⁰ Among biologists critical of the neo-Darwinian synthesis, see for example E. S. Russell, 'The Diversity of Animals', *Acta Biotheor.* 13 (Suppl. 1) (1962), 1–151. Cf. A. Vandel, *L'Homme et L'Evolution*, Paris, 1949. Among recent philosophers, see M. Polanyi, *Personal Knowledge*, Chicago and London, 1958 (Torchbook edition: New York, 1962), Ch. 13.

²¹ A. Vandel, *op. cit.*

²² Cf. my analysis of R. A. Fisher's 'Genetical Theoretical Theory of Natural Selection', in *The Knower and the Known*, pp. 253–66; Chapter VIII of this volume.

²³ Cf. T. Dobzhansky, 'On Some Fundamental Concepts of Darwinian Biology', *Evol. Biol.* 2 (1968), 1–33, where efforts are made, not wholly successfully, to disentangle some of these concepts.

²⁴ See my account in *Approaches to a Philosophical Biology*, New York 1969, Ch. 2, pp. 74–75, and this volume, Ch. XVIII.

²⁵ R. R. Sokal, 'Typology and Empiricism in Taxonomy', *J. Theoret. Biol.*, 3 (1962), 230–67.

²⁶ For a detailed account of Aristotle's usage in his own biological writings the reader should consult Professor David Balme's definitive treatment, as well as A. L. Peck's notes in the Introduction to his edition of the *Hist. Anim.* D M. Balme, 'ΓΕΝΟΣ and

ΕΙΔΟΣ in Aristotle's Biology', *Class. Quart.* **12** (1962), 81–98; A. L. Peck, Introduction, in Aristotle, *Hist. Anim.*, volume I, Cambridge, Mass., 1965, esp. notes 5–11. Cf. also D. M. Balme, 'Aristotle's Use of Differentiae in Zoology', in *Aristote et les Problèmes de Méthode*, Louvain 1960, pp. 195–212.

[27] The two aspects I am separating are, admittedly, brought very close together by Aristotle himself, not only in *Post Anal.* (94 a 20ff.), where *genos* is given a place corresponding to that of 'material cause' in the *Physics*, but also in *Metaphysics* Δ 1024b 8, Z 1038 a 6, H 1045a 23 f. and I 1058 a 23, where *genos* is identified with *hyle*. I am grateful to Professor Balme for calling my attention to the latter passages, and confess that I might not have made the distinction between the two pairs of terms as flatly as I tried to do, had I read his papers (just referred to) before writing this essay. No one interested in Aristotle's biology and its relation to his philosophy of science can afford to neglect Professor Balme's careful and illuminating work.

[28] G. L. Stebbins, *Basis of Progressive Evolution*, Chapel Hill, N.C., 1969, pp. , 5–6.

[29] The *locus classicus* for the application of information theory in science is *Science and Information Theory*, Leon Brillouin, New York, 1956. The application of information theory to biology has been discussed in a number of places, notably by Henry Quastler, see *The Emergence of Biological Organization*, New Haven, 1964. The point that the distinction between living and non-living systems with respect to information is quantitative – though great enough to appear qualitative – I owe to Dr. Thomas Ragland of the University of California, Davis, who has lectured on this subject to my class in the philosophy of biology. Michael Polanyi, in 'Life's Irreducible Structure' (in *Knowing and Being*, Chicago and London, 1969, pp. 225–39), argues, on the contrary, that the distinction between the two kinds of systems is logical and qualitative; yet he admits, in terms of evolution, a continuous transition from one to the other.

[30] *The Knower and the Known*, p. 233.

[31] Hilary Putnam, 'The Mental Life of Some Machines', *Intentionality, Minds and Perception*, Detroit, 1967, pp. 177–200; Gilbert Ryle, *The Concept of Mind*, London 1949; Michael Polanyi, *op. cit.*

[32] M. Polanyi, 'Logic and Psychology', *Amer. Psychologist*, **23** (1968), 39–40.

[33] *Loc. cit.*

[34] Cf. for example, the obscure but thought-provoking argument of David Hawkins in *The Language of Nature*, New York, 1967.

[35] G. G. Simpson, *op. cit.*, 114.

[36] Darwin, *Origin of Species*, Ch. XV ('species are only well-marked varieties'); Paul R. Ehrlich and Richard W. Holm, 'Patterns & Populations', *Science* **137** (1962), 652–57.

[37] See Ernst Mayr, *Animal Species and Evolution*, Cambridge, Mass., 1969, pp. 18–20.

[38] See e.g., Simpson, *op. cit.*

[39] See *Portrait of Aristotle*, esp. Ch. III, and Wieland, *op. cit.*

[40] *Post. Anal.* II, 13, 96a33ff. and *Part. Anim.* 634b 29ff.

[41] *Part. Anim., loc. cit.*

[42] For the complexities, e.g., of bacterial taxonomy, see Mortimer P. Starr and Helen Heise, 'Discussion', *Systematic Biology*, *Nat. Acad. Sci.* publication 1962 (1969), 92–99.

[43] Professor Dennis Barrett of the University of California, Davis Zoology department, while denying that he is a 'zoologist', admits sadly that he has to know 'his' organism, the sea urchin, in order to study in it the development of the fertilization membrane.

[44] Such founders of modern science as Harvey and Newton, indeed, *thought* they

derived their great discoveries very directly from experience; we, with three centuries of hindsight, know they were more daringly imaginative than they believed.

45 C. F. A. Pantin, 'The Recognition of Species', *Science Progress* **42** (1954), 587-98; cf. his posthumous Tarner lectures, *The Relations between the Sciences*, Cambridge 1965.

46 'The Recognition of Species', p. 587.

47 D. Hull, 'Contemporary Systematic Philosophies'. *Annual Review of Ecology and Systematics*, I (1970), pp. 19–54 and M. Starr and H. Heise, *op. cit.*

48 J. S. L. Gilmour, 'Taxonomy', *Modern Botanical Thinking*, Edinburgh 1961, pp. 27–45.

49 N. R. Hanson, *Patterns of Discovery*, Cambridge, 1965. Cf. also his posthumously published *Perception and Discovery*, San Francisco, 1969.

50 M. Merleau-Ponty, *La Phénoménologie de la Perception*, Paris 1945; *The Phenomenology of Perception* (transl. by Colin Smith), New York and London, 1962.

51 See E. W. Straus, *The Primary World of Senses* (trans. by J. Needleman), New York 1963 and *Phenomenological Psychology* (trans. in part by Erling Eng), New York 1966, and this volume, Ch. 17.

52 M. Polanyi, *op. cit.*, also *The Tacit Dimension*, New York, 1966. Cf. William T. Scott, 'Tacit Knowing and The Concept of Mind', *Phil. Quart.* **21** (1971), 22–35.

53 Cf. 'What-is-being', the rendering of Joseph Owens in his *Doctrine of Being in the Aristotelian Metaphysics*, Toronto, 1957.

54 The 'logical' is one of the aspects of the τί ἦν εἶναι distinguished by C. Arpe in his dissertation, *Das* τί ἦν εἶναι *bei Aristoteles* (Hamburg 1937). Cf. E. Tugendhat, TI KATA TINOΣ (Freiburg/München 1958), 18, n. 18: 'Erst Arpe hat, anknüpfend an Natorp (*Platons Ideenlehre*, S.2) gezeigt, dass die Form des Ausdrucks nur aus der Definitionssituation verstanden werden kann und sich, wie am besten aus der Topik zu ersehen ist, auf jede beliebige Kategorie bezieht'. But cf. also Wieland, *op. cit.*, 174; the 'logical' here is perhaps closer to what I am calling the 'methodological' interpretation. It is closer to heuristics – the search for principles – than to demonstration *from* them. The deductive aspect is stressed by Prof. Moravcsik in his reading of the τηε (oral communication).

55 Wieland, *op. cit.*, pp. 174–75. Wieland emphasizes the methodological function of the τηε (which he identifies, by implication, with Arpe's 'logical' function) to the exclusion of its other aspects, esp. what Arpe calls the 'physical' or 'teleological' and the 'ontological'.

56 See Arpe, *op. cit.* For a clear summary of the traditional view, see Bernard J. Lonergan, S. J., *Verbum – World and Idea in Aquinas*, Notre Dame 1967, pp. 16–25.

57 In a discussion of a similar argument, Professor Günther Patzig has pointed out to me that the principle of the plurality of science is non-, even anti-Kantian, if by it we mean to espouse a plurality of scientific methods. For Kant the method of science was indeed one. What *is* Kantian, however, is the denial that we can have one unified, finished system of knowledge for the *whole* of nature. If we could have such a system, we could *not* have diversified sciences with diversified methods. If we cannot have such a system, on the other hand, then a plurality of fields, and of methods, is at least logically possible, and on Kantian grounds.

CHAPTER VI

IS GENUS TO SPECIES AS MATTER TO FORM?
ARISTOTLE AND TAXONOMY*

Whenever I teach the philosophy of biology, and especially any part of it to do with the species problem, and whenever I read, or hear, current discussions in philosophy of science, which await, it seems to me, some new resolution of the problem of scientific realism, I suffer from an unhappy consciousness that Aristotle could somehow tell us something on these matters if only we could read his message right. If we want to assert that there are real kinds in nature which the scientist – in particular the taxonomist – is trying to order and understand, surely Aristotle has something to tell us. But what?

Before I try to take a few first steps toward answering this question let me make a pair of introductory remarks to put our inquiry in perspective.

First we must set aside two errors made by some modern biologists who refer to Aristotle. One, he is not a classifier in anything like the Linnean sense; his actual classifications simply cannot be put into any large all-embracing arrangement and even his theoretical discussions differ, depending – like (almost?) everything in the corpus – on the context. Nor, secondly, is he a 'typologist' – for if that epithet means anything (though I sometimes think its invocation constitutes a pure case of what was formerly called 'the emotive use of language'), it means slapping a prioristic divisions onto nature by applying *one* character – truly an anti-biological proceeding, as Aristotle was well aware. The discussions of division both in *Part. Anim.* I (643a 27ff., 643b 14–15) and in Zeta XII (1037b 28ff.) emphatically insist that biological classification demands *many* differentiae. With these misapprehensions out of the way, let us ask what Aristotle is doing in his own use of what look to us like taxonomic terms.

I shall try to approach this difficult question in two steps. First, by looking at the main differences in the use of some key terms in various parts of the corpus; then by looking a little more closely at the main import – if we can find one – of Aristotle's usage in the logic and metaphysics.

The key terms we have to ask about, obviously, are *eidos*, and in the context of taxonomy, *genos*. But if only for the reason that *eidos* is paired with *hyle* as well as with *genos*, the concept of matter will have to be considered at least along the way. *Ousia* will of course also prove to be a key concept in our analysis, and I think the τί ἐστι ('the what-it-is') and especially the τί ἦν εἶναι ('the being what-it-is') ought to be so also, though we shall hardly get to them on this occasion.

First let us catalogue the use of the obviously taxonomic terms, *eidos* and *genos*, in the corpus as a whole. Here we have to distinguish three different areas: first, the biological works themselves (except for the *De Anima* and the introductory parts of *PA* and *HA*), second, the two last-named, which deal explicitly with biological methodology, and third, the logic and metaphysics (as well as some passages in the *Physics* and the psychological works). First, then, biology. While I still believe, with D'Arcy Thompson and others, that Aristotle's ontology, and the ontological foundation of his logic, clearly reflect his biological interests, I must admit that his usage in his own biology strikingly contradicts the biologically grounded metaphysics he seems to have articulated elsewhere. From the point of view of any coherent concept of an *eidos/genos* distinction, his usage is simply chaotic: an *eidos* can be as wide as you like and a *genos* as narrow. David Balme, in his paper, 'ΓΕΝΟΣ and ΕΙΔΟΣ in Aristotle's Biology'[1], in which he examined 413 occurrences of *genos* and 96 of *eidos* in the biological writings (except for *PA* I & *De An*), has conclusively shown this to be the case (indeed Bonitz rather sadly recognized it too). In fact, Balme points out, of the 96 occurrences of *eidos* only 13 refer to particular species. In terms of reference, in other words, there is simply no distinction made between species and genus in the biological works. One passage may stand for many: at 680a 15–16, Aristotle says: 'there are many genera, for there is no single species of sea-urchins' (ὄντων δὲ πλειόνων γενῶν (οὐ γάρ ἓν εἶδος τῶν ἐχίνων πάντων ἐστί)) At the same time, Balme remarks, one can usually render *genos* as 'kind' and *eidos* as 'form': so the *sense* of the two terms is constant – only the distinction between them seems to have, at least at first sight, little if any relation to the technical distinctions we are used to either in the *Categories* or the *Topics* – in the former case, at least, we would suppose, a clearly taxonomic distinction. Certainly if taxonomy is supposed to establish classes, then Aristotle in his own biology is not using '*genos*' and '*eidos*' as taxonomic tools at all. He is just using the ordinary words corresponding to our 'kind' and 'form' as any amateur naturalist might: 'Is this the same kind of wasp?' 'The California jay is a different form from the Eastern.'

Secondly, there are explicitly methodological passages in the biological works where something like the technical distinction is applied. Balme finds four such passages in *HA*, all in introductory sections. In addition, there is the treatment in *PA* I, which is clearly a discussion of the philosophy of biology intended as an introduction to the biological lectures themselves. Here we have to consider both the discussion of division (*PA* I, II and 3, 64 b 21ff.) and the explicit treatment of the relation between individual, species and genus. The division passage, in Balme's analysis[2], is entirely consonant with the treatment of division in Zeta and I shall return to it in that context. What of the account of likeness and difference in *eidos* and *genos*?[3] Things alike in *eidos*, Aristotle says, differ only in number, and things differ in genus only analogically, while things in the same genus differ 'by excess and defect'. In terms of Aristotelian biology the last distinction is at first sight a bit puzzling. Is 'two-footed' an excess or deficiency of feet? Is 'fish' an excess or deficiency of 'animal' or 'catfish' of 'fish'? But I think this point too will fit in to our reading of key passages in the *Metaphysics* and I shall return to it too there.

For there is, thirdly, a range of uses of the key terms in the logical and metaphysical writings which partly contrasts, but partly harmonizes, with the two sets of texts already noticed. Let us turn, then, to the use(s) of *eidos* and *genos* in the rest of the corpus, chiefly in the *Metaphysics* – and chiefly in fact the uses of *genos*.

I have to deal chiefly with *genos* for several reasons. First, despite the fact that for the late Latin and modern tradition, *species* and *forma* are two concepts, not one, and in disagreement with Professor Balme, I am still convinced that somehow *eidos* for Aristotle has one overarching meaning, while for *genos* this is, if at all, much less clearly the case. But, secondly, if I want to maintain that there is one Aristotelian concept of *eidos*, it would seem to be helpful if we could reduce to unity the welter of meanings of *genos*. Some people *do* try this, notably by equating *genus* with *matter*. Professor Balme seems to favor this view; so it seems, if I understand him, does A. C. Lloyd in his papers on what he calls '*P*-series' and on individuation[4] and so does Richard Rorty in his paper on 'Genus and Matter'.[5] The Balme-Lloyd-Rorty thesis – or the B-L-R thesis –, therefore, must be considered in some detail.

But let me start with *eidos*. That it has one sense, applicable in many contexts, rather than many senses, is suggested, at least, by the absence of a chapter on *eidos* in Delta. Moreover, we can get, I think, some hint of what such a unified sense might be, if we consider the English expressions 'the form of a thing' – as distinct from its components – or 'a certain form of thing' – as distinct from other forms included in some larger class. 'Form' has, it seems to me, still *one* clear sense in both these contexts. This is as it should be. For plainly *eidos*, with the related concepts *ousia* and τί ἦν εἶναι, occupies a central place – and *one* central place – in Aristotle's account of his own scientific enterprise. It is the ἄτομον εἶδος that Aristotelian differentiation – as, in his view, distinct from Platonic division – is after. In *PA* I (643a 24), Aristotle tells us ' ''Εστι δ'ἡ διαφορὰ ἐν τῇ ὕλῃ τὸ εἶδος'. That is what (on Balme's interpretation) the division passage that follows (644a 24ff.) is about: it is an attack, not on division as such, but on dichotomy and single differentiae, because that method is impotent to capture the *atomon eidos*; it does not lead us, in the terms of Z XII (1038a 19–20) to 'the last differentia' which is 'the substance of the thing'.

This objective of Aristotelian division – and of Aristotelian science – is elucidated in Lloyd's 'ordered P-series' paper. It is a very complicated argument, and I could not possibly disentangle it all; besides, I shall refer to it again later. But I think it is important to acknowledge one aspect of Lloyd's thesis at this juncture. In contrast to series like colors, figures or souls, where the genus is nothing *para ta eide*, he points out, a biological series of differentiae proceeds differently, just because it has a unique terminus in the definition of a substance, and so carries earlier differentiae along (footed into two-footed) in a different fashion from other sorts of divisions. Aristotle does indeed admit division in a variety of contexts; yet he insists that the logical process of differentiation which terminates in the definition of an *atomon eidos* is unique. Iota 7–9, for instance, starts by discussing contraries and intermediates in the case of color (which is, in Lloyd's term, a P-series) and concludes by contrasting another such case – the case of male and female – with that of substantial differentiation (the latter indeed, Lloyd admits, contains P-series, but is not one). In short, if we are to practise Aristotelian science, we have to single out the kind of division that zeroes in on an *atomon eidos* from other classificatory series. Aristotle himself contrasts his method with that of his predecessors in *PA* I (642a 24): 'what they lacked was the τί ἦν εἶναι of each kind of thing and the definition of substance'. But the 'definition of substance' is the statement of what its *eidos* is in terms of genus and difference.

If, then, we are to take *eidos* as in some such way a single concept, what are we to do about *genos*? Here things get very sticky. Let us start from Delta 28 and see what Aristotle himself has to say about genos as a πολλαχῶς λεγόμενον:

Γένος λέγεται τό μὲν ἐὰν ᾖ ἡ γένεσις συνεχὴς τῶν τὸ εἶδος ἐχόντων τὸ αὐτό, οἷον λέγεται ἕως ἂν ἀνθρώπων γένος ᾖ, ὅτι ἕως ἂν ᾖ ἡ γένεσις συνεχὴς αὐτῶν· τὸ δὲ ἀφ' οὗ ἂν ὦσι πρώτου κινήσαντος εἰς τὸ εἶναι· οὕτω γὰρ λέγονται "Ελληνες τό γένος οἱ δὲ "Ιωνες, τῷ οἱ μὲν ἀπὸ "Ελληνος οἱ δὲ ἀπὸ "Ιωνος εἶναι πρώτου γεννήσαντος· καὶ μᾶλλον οἱ ἀπὸ 'τοῦ γεννήσαντος ἢ τῆς ὕλης (λέγονται γὰρ καὶ ἀπὸ τοῦ θήλεος τὸ γένος, οἷον οἱ ἀπὸ Πύρρας). ἔτι δὲ ὡς τὸ ἐπίπεδον τῶν σχημάτων γένος τῶν ἐπιπέδων καὶ τὸ στερεὸν τῶν στερεῶν· ἕκαστον γὰρ τῶν σχημάτων τὸ μὲν ἐπίπεδον τοιονδὶ τὸ δὲ στερεόν ἐστι τοιονδί· τοῦτο δ' ἐστὶ τὸ ὑποκείμενον ταῖς διαφοραῖς. ἔτι ὡς ἐν τοῖς λόγοις τὸ πρῶτον ἐνυπάρχον, ὃ λέγεται ἐν τῷ τί ἐστι· τοῦτο ⟨γὰρ⟩ γένος, οὗ διαφοραὶ λέγονται αἱ ποιότητες. τὸ μὲν οὖν γένος τοσαυταχῶς λέγεται, τὸ μὲν κατὰ γένεσιν συνεχῆ τοῦ αὐτοῦ εἴδους, τὸ δὲ κατὰ τὸ πρῶτον κινῆσαν ὁμοειδές, τὸ δ' ὡς ὕλη· οὗ γὰρ ἡ διαφορὰ καὶ ἡ ποιότης ἐστί, τοῦτ' ἔστι τὸ ὑποκείμενον, ὃ λέγομεν ὕλην. (1024a 29 - b 9).

The term 'genus' (or 'race') is used: (a) When there is a continuous generation of things of the same type; *e.g.*, 'as long as the human *race* exists' means 'as long as the generation of human beings is continuous'. (b) Of anything from which things derive their being as the prime mover of them into being. Thus some are called Hellenes by race, and others Ionians, because some have Hellen and others Ion as their first ancestor. (Races are called after the male ancestor rather than after the material. Some derive their race from the female as well; *e.g.* 'the descendants of Pyrrha'.) (c) In the sense that the plane is the 'genus' of plane figures, and the solid of solids (for each one of the figures is either a particular plane or a particular solid); *i.e.*, that which underlies the differentiae. (d) In the sense that in formulae the first component, which is stated as part of the essence, is the genus, and the qualities are said to be its differentiae. The term 'genus', then, is used in all these senses – (a) in respect of continuous generation of the same type; (b) in respect of the first mover of the same type as the things which it moves; (c) in the sense of material. For that to which the differentia or quality belongs in the substrate, which we call material. (Loeb transl.)

The first two meanings here agree with the usage in the biological works. They present no special puzzles, nor have they any bearing on our problem.

The other two meanings are trickier. Since they are summed up together as equivalent to 'matter', we seem to have here definitive support for the B-L-R position. But there are difficulties. First, in two of the later passages where Aristotle links *hyle* and *genos*, he contrasts this equivalence with some other use of *genos* – at 1058a 7, he refers to *genos* 'whether as matter or otherwise', and at 1038a 6 (of which more later) he says 'genos is *either* nothing πτε or matter', etc. Putting these two passages together, we get a division of senses of *genos*: either it belongs to a P-series (is nothing πτε) or it means matter. Now look at case 3 in Delta: figures are notoriously P-series. So this *should* be one of those cases where *genos* does *not* mean matter. This case, moreover, is mathematical, not physical; if *genos* is here to be equated with *hyle*, it is not, it would seem at first sight, the *hyle* of which natural substances are made, the *hyle* of generation, that is in question. So *genos* does not mean *hyle* simply. The fourth case, indeed, is purely logical: *genos* here is intelligible matter, and surely this is only analogical to what gets worked up through all four kinds of change into Socrates or Coriscus or Fido or Shep. Yet, admittedly, at 1058a 23 Aristotle does equate the meaning of *genos* as *hyle* with its use 'en te physei'. So perhaps it *is hyle* in general – including the matter from which generation occurs – with which *genos* is here identified.

But again it should be noticed, two lines below the seeming identification of *hyle* and *genos*, in Delta, in the discussion of 'difference in genus', they are pulled apart again:

ἕτερα δὲ τῷ γένει λέγεται ὧν τε ἕτερον τὸ πρῶτον ὑποκείμενον καὶ μὴ ἀναλύεται θάτερον εἰς θάτερον μηδ' ἄμφω εἰς ταὐτόν, οἷον τὸ εἶδος καὶ ἡ ὕλη ἕτερον τῷ γένει, καὶ ὅσα καθ' ἕτερον σχῆμα κατηγορίας τοῦ ὄντος λέγεται (τὰ μὲν γὰρ τί ἐστι σημαίνει τῶν ὄντων τὰ δὲ ποιόν τι τὰ δ' ὡς διήρηται πρότερον)· οὐδὲ γὰρ ταῦτα ἀναλύεται οὔτ' εἰς ἄλληλα οὔτ' εἰς ἕν τι. (1024b 9-16.)

Things are called 'generically different' whose immediate substrates are different and cannot be resolved one into the other or both into the same thing. E.g., form and matter are generically different, and all things which belong to different categories of being; for some of the things of which being is predicated denote the essence, others a quality, and others the various other things which have already been distinguished. For these also cannot be resolved either into each other or into any one thing. (Loeb transl.)

Eidos and *hyle* are here said to 'differ in genus': so plainly the terms '*hyle*' and '*genos*' are not constantly used as synonyms, not even constantly in the chapter in which the alleged identity has been introduced.

What can we make of this tangle? Let me first call attention to some of the passages that testify *pro* and *contra* the B-L-R thesis and then propose a general approach to the interpretation of *genos* which may perhaps solve some of the problems which these and other passages raise.

Apart from Delta (1024a 8), the chief texts *pro* the B-L-R thesis are:

τοῖν δυοῖν δὲ τὸ μὲν διαφορὰ τὸ δὲ γένος, οἷον τοῦ ζῷον δίπουν τὸ μὲν ζῷον γένος διαφορὰ δὲ θάτερον. εἰ οὖν τὸ γένος ἁπλῶς μὴ ἔστι παρὰ τὰ ὡς γένους εἴδη, ἢ εἰ ἔστι μὲν ὡς ὕλη δ᾽ ἐστίν (ἡ μὲν γὰρ φωνὴ γένος καὶ ὕλη, αἱ δὲ διαφοραὶ τὰ εἴδη καὶ τὰ στοιχεῖα ἐκ ταύτης ποιοῦσιν), φανερὸν ὅτι ὁ ὁρισμός ἐστιν ὁ ἐκ τῶν διαφορῶν λόγος. (1038a 3–9).

In general it does not matter whether it contains many or few terms, nor, therefore, whether it contains few or two. Of the two one is differentia and the other genus; *e.g.*, in 'two-footed animal' 'animal' is genus, and the other term differentia. If, then, the genus absolutely does not exist apart from the species which it includes, or if it exists, but only as matter (for speech is genus and matter, and the differentiae make the species, *i.e.* the letters, out of it), obviously the definition is the formula composed of the differentiae. (Loeb transl.)

supported (by implication) in 1045a 14–25:

τί οὖν ἐστιν ὃ ποιεῖ ἓν τὸν ἄνθρωπον, καὶ διὰ τί ἓν ἀλλ᾽ οὐ πολλά, οἷον τό τε ζῷον καὶ τὸ δίπουν, ἄλλως τε δὴ καὶ εἰ ἔστιν, ὥσπερ φασί τινες, αὐτό τι ζῷον καὶ αὐτὸ δίπουν; διὰ τί γὰρ οὐκ ἐκεῖνα αὐτὰ ὁ ἄνθρωπός ἐστι, καὶ ἔσονται κατὰ μέθεξιν οἱ ἄνθρωποι οὐκ ἀνθρώπου οὐδ᾽ ἑνὸς ἀλλὰ δυοῖν, ζῴου καὶ δίποδος, καὶ ὅλως δὴ οὐκ ἂν εἴη ὁ ἄνθρωπος ἓν ἀλλὰ πλείω, ζῷον καὶ δίπουν; φανερὸν δὴ ὅτι οὕτω μὲν μετιοῦσιν ὡς εἰώθασιν ὁρίζεσθαι καὶ λέγειν, οὐκ ἐνδέχεται ἀποδοῦναι καὶ λῦσαι τὴν ἀπορίαν· εἰ δ᾽ ἐστίν, ὥσπερ λέγομεν, τὸ μὲν ὕλη τὸ δὲ μορφή, καὶ τὸ μὲν δυνάμει τὸ δὲ ἐνεργείᾳ, οὐκέτι ἀπορία δόξειεν ἂν εἶναι τὸ ζητούμενον.

What is it, then, that makes 'man' one thing, and why does it make him one thing and not many, *e.g.* 'animal' and 'two-footed', especially if, as some say, there is an Idea of 'animal' and an Idea of 'two-footed'? Why are not these Ideas 'man', and why should not man exist by participation, not in any 'man' but in two Ideas, those of 'animal' and 'two-footed'? And in general 'man' will be not one, but two things – 'animal' and 'two-footed'. Evidently if we proceed in this way, as it is usual to define and explain, it will be impossible to answer and solve the difficulty. But if, as we maintain, man is part matter and part form – the matter being potentially, and the form actually man –, the point which we are investigating will no longer seem to be a difficulty. (Loeb transl.)

and 1058a 21–6:

ὥστε φανερὸν ὅτι πρὸς τὸ καλούμενον γένος οὔτε ταὐτὸν οὔτε ἕτερον τῷ εἴδει οὐθέν ἐστι τῶν ὡς γένους εἰδῶν, προσηκόντως· ἡ γὰρ ὕλη ἀποφάσει δηλοῦται, τὸ δὲ γένος ὕλη οὗ λέγεται γένος — μὴ ὡς τὸ τῶν Ἡρακλειδῶν ἀλλ' ὡς τὸ ἐν τῇ φύσει, οὐδὲ πρὸς τὰ μὴ ἐν ταὐτῷ γένει. ἀλλὰ διοίσει τῷ γένει ἐκείνων, εἴδει δὲ τῶν ἐν ταὐτῷ γένει.

Thus it is evident that in relation to what is called genus no species is either the same or other in species (and this is as it should be, for the matter is disclosed by negation, and the genus is the matter of that of which it is predicated as genus; not in the sense in which we speak of the genus or clan of the Heraclidae, but as we speak of a genus in nature); nor yet in relation to things which are not in the same genus. From the latter it will differ in genus, but in species from things which are in the same genus. For the difference of things which differ in species must be a contrariety; and this belongs only to things which are in the same genus. (Loeb transl.)

I have discussed the first; the other three will, I hope, be fairly assimilated to my own interpretation of *genos*, which follows. Bonitz includes 1033*a* 3–6 as a supporting passage:

ἀμφοτέρως οὗ λέγομεν τοὺς χαλκοῦς κύκλους τί εἰσι, καὶ τὴν ὕλην λέγοντες ὅτε χαλκός, καὶ τό εἶδος ὅτι σχῆμα τοιόνδε, καὶ τοῦτό ἐστι τὸ γένος εἰς ὃ πρῶτον τίθεται ὁ δὴ χαλκοῦς κύκλος ἔχει ἐν τῷ λογῳ τὴν ὕλην.

But then is matter part of the formula? Well, we define bronze circles in both ways; we describe the matter as bronze, and the form as such-and-such a shape; and this shape is the proximate genus in which the circle is placed. The bronze circle, then, has its matter in its formula. (Loeb. transl.)

But I think Ross's interpretation, which comfortingly takes the 'kai touto' clause as a gloss, makes better sense of the whole chapter than would its retention and its B-L-R-like interpretation. Besides, neither Balme nor

Wieland in his brief treatment of this question include it.[6] Bonitz also
lists *De Caelo* 342b 7, but here *genos* so clearly means 'kind' in a very
general sense – 'what *can* be hot' – that its implications if any for the
relatively technical meaning seem slight.

Contra the B-L-R thesis, on the other hand, there is weighty evidence.
The difficulties in the context of 1024b 8, as well as in 1024b 10, I have
already noted. A little earlier, at 1015b 33 (on the term 'one'), the ways
in which 'man' and 'cultured' are 'in' Coriscus are contrasted as on the
one hand the *genus* 'in' the substance and on the other a 'hexis or pathos'
of the substance – where the latter, so to speak, accidental accident would
be, in Aristotle's sense, 'material', the former surely not so. At 1016a
24–6, further, a consideration of 'genos' is said to be *similar* to that of
matter: so they are clearly not identical. At 1071a 37 Aristotle considers
problems about 'genos' and predication, and then goes on to 'matter' as
a *further* consideration: again, they are clearly not the *same* concept.
Moreover, at 1058b 20ff. – again the continuation, without a break of
context, of one of the major *pro* passages, Aristotle explains that male and
female do not constitute specific differentiae because they differ 'only in
matter and body'.

τὸ δὲ ἄρρεν καὶ θῆλυ τοῦ ζῴου οἰκεῖα μὲν πάθη, ἀλλ' οὐ κατὰ τὴν
οὐσίαν ἀλλ' ἐν τῇ ὕλῃ καὶ τῷ σώματι, διὸ τὸ αὐτὸ σπέρμα θῆλυ ἢ
ἄρρεν γίγνεται παθόν τι πάθος. (1058b 22–6.)

'Male' and 'female' are attributes peculiar to the animal, but not in virtue of its substance; they are material or physical. Hence the same semen may, as the result of some
modification, become either female or male.
We have now stated what 'to be other in species' means, and why some things differ
in species and others do not. (Loeb transl.)

But of course if they don't even differ in species, so much the less do they
differ in genus; so, again, genus ≠ matter! The *coup de grâce* is given,
however, again in Delta, 1016 B 33, still in the chapter on unity. Those
things are one in number that are one in *matter*; those things are one in
genus that belong to the same *category*. In between are things one in *eidos*
(forma = species); thus their material differences individuate them from
one another, their generic unity puts them together, not even into one
biological genus, but into one σχῆμα τῆς κατηγορίας.

Ἔτι δὲ τὰ μὲν κατ᾽ ἀριθμὸν ἐστιν ἕν, τὰ δὲ κατ᾽ εἶδος, τὰ δὲ κατὰ γένος, τὰ δὲ κατ᾽ ἀναλογίαν, ἀριθμῷ μὲν ὧν ἡ ὕλη μία, εἴδει δ᾽ ὧν ὁ λόγος εἷς, γένει δ᾽ ὧν τὸ αὐτὸ σχῆμα τῆς κατηγορίας, κατ᾽ ἀναλογίαν δὲ ὅσα ἔχει ὡς ἄλλο πρὸς ἄλλο. ἀεὶ δὲ τὰ ὕστερα τοῖς ἔμπροσθεν ἀκολουθεῖ, οἷον ὅσα ἀριθμῷ καὶ εἴδει ἕν, ὅσα δ᾽ εἴδει οὐ πάντα ἀριθμῷ· ἀλλὰ γένει πάντα ἕν ὅσαπερ καὶ εἴδει· ὅσα δὲ γένει οὐ πάντα εἴδει ἀλλ᾽ ἀναλογίᾳ· ὅσα δὲ ἕν ἀναλογίᾳ, οὐ πάντα γένει.

Φανερὸν δὲ καὶ ὅτι τὰ πολλὰ ἀντικειμένως λεχθήσεται τῷ ἑνί· τὰ μὲν γὰρ τῷ μὴ συνεχῆ εἶναι, τὰ δὲ τῷ διαιρετὴν ἔχειν τὴν ὕλην κατὰ τὸ εἶδος, ἢ τὴν πρώτην ἢ τὴν τελευταίαν, τὰ δὲ τῷ τοὺς λόγους πλείους τοὺς τί ἦν εἶναι λέγοντας.

Again, some things are one numerically, others formally, others generically, and others analogically; numerically, those whose matter is one; formally, those whose definition is one; generically, those which belong to the same category; and analogically, those which have the same relation as something else to some third object. In every case the latter types of unity are implied in the former : e.g., all things which are one numerically are also one formally, but not all which are one formally are one numerically; and all are one generically which are one formally, but such as are one generically are not all one formally, although they are one analogically; and such as are one analogically are not all generically.
It is obvious also that 'many' will have the opposite meanings to 'one'. Some things are called 'many' because they are not continuous; others because their matter (either primary or ultimate) is formally divisible; others because the definitions of their essence are more than one. (Loeb transl.)

Now clearly on the face of it, matter as principle of individuation is not identical with genus as category. Thus for example, things that are *not* one in genus in terms of this analysis differ only analogically, while plainly Socrates and Coriscus differ not analogically but as literally and concretely as possible: Socrates is one individual differing in number *because* in matter, from Coriscus. (See for corroboration 1034a 5ff. – Callias vs. Socrates – on matter as principle of individuation.[7] But *Socrates*, on the other hand, is of course the same as Coriscus in *eidos*, and *a fortiori* in *genos*: he is a man and a substance. Putting aside for the moment the significant fact that *genos* here refers to category, not to biological genus (I shall return to that shortly), let us consider briefly what is implied for Aristotelian metaphysics if we take the *hyle = genos* identity seriously here, as at least the *R* of the B-L-R axis (if I understand him correctly) suggests that we do. We have then to say that the individual differences of members of a species are individuations not within eidetic possibilities, but (because material) within the genus. Thus – and this I gather from conversation with him Rorty accepts – the difference of Socrates's snub nose from Coriscus's straight one is a difference in animality, not in humanity. Now of course no two men differ *in* humanity: *eidos* is one and indivisible. But they do differ, it seems clear, in the matter appropriate to the generation of a *man*, not of any old animal. Snub is a shape of nose, and noses are *human* organs as distinct from snouts or probosces in general. That's why there must be a reference to a *kind* of matter in the definition – all right, that's the genus; but as Aristotle insists in Iota VIII, two individuals that differ in species within a genus are neither the same nor different in species in relation to the genus – for the genus is transformed when specified: the animality needed to make a man is already human animality. Iota IX, already referred to, further confirms this: Callias is definition *and* matter (1058b 10–11). But the definition is already *genus* and specific difference, which expresses 'to eschaton atomon'. 'Matter' is something *else*! The thesis that the matter worked up in the natural world is not already anthropoid matter, but each time again mere animality waiting to take shape as man or oak tree: that thesis is not only contradicted by such texts as these just referred to; it is simply too counterintuitive biologically to be true. As Wieland puts it, *genos cannot* be matter as ἐξ ὅυ; even if the concept *genos* has some matter-like function sometimes in the analysis of nature, the two cannot be simply identified.

After that tirade, let me back off a bit and see what we can make of the Aristotelian language game with *genos* (for there *is*, I believe, a family of uses here that does make sense), and I'll return with a somewhat politer bow to B, L and R at the close of my account.

Let us look, then, at the main uses of the term *genos* outside the biological writings. There are, I think, five members of the family to be distinguished, which we can relate to one another in terms of Aristotle's dialectical approach to the problem of Being and Substance.

First, *genos* in Aristotle's argument often refers to Platonic genera, especially to the question whether Being and Unity are genera. This usage is frequent, of course, in Beta. It recurs, for example, at 1059b 27, in the introduction to Theta. It looks as if every time Aristotle starts on a new approach of his own to Being *qua* Being, he wrestles first with the Platonic conception. The importance of this context – the need to come to terms with the Platonists, and to vindicate his own method and doctrine in the context of *their* problems – can hardly be exaggerated. In short, the use of 'genos' to refer to such Platonic genera as Being and Unity not only provides the first in our series of uses; it may serve to call our attention to the pervasive or at least recurrent presence in the *Metaphysics* of the context of Platonic discussion: we will need to refer to it again in the course of our interpretation, especially of 1045a 14ff. and indirectly therefore of 1038a 6.

If, then, it was the snares of *Platonic* ontology from which Aristotle was devising an escape, he did it, secondly, by stabilizing the threatened homonymity of 'being' through the categoreal schemata: substance and the other nine. There is no genus of Being as such, but there 'are' the different senses of 'being', the genera substance, quality, quantity, and so on. There is a first answer to the question 'what is it' along a fixed plurality of lines which are only comparable (by analogy) with one another. This usage – genos = category – is very common.[8] We noticed it already in Delta 6 (1016b 33): things with the same matter, we were told, are the same in number; things with the same form but different matter differ in number; things with different forms in the same genus (or category) differ by more or less, or by excess and deficiency: more or less tall, or footed, or rash, or long-lasting. There may seem to be a difficulty in this interpretation, of course. It works well for color, for example, as in Iota, or perhaps for any accidental category, but how substances can differ by more or less appears at first to be a puzzle. It should be remembered, however, that if one is dealing scientifically with substances, in biology, for instance, with their parts, their generation and so on, not as such with other categories, it is only by applying terms from other categories as differentiae that we can ever achieve knowledge of substances.[9] And in this way the distinction of more and less does enter indirectly into the differentiation of substances, even though there is no more and less in substance itself.

Let us then take the categories in general as an Aristotelian and systematically homonymous solution to the Platonic problem of Being or Unity as genera (as they are also at 1042b 25 to the problems of Democritean metaphysic). So then the technique of working with the right more-and-lesses within a given (accidental) category *is* the technique that is needed to get at what Aristotle believed he could provide as his predecessors could not: the being-what-it-is of each kind of thing and the definition of substance. We must note, further, two important corollaries of this thesis: first, there *is* a connection between matter and genus in the categoreal sense. In Iota 4 at 1055a 30 (and cf. also 1055a 6), Aristotle notes that contraries (sc. in the same genus = category) have the same matter and *a contrario* things in different categories have different matter. Now clearly this is a *very* general sense of 'matter': the 'matter' that is common substrate for change of some kind. Thus anything that can *be* colored can change in color – but what is in one or another position doesn't change in color, it changes in position. Secondly, it is this very diffuse, or analogical, sense of matter, I submit, that supplies the context for 1058a 23, the B-L-R passage in which we are told that 'genos ἔν τῇ φύσει' means 'matter' – and this is the use also to which, I believe, Aristotle is referring at 1058a 7 (where, as at 1038a 6, he contrasts with cases where the genus is *nothing* πτε those where it is 'matter'). What he means, I think, is something like this: either, as the Platonists admit, we classify colors or figures with the 'genus' individuated into blue or circle, and no ontological residue remaining (as it seems – according to Lloyd – academic argument had already admitted) – or we look at the genus (that is, the schema of predication, the category) as 'matter' for specification by degrees of contrariety. In neither case does one have separated genera as the Platonists would have liked to have them. The interpretation of *genos* (i.e. category) as 'material', however, is pertinent to the study of nature – and this is our use ἐν τῇ φύσει – in one of two senses. Either – and I think, in the context of Iota as a whole, from 4–9 at least, this is the more likely reading – 'matter' is simply the unity of categoreal context open for distinctions of more or less – a very metaphorical meaning, which can *not* be identified with the concrete, worked up matter 'out of which' the sculptor makes his bronze statue or the father his son. Or, as I *think* Lloyd would have it, the concept of genus as matter, as alternative to the P-series case, is appropriate to physics because it is here that the kind of matter workable into a statue or a man in fact – technically or physically – comes into play and so saves technical or biological differentiation from *being* the P-series used to implement it. I shall return to this kind of distinction shortly in connection with 1038a 6; there it seems to me it works, but here the context suggests the weaker reading.

The Aristotelian genera, or categories, then, have saved the meanings of Being from the total equivocity inherent in the Platonic search for ultimate, self-existent genera. Being is not a genus, and neither is One, but there *are* the genera within each of which we can distinguish kinds through applying distinctions of degree within the genus in question. Given this technique, further (thirdly in my present enumeration), we can apply our what-is-it question within a given category at a number of levels of particularity or generality. In this context we can use the concepts *genos* and *eidos* relatively to one another in any category. This is the harmless logical use of the *Topics* and the comparatively non-technical use of many other passages, e.g. at *GA* 767b 33, where 'genus' means simply the kind as distinct from the individual, in this case man as against Coriscus. 'Genus' is used simply to sort out any wider from some narrower class (or even from the individual), or a wider or narrower property used to specify a class. This use, moreover, appears to derive, by a narrowing of reference, from the 'category' meaning. Thus, for example, in *PA* I and *HA* I Aristotle uses the very same criteria of oneness and difference (in number, *eidos* and *genos*) that he uses of the 'category' sense in Delta. We have:

numerical identity	eidetic identity	generic identity	no identity
identity of matter	identity of formula, difference of matter	difference by more and less or excess and defect	difference in genos, likeness only by analogy

Although he is here speaking of different kinds of animals, not different kinds of being (or of predication), the arrangements of identities and differences are identical; the sense of 'genus' here, therefore, must surely be systematically related to the category sense. It seems that once the 'schemes of predication' have stabilized the total equivocity of Being in the Platonic sense, we can use this fundamental sense, or one continuous with it, along a scale of wider and narrower reference.

But now, fourthly, we come to a special kind of case, the case considered in Zeta 12 and in *PA* I, 2–3, where we are examining, not classification by more or less in any category, but Aristotelian division, which is meant to lead us – uniquely – to the *atomon eidos*. Here the genus is not any class wider than some other, but that kind of class which can be properly differentiated into the kind of *eide* that will give us definitions of *ousiai*. And again, we have here one of the key B-L-R passages (1038a 3–8)

where *genos* appears to be identified with *hyle*: we have here again a
contrast with the cases where the *genos* simply is not p.t.e. – and the case
where it 'is matter'. But if we look at this discussion in its own context,
and also alongside 1045a 23, with which we are asked to compare it, some
puzzling points emerge. On the one hand, admittedly, it's true, it is the
segregation of *eide* out of a *genos* (as 'matter') that distinguishes the true
atomon eidos case from any other differentiation by contraries within any
non-substantial category. But on the other hand, look at the way Aristotle
puts it: if the genus isn't just nothing apart from the species, it's *only*
matter. In other words, you Platonists yourselves admit that for colors,
for instance, there is no residual separable genus over and above the
differentiated kinds of colors – or if you want me to find the genus as
something more than nothing, well, it's only matter – and that's cold
comfort to a Platonist. Hence, perhaps, the ambiguous case of 'speech' – a
universal that *can* be worked up into one letter ('phone') or another, or an
indifferent material voicing, informed now one way or another as this or
that sound. The emphasis here, moreover, is on the unity of the definition,
in which the genus lingers only as it is differentiated by the difference.
And, again, in the context of definition, the task of genus, while it can be
seen as 'matter-like', cannot *be* matter. If speech is that out of which
letters are made it is material but not generic; if it is generic it is analogous
to matter, but *is* not material. Wieland is right in emphasizing that both
genos and *eidos* in Aristotle's usage (and he cites this passage as well as
1058a 23) are *Reflexionsbegriffe*. They serve to guide us, analytically, in
our search for the *atomon eidos*. But they cannot be identified with one
another just because they both guide us in the same direction. To put it
another way: they are intentional concepts; substitution may be sought
here only at one's own risk.

1045a 23, moreover, applying 'matter' and 'form' to the same example,
sets the discussion even more explicitly in the framework of an anti-
Platonist polemic. Here Aristotle has put the question of the unity of
definition in the context of the question: is there one Idea of two-footed
and another of animal? The whole inquiry, therefore, (1) is directed to the
problem of Platonic genera, (2) deals with the logical, not physical,
genos-eidos relation and (3) suggests a kinship of *genos* and *hyle* as an
anti-Platonist move: your 'genera'? They're either nothing, or only
matter. So man, even as defined – as intelligible – is partly form and

partly matter. His unity as defined is possible only when the genus is read as matter for specific form, not as some independent genus off on its own. The passage concludes, moreover, with the explicit distinction between *intelligible* and sensible matter – and *genos* is clearly, in this context, matter in the former, not the latter, sense.

We have then, so far, four *topoi* for the use of 'genus': (1) Platonic, (2) categoreal, (3) logical (and relative), (4) in relation to the search for the *atomon eidos*. And fifthly and finally, at long last, within the category of substance itself there are the secondary substances, genera, which embrace a number of *eide*. This is the 'technical' sense which one expects but does not find in Aristotle's own biological research, nor, if I am right, in his programmatic statement about *genos* and *eidos* in *PA* I, nor, indeed, in most of the uses of that term in the *Metaphysics* and elsewhere. What is the relation of *genos* to *hyle* here?

Very briefly, I think it is twofold. First, following a number of eminent authorities, from Wieland back to St. Thomas, I believe we must accept a parallel, but not an identity, between genus in the definition and matter in the concrete thing. We may use our grasp of intelligible and of sensible matter to further our grasp of what substances are, but we must not identify them. Is there, though, in some other area, a closer union of *genos* and *hyle*? Is there, as I *think* the proponents of the B-L-R thesis want us to believe, a causal role of genus which *is* the causal role of matter – or, as Lloyd puts it, of a kind of matter? Perhaps, if we take 'kind' as limited by the identification of species and form. Relatively, a kind of matter can be informed in such or such a way. But, I have argued, for the generation of a man this is hominoid matter, potential for this species, not just generic. The seed may not develop, but if it does, it will be a man, not a mouse. Yet surely, Aristotle says, 1050a 15–16, the matter exists potentially because it may take on form, and when it exists actually it is in the form (ἔτι ἡ ὕλη ἔστι δυνάμει ὅτι ἔλθοι ἂν εἰς τὸ εἶδος· ὅταν δέ γε ἐνεργείᾳ ᾖ, τότε ἐν τῷ εἴδει ἐστίν.) and that is exactly the role of genus: animal may be four-footed or two-footed; when it is four-footed, the animality is in the four-footedness, since the genus is differentiated in the species. The only way it can be at all, p.t.e., is the way it precipitates at the close of an Aristotelian division into least species or the members of least species. Now I admit there is a meeting here of a number of lines of Aristotelian analysis in one terminal point: the generation and the existence of the concrete this-such with its unified, defining essence. But what I think the B-L-R interpretation, as distinct, say, from that of Wieland, is inclined to do, is to run together too quickly separate tracks of analysis which cannot in fact be so amalgamated. If we took the logical analysis of definition, the physical analysis of natural substances in terms of matter and form, and the *alternative* analysis of substance in terms of act and potency (Aristotle quite plainly lists the last two as alternatives both in Delta and in Physics II) – if we took all these as one, then we would indeed come out where B-L-R want us to. But then, I fear, we would be gravely misusing Aristotelian methods. At the same time – and here is my polite bow – I think there is perhaps a point where the categoreal plurality of the Zeta approach meets the act-potency approach of Theta – there perhaps where we look at the before and after of substantial change – where matter behaves rather like genus. Yet – I must partly retract my concession – the genus-like character in question *is* not the technical genus which stands as mere animal to the man who is both human form and animal matter, it is *humanly* animal matter, not (unless he be a mere monster) just animal as such. Even here, I must hold in conclusion, the relation of genus and matter is more analogical than literal.

I have said much more, alas, about the plurality of '*genos*' than about the unity of '*eidos*'. To defend my thesis about the latter – or rather to develop my hunch – I would have to go farther afield, into the question of the τί ἦν εἶναι and the Aristotelian theory of perception and its role in knowledge. After all that, there would still be the question of the bearing of Aristotle's taxonomic concepts on the modern species problem. Despite Aristotle's anti-evolutionism and despite the deep differences between Aristotelian and modern science, I am confident that somehow Aristotle has something to say to us here. But I have not come near raising that problem, let alone resolving it. I can only conclude, with Hume: 'Others, perhaps, or myself, upon more mature reflexions, may discover some hypothesis, that will reconcile those contradictions'.[10]

NOTES

* This paper is – as A. L. Peck would put it – 'zetetic rather than expository'. Indeed, I could not have searched even this far without the help of a number of colleagues,

especially of Professors Michael Frede, Dorothea Frede and David Balme. My confusions, of course, are still my own, but if I have achieved any clarity on the difficult problem before us, I have them to thank for putting me on the road to it.

[1] *Classical Quarterly* **12** (1962), 81–98.

[2] *Proc. Cambr. Phil. Soc. N. S.* **16** (1970), 12–21; cf. *Aristotle's De Partibus Animalium I and De Generatione Animalium I* (translated with Notes by D. M. Balme), Clarendon Press, Oxford, 1972, Notes on PA 1, 2 and 3.

[3] PA I, 4, 644a 12ff.; cf. *e.g.* HA I, 1, 486a 15ff.

[4] 'Genus, Species and Ordered Series in Aristotle', *Phronesis* **7** (1962), 67-90; 'Aristotle's Principle of Individuation', *Mind* **59** (1970), 519–529.

[5] 'Genus as Matter', in *Exegesis and Argument*, New York, 1973, pp. 393–420.

[6] *Die Aristotelische Physik*, Göttingen, 1962, pp. 209–10.

[7] 1034a 8ff.

[8] See the *De Anima* for example.

[9] It has been argued by Professor Montgomery Furth that Aristotle's discussion of differentiae in *Cats.* V in fact assimilates these to the category of substance. I do not believe that the text quite bears this out, and if it did it would make nonsense of Aristotle's discussion of problems connected with other categories – as in Iota, for example – as well as of his treatment of division and of his scientific method as such.

[10] Professor Balme has kindly commented on my argument; his remarks bring out a number of points which merit, and will, I hope, receive attention on some other occasion. He writes:'If I define a genus in my way of thinking, I have to state a group of possibilities; e.g. Bird is Animal having two long/short thick/thin legs, long/short feathers, etc. These are the differences possible within the generic similarity. Therefore the definition of a genus cannot be a list of determinations (as the species is) but a list of alternative possibilities which can be more or less vague and far-reaching according to the taxonomic level at which you put the genus (and this is only a matter of convenience, again unlike the species). The proximate matter of a bird is stuff capable of these differences; it may be merely stuff capable of developing wings rather than forelegs, or it may be worked up to the point where it is capable of long or short wings but not this that or the other – i.e. it is nearer the specific determination. So far, the definition of genos at each level of greater or less precision will be the same as the definition of the proximate matter at corresponding levels. But the proximate matter also produces differences in the individual which are not in the specific definition – snub nose, blue eyes. They are due to irregularities in the matter, and to mutual interference between movements in the matter (GA 778a 6). They are excluded from the specific definition because they occur in fewer than the majority of individuals, i.e. they are unpredictable. But they are still within the generic possibility, for otherwise they could not be present. (Matter qua matter has of course no formal attributes, while the snub nose is a shape. Just why it is unpredictable makes an interesting question.) Therefore we can say that Socrates and Callias are one in genus in spite of their differences, because their genus is not a hard list but a statement of alternative possibilities; it is a statement of the overarching similarity of these possibilities. They are separate in number and in matter, because each is a separate instance of the genos – i.e. a separate collection of generic attributes housed together, and that *is* a separate bit of promimate matter. They are the same in species for quite a different reason, which I think you are hinting at in your unfulfilled program – the specific definition is confined to stating the really essential *ti en einai*, which does not include the blue eyes. This way of looking at it depends on defining the genus by multiple differentiae. Then the puzzle about excess and defect becomes

tractable: the catfish's attributes are determinations of fish-possibilities – *such* fins, *such* organs, all being quantitative variations of given structures. *MA* I 486a 22–b 17 is quite helpful. Wieland seems to me wrong in saying that the genus cannot be the *out-of-which*. The specific difference is not something added to the generic definition, but is a determination of potentialities already given in the genus. I think he forgets that 'Bird' is the *name* of a genus, not its definition.'

Balme's contrast between Socrates' and Callias' unity in genus and in species is especially important: he is clearly using 'genus' here, as Aristotle does, to mean 'kind' at *any* range of generality, while the specific form, which is unique, plays a special role both in nature and in our knowledge of it. It is this distinction, I believe, which is neglected in Professor Furth's illuminating paper given at the same symposium at which my paper was read. Even in generation, the controlling role of the specific form throughout the process needs to be kept sight of. Thus Furth's statement that 'the process that takes matter into specific form and the process that takes mass into countable individuals are one and the same process' needs, in my view, to be carefully qualified if it is not to misrepresent Aristotle's account. Even the mass contributed by the mother is 'almost semen'; a woman is (in Aristotle's view) a deformed *man*, not a failed lion or fish or bird. Again, it is *human*, not just animal, potentiality that submits to form to produce this blue-eyed, snubnosed baby instead of this brown-eyed, straight-nosed one.

CHAPTER VII

TWO EVOLUTIONARY THEORIES

I. THE TWO THEORIES

1959 will find the majority of Western biologists assenting with renewed enthusiasm to the basic principles enunciated a century ago in *The Origin of Species*. But there continues to be an heretical minority, and, moreover, the heretics speak with authority and vigour. When, in turn, the orthodox pause to answer the dissenters' arguments, they offer an illuminating paradigm of scientific controversy: illuminating both for a study of the nature of scientific disputes in general and for the epistemological problems inherent in evolutionary controversy in particular.

I propose to limit myself here to the analysis of one such controversy: the disagreement between the theories of evolution stated by two palaeontologists, G. G. Simpson in *Major Features of Evolution* (1953) and O. H. Schindewolf in *Grundfragen der Paläontologie* (1950).[1]

Professor Simpson is the principal American spokesman of neo-Darwinism, or, as he now prefers to call it, the 'synthetic' theory, and his *Major Features* is certainly one of the most carefully elaborated statements of this position.[2] The very fact that he speaks *as* a palaeontologist lends weight to that statement. Darwin himself had trouble with palaeontology;[3] and there have been palaeontologists ever since who have failed to adhere to the Darwinian position even when it had become an orthodoxy. Darwinism – and in this neo-Darwinism is no different – presents a picture of life streaming endlessly forward, confined in this or that channel only by the combined effect of environmental change and natural selection, which selects the useful by breeding out the less useful. This picture, however, has seemed to some palaeontologists a distortion of the fossil record, and particularly in two ways.[4] First, the record itself is full of gaps; forms appear and disappear suddenly. There are a few famous transitional forms, like *Archaeopteryx*, but not nearly all there should be if evolution really happened as gradually as the Darwinians think. Secondly, such continuities as the fossil record does exhibit often seem to be unrelated to utility (or in modern language to adaptation) and hence to lie beyond the control of natural selection. The stock examples of this are found in the sabretooth tiger and the Irish elk. Respectively they have teeth or antlers that go on and on getting bigger and bigger even after they become a nuisance, and this was said to illustrate the phenomenon of 'orthogenesis'. This concept is still used by some scientists (including Schindewolf), but the problem it raises really forms part of a larger question: are there important phenomena in evolution which appear to be non-adaptive, and which, therefore, cannot be reasonably interpreted under the aegis of selection? Up to recent years, at least, many palaeontologists thought there were such phenomena: they thought, in other words, that the concepts of adaptation and selection were not adequate for the interpretation of their subject matter.

Against these opinions we have Simpson, himself a most distinguished palaeontologist, the foremost living authority on equid history and a pioneer in work on rates of evolution, lending the weight of his authority to the neo-Darwinian view. He sees evolution as a continuous series of minute changes in innumerable directions, in which all alterations of any significance, larger as well as smaller, quicker as well as slower, are determined by the great co-operating 'pressures' of mutation, geographical isolation, and selection, with adaptation as the universal effect, and criterion, of systematic change. The basic concept, ultimately, is variation in the occurrence of genes; out of such variations all the systematic relations of living things have been gradually evolved.

There are a number of prominent biologists who are dissatisfied with this view;[5] Simpson, in *Major Features*, takes special pains to reply to the critique of one of them: the German palaeontologist Schindewolf; and it is with their disagreement that I propose to deal.

Schindewolf finds the fossil record full of striking discontinuities, even where there can be no serious question of gaps in the record; new structures just suddenly turn up. Such new structures, moreover – for instance, the limbs of tetrapods – were not, he believes, in themselves adaptive. Land dwellers *might* have started with two legs, or half a dozen, or, octopus-like, with a ring of tentacles. How can we say that four is the best-adapted number, or that the pentadactyl ground-plan is the best-adapted pattern for such limbs in all the great variety of circumstances in which they have come to be used? This happens to be the characteristic tetrapod 'Bauplan', but there is no conceivable reason of utility why this number and no other, this blueprint and no other[6] should be found. Once present, of course, these four pentadactyl limbs did indeed develop in specialised, adaptive ways: for running, digging, jumping, etc. Their common structure, however, was not in itself, in Schindewolf's view, an adaptation, it was simply a new type of organisation, four-footedness, capable of numerous and varied adaptations.

Furthermore, Schindewolf agrees with the older palaeontologists that within each type, once it has appeared, there is a progressive, orthogenetic development. In fact, there is a rhythm analogous to that of birth, maturation, and senescence: the sudden appearance of a new type, its orthogenetic advance, and finally a stage of the breaking up of types which usually leads to extinction. [7] The choice between a paraxonic and mesaxonic limb, for example, appears to be, from the point of view of adaptation, a matter of indifference. For speed of locomotion, either will do. Once the choice was made, however, there was a progressive, orthogenetic trend within each form, to the horse on the one hand, and the camel on the other. The major type – i.e. the structural plan of artiodactyl as against perissodactyl limb – had to come first, and within this the development of specialised forms occurs – e.g. cows, pigs, and camels as against horses, rhinos, and elephants.

Notice, moreover, that in each case the more general category appears first, and specialisation is specialisation *within* it. Whereas for Simpson it is a question of particular small changes adding up to big ones, for Schindewolf the big changes in structure come first, the new general structures breaking down thereafter into more special adaptations. This is related, as we shall see, to the fact that Schindewolf looks at structures first, putting morphology ahead of phylogeny, while Simpson, like all Darwinians, wants to make phylogeny – the history of life – the one foundation on which all the rest of biology is built.

II. PARALLELISM OF THE TWO THEORIES

We have here, then, a nicely balanced pair of theories: one stressing continuity and the adaptive character of all evolutionary change, the other stressing discontinuity (novelty of forms) and the non-adaptive character of major changes. The two writers disagree seldom, if at all, about the 'facts', yet each rejects emphatically the other's inferences from the facts and is at pains to refute them. Consider briefly how the two views balance one another in this respect.

(a) Sometimes the very same matter of fact seems to the two of them to provide evidence for diametrically opposite conclusions. For instance, the fact that some very old genera, like *Lingula*, still survive, is proof, for Simpson, that evolution does not exhibit the general pattern of racial life and death: it cannot be said to go on in a cyclical fashion if some old survivors are left happily in their ancient and well-established niches.[8] The same sort of fact is proof, for Schindewolf, that selection is by no means all-powerful, since if it were it would wipe out all simpler forms in favour of the more advanced, i.e. more efficient.[9]

(b) Each one accuses the other of committing the same or similar fallacies. About the question whether higher categories – orders, classes, phyla – originate before or after species, each one accuses the other, in nearly the same words, of arguing retrospectively, and reading into the record what he himself wants to put there.[10] Did the class of birds originate, in one step, with *Archaeopteryx*, before the innumerable species and sub-species and varieties of birds had developed? Simpson says *Archaeopteryx* was a species like any other, originating by normal speciation from other reptilian species; only when we look back over the whole vista of evolution do we say, this particular species was the first of what has turned out to be a new class.[11] Schindewolf says, *Archaeopteryx* is a bird: i.e. an animal that flies by means of feathers. There were never any before and there have been a great many since – but the transition, from class Reptilia to class Aves, happened then and there.[12] The principle, flight by means of feathers, was genuinely new. Only the neo-Darwinians, late in time, come and conjure away its essential novelty.

Or, again, each accuses the other of unnecessary and mystifying assumptions. Simpson considers that orthogenesis smacks of 'purposiveness'[13] – the worst possible name-calling in evolutionary controversy; and the division of the history of a type into stages analogous to birth, maturity, and old age, he considers viciously anthropomorphic and unjustified by the facts.[14] Schindewolf, on the other hand, insists that while his use of orthogenesis and of the type concept is purely descriptive,[15] nothing could be more mystifying than the prophetic vision of selection which the neo-Darwinians implicitly assume: the idea that selection selects what *will* be advantageous in another ten million years, and thus foresees advantages that will accrue to the remote descendants of the forms selected.[16]

True, a neo-Darwinian, accused thus baldly, would simply deny that he makes any such preposterous assumption. Selection, he says, selects what is useful in *each* generation, otherwise it would not be selected. There is no 'foresight' involved, but only the automatic and immediate control of variation through utility. Yet there is nevertheless some plausibility in Schindewolf's criticism; for there are many cases in which Darwinians are forced to relax this basic principle, and to loosen the connection between adaptation and selection. It seems highly unlikely, for example, that the very minutest beginnings of ultimately co-adaptive structures, like the appendages of crustaceans, should have been of immediate and constant advantage to their bearers.[17] What does neo-Darwinism make of such cases? Sir Julian Huxley tells us that *very* slight selective values can now be seen to be effective, because *evolutionary periods are so long*.[18] But can an incomprehensible happening become comprehensible by lasting a long time? Either one is simply describing the direction of a trend that has already happened, and 'selection' is just shorthand for such a description, rather than an explanatory concept; or 'selection' stands for some mysterious force which controlled the whole process, in Dobzhansky's phrase (to quote another eminent neo-Darwinian), *sub specie aeternitatis*.[19] This sounds suspiciously like Schindewolf's foresight of selection, as does Simpson's statement: 'it is certain that if we can see any advantage whatever in a small variation (and sometimes even if we cannot), selection sees more'.[20]

To return to Schindewolf: as against the highly abstract and hypothetical Darwinian theory, he makes the assumption: *that life can originate novelty*.[21] This is, he says, a preferable assumption because it is simpler. He does not pretend to 'explain' this proposition, and in that sense it may be 'mysterious', but no more mysterious, he says, then physical concepts like 'force' or 'gravitation' which everyone is prepared to accept as 'explanatory' – and not nearly so mysterious, he believes, as the whole nexus of assumptions implicit in the concept of 'natural selection' as the neo-Darwinians use it.[22]

(c) Finally, and most neatly characteristic of this type of situation, the two scientists argue in exactly opposite directions, each taking as his premiss the negation of the other's conclusion. Simpson, wedding palae-ontology to the statistical methods of population genetics, sees a gradual change in populations such that the sharp divisions of traditional mor-phology become false.[23] Schindewolf, basing his theory on the logical priority of morphology, concludes that the gradualist, statistical picture of neo-Darwinism is false.[24] To put it very schematically; Simpson argues: the neo-Darwinian theory is true; morphology implies that neo-Dar-winism is not true; therefore morphology is wrong. Schindewolf argues: morphology must first be accepted as true; morphology implies that the neo-Darwinian theory is wrong; therefore the neo-Darwinian theory is mistaken. Or to put the matter another way, they agree on their major premiss: *traditional morphology and neo-Darwinism are incompatible.* One says: Darwinism, therefore not morphology; the other says: morphology, therefore not Darwinism.

In each book this argument – that is, in effect, the argument on the relative priority of morphology and phylogeny – comes late; but in each book it is a most decisive argument, which has been in fact moulding the shape each theory has taken all along. Why is it so decisive? In neither case has a crucial experiment falsified the other theory. Nor is it merely that the two writers start from different premisses. But these different premisses, leading each to a conclusion directly contradicting the alter-native premiss, are symptoms of a more pervasive difference, a difference in ways of thinking. To put it very simply, the two writers look at their material differently. It is this difference in 'looking' that I want to analyse and in part to evaluate.

III.   PLANES OF DISAGREEMENT

When we consider the two theories a little more closely, we find that the differences between them are themselves of different kinds. Perhaps we may say that our two scientists disagree on different *levels* or *planes* of *thought*. There is, first, the *verbal* plane. They differ at many points simply in the formulation of observed phenomena or of their generalisations from them. Secondly, there is the *visual* plane. They use different sorts of visual imagery to support their theories; the models on which they rely are different in kind. Third, the plane of *attention*. They pay heed to different aspects of the phenomena. Different *kinds* of question interest them. Finally, and most fundamentally, we come to the *conceptual* plane. Their 'outlook', the interpretive framework *out of which* they are looking at their material, is different.[25] Here, and indeed on our second and third levels, we will have moved beyond the range of differences specifiable in terms of vocabulary and syntax, to the more pervasive *tacit* component of theoretical knowledge.[26]

IV. THE VERBAL PLANE

Consider, then, first, the verbal differences between the two theories – those differences which *are* specifiable in terms of vocabulary and syntax. The interesting thing about these differences is that when one looks at them hard, they disappear. It looks as if a great many statements in the two theories could be translated without loss of meaning one into the other. This does not mean that their disagreement is 'merely verbal'. On the contrary, if their statements were all translated into a common terminology, they would agree verbally but would still disagree.

Let me mention briefly two examples of such verbal disputes. Simpson objects to 'orthogenesis' but speaks of 'essentially rectilinear evolution',[27] which is all that Schindewolf's 'orthogenesis' is shorthand for. Schindewolf insists that 'adaptation and genuine evolution are two quite different things', yet in speaking of the origin of mammals he says this was a 'general improvement' as distinct from the adaptations of particular groups of mammals to particular environments – as the limbs of bats, whales, etc.[28] But Simpson also distinguishes between adaptations of a general and variable kind, and more limited and specialised adaptations – like the rodent's tooth.[29] Why couldn't Schindewolf speak of 'general adaptations' instead of 'general improvement'? But he would be very angry at this suggestion, as Simpson would be if told that he might as well say 'orthogenesis' instead of 'essentially rectilinear evolution'.[30] Obviously the verbal disagreement, which, in these and a number of other cases, could be easily eliminated, is relatively unimportant. It is the difference that would remain in spite of verbal agreement that matters.

V. THE VISUAL PLANE

Let us try, then, to look behind the verbal level of the dispute and consider the differences that underlie it. The two scientists, we said, *look* at the facts differently. They differ in the visual imagery by which their theories are supported. In Simpson's case the visualisation is at a relatively abstract level. Sometimes he uses graphs illustrating complex statistical relationships, sometimes models in the sense of cognitive maps. In Schindewolf's case it is usually, more directly, the actual shape of fossils that we are asked to envisage: these shapes and differences in shape illustrate the relationships characteristic for evolution.

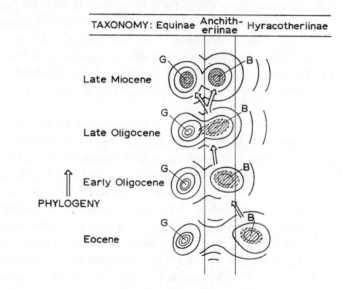

Fig. 1.

Consider an example from each. Simpson (Figure 1) summarises the development of browsing and grazing types in equid phylogeny in terms of the 'selection landscape' technique introduced by Sewall Wright.[31] Selection landscapes are constructed by applying abstract topological techniques to a wide range of data already subject to elaborate statistical manipulations. The result is a series of 'peaks' and 'valleys', each peak representing an adaptive optimum for the characters and groups in question.[32] In the case of horse evolution, we have two adaptive peaks, a browsing peak (B) and a grazing peak (G), with periods of shifting between them, and finally (not shown on this diagram) the disappearance of the browsing peak.

Compare with this a diagram of Schindewolf's (Figure 2),[33] showing the development of one type of modern coral (f–h) from an early onto-genetic stage (b) of an ancient coral (a–c). It seems unlikely, if we look at (c) and (h), that a transition between them could have occurred by small cumulative steps. We could imagine, however, that in a young Pterocoral, the third pair of protosepta (c III) failed to develop, and instead (as is the case with modern Heterocorals), the second (f II) and then the first pair (g I) divided laterally, producing the four divisions (and other novelties in the arrangement of the subdivisions) typical of the new kind of animal. This is something one sees intuitively, by looking at the actual shapes of fossil and recent corals at early and later stages of growth.

VI. THE PLANE OF ATTENTION

The difference in visualisation is symptomatic, further, of a difference of attention. The two are looking at different aspects of their common subject matter. What is central for one is peripheral for the other, and vice versa. Simpson is applying statistical methods in palaeontology, in close reliance on the theories of population genetics, which are also statistical. He is dealing, not with individuals, which do not evolve, but with populations, which *do*. Schindewolf is looking rather at individuals of different shapes and sizes, and the problem is to relate them phylo-genetically; this can best be done in reliance on analogies drawn from embryology. From these different perspectives the same 'facts' look different. For example, Schindewolf uses the development of artiodactyl and perissodactyl limbs as evidence that major steps in evolution occur discontinuously – and that therefore his theory rather than the neo-Dar-winian is correct.[34] Simpson cites the development of the artiodactyl limb as a case of what he calls quantum evolution, that is, evolution by an all-or-none transition.[35] Here the two are even in verbal agreement, for Schindewolf describes his transitions from type to type as 'quantenhaft'. Moreover, both agree that the transition in this case constitutes a sharp dividing line. Yet the different field of attention in each case puts such divergences into a very different context. Simpson is looking at gradual transitions in the genetical make-up of populations, while Schindewolf is looking at changes in the structure of individuals, which often vary slightly though sharply, and between two such sharply divided varieties organisms fall easily into 'types'. Thus Simpson sees a sequence of particular changes building up to a generalised trend, Schindewolf a general pattern with lesser changes within it.

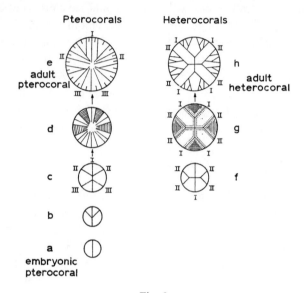

Fig. 2.

Let us look more closely at Simpson's account. Populations, from the point of view of population genetics, are aggregates, not even of individual organisms, but of gene-complexes, tending to vary this way or that. A population is conceived in terms of its 'gene pool', that is, the sum total of the genes of its members' germ cells. 'Genetical selection' is the concept used to describe the systematic direction of the variation of genes, as distinct from drift, a minor factor, chiefly in small populations, which consists in fluctuations that fail to establish well-defined trends. Concepts such as 'threshold' or 'break-through' serve to facilitate the understanding of relatively abrupt shifts in such gradually changing trends; but what is of primary interest is the gradual change in the overall distribution of genes in interbreeding populations. That there was a first *Eohippus* or *Mesohippus* or *Equus* is not nearly so interesting as the fact that there were gradually more.

Attention to overall population trends, moreover, is supported by attention to the shifting organism/environment relations – strictly speaking, gene-pool/environment relations – which underlie them. Individual wholes are not of interest in this vision of pattern growing out of flux. And of course, it is primarily the 'mechanism' of these innumerable small changes that population genetics has so successfully interpreted. Minute alterations in the material of heredity do tend to appear in experimental populations, and to perpetuate themselves. Such systematic change in hereditary material is called 'genetical selection', and selection is said to be the 'mechanism' by which such change happens. It is not, however, a 'mechanism' in terms of a cause-and-effect relationship, for the occurrences it correlates are random. Nor are internal laws of structure or development of interest to this kind of evolutionary thinking. Selection in the sense of population genetics is simply any systematic change, regardless of cause, in the make-up of the gene-pool of an actively interbreeding population. The occurrence of mutations being taken as random, selection is the settling of these random occurrences into statistically manageable averages. It is represented by the curve established by reading averages of gene frequency over a number of generations. Except in direct relation to reproduction, therefore, selection, in the sense of genetical selection, has no *intrinsic* relation even to adaptation. Simpson himself makes this quite plain. That such systematic change produces adaptation can be stated only on empirical evidence, or as a consequence of relaxing the strict genetical definition.[36]

Schindewolf, on the other hand, is looking at the difference, say, in the shells of cephalopods or the structure of corals. The interesting point is not that the number of ammonites or of heterocorals came gradually to a peak, but that there were first ones, abruptly different from anything appearing in lower layers of fossil deposits. These changes cannot be interpreted as crystallising out of fluid gene-pool/environment relations, since they often appear when there is no evidence of significant environmental change. In fact, in Schindewolf's view, it is by the change in fossils that the succession of strata must be ordered, not vice versa. Moreover, the new types often show no perceptible difference in adaptive

relations. In the history of corals, the change from six to four divisions
is not a change in adaptation. It is just a change in plan. What we are
looking at is a set of forms, macroscopic structures, Gestalten, which are
in their nature discrete. The problem, for example, for the two types of
coral, is to account for the continuity between them: and here the concept
of early embryonic change (proterogenesis, or in English usage paedo-
morphosis) comes to our aid. Such change is also mentioned by Simpson,
but concerns him little; for what interests him is the statistical trends in
populations, not the life history of the individual embryo. Thus, though it
is Simpson who seems to have moved beyond a succession of stable fossil
forms to the genetic basis of evolution, it is Schindewolf who sees these
stable forms as relics of living individuals and connects them, in imagination, through concrete, developmental stages rather than through abstract,
statistical change. It is true that populations, not individuals, evolve;
but populations are populations of individuals, and something that
happened in individual development must have underlain the abstract
statistical relationships which population geneticists construct. It is this
underlying *real* change that Schindewolf is trying to envisage. In other
words, where Simpson interprets the palaeontological data in conjunction
with the abstract generalisations of population genetics, Schindewolf relies
on the embryologist's conception of individual development.[37]

In short, looking at populations throughout evolutionary history, we
see a series of ever-branching particular directions establishing relatively
stable interplay of gene-complex/environment relations. Both these
variables change continuously, and so does the result, statistically viewed,
of the outcome of their interactions. Looking at individuals, we see a
multiplicity of types, which, once there, adapt themselves to particular
variations in environment, and of course die when they become mal-
adapted; but the initial divergence in types is the fundamental, and irreducible, starting-point. For Simpson, evolutionary change consists in the
'break-through' of a population into a new 'zone'. This may occasionally
happen as an all or none reaction, but not generally. The so-called explo-
sive phase of evolution is usually just a speed-up of normal speciation
'due' to intensive selection pressure in a new 'zone'. For Schindewolf,
evolutionary change occurs when a youthful animal takes a new turning
on its ontogenetic course, and so, as Hardy puts it, 'escapes from specialization',[38] with the result that something new – a new ground plan –

appears which is withdrawn momentarily from the confining influence of selection, and proliferates, out of sheer exuberance, in all sorts of directions.

VII. THE CONCEPTUAL PLANE

Our two scientists, then, not only lean on different kinds of visual images as models, but look at different aspects of the phenomena. Why these differences? Why do they use such different visual aids? Why is their attention so differently directed? The answer can be given only out of a third sense of 'looking': what they are looking *for*, the general perspective out of which they look, is different. They get different answers from nature, in Kantian terms, because they are putting different questions: and they put different questions because they stand at different places in relation to their subject-matter: their 'outlook' is different. This may sound like playing with words: one can only see what one is looking at, and one only looks at what one is looking for. But consider our two cases in terms of this question: what are they looking *for*? What is their outlook, their way of seeing? That is really the most fundamental kind of difference, on which the differences in verbal usage, in visualisation, and in attention, all depend.

Clearly, as I said at the beginning, Darwinism (and following it, neo-Darwinism) takes its stand firmly on the ground of continuity.[39] It sees life, and ultimately matter, as continuous and therefore explicable, in the last analysis, in terms of a *single* set of principles. Schindewolf's sort of theory envisages a variety of types, a set of discrete wholes or patterns, in step-like arrangement. This simple alternative may seem to be contradicted by the statement of Dobzhansky (certainly among the most authoritative neo-Darwinians) that the most striking phenomenon calling for explanation by the biologist is *dis*continuity.[40] But actually Dobzhansky's statement confirms my interpretation: continuity is the fundamental framework within which neo-Darwinian biologists are thinking, and discontinuity is the surprise they must explain away. They have to show how out of continuity the actual existent discontinuity – the biological discreteness of species, genera, etc. – could have arisen. Schindewolf, starting with the discreteness of different morphological types, has to explain their continuity. For him the panorama of forms comes first, and since these do appear, in a series of strata, with increasingly complex novelties succeeding one another, it becomes necessary to explain how transitions could have happened, as Schindewolf certainly agrees they did, between such basically different forms. Hence, as we have seen, his interest in early embryological change, for in this kind of change one can envisage an abrupt alteration in structure which nevertheless takes account of the actual continuity in the successive generations of living things.

Their disagreement about the relative priority of morphology and phylogeny lies close to the heart of the matter. Schindewolf looks first at structure, and can't understand how anyone can look at phylogenetic relations without having first looked at structure. But the great vision of Darwinism is precisely the dissolution of stable forms into a continuum. What Darwin did was to unsettle the fixity of species, to make nature flow. If he threw Paley's watch into the middle of the ocean, life itself became that ocean; in Simpson's image of selection landscape: 'rising, falling, merging, separating, moving laterally, at times more like a choppy sea than like a static landscape'.[41] And once life is seen as flux, change in its 'forms' is reduced to nearly imperceptible transition, the gradual elongation of the elephant's trunk or the gradual disappearance of the equid's extra toes, and so on.

If, however, the stress on continuity is essential for neo-Darwinism, that conception alone is by no means definitive for the neo-Darwinian outlook. A Lamarckian theory would also stress continuity, through the gradual establishment of new techniques to meet new needs. A Lamarckian theory would also, of course, stress adaptation, which the Darwinian view, as against Schindewolf's, also emphasises. What differentiates the neo-Darwinian theory is the peculiar combination of three basic concepts: continuity, particulate inheritance (and mutation), and adaptation. What is this peculiar combination? How do these three ideas hang together in the neo-Darwinian conception of evolution?

First, let us examine the relation between the particulate mechanisms of heredity and changes in heredity. This at first sight looks incompatible with the emphasis on continuity and gradualness of change and, of course, mutationism was for a while thought to contradict the basic Darwinian view. But as Fisher declared twenty-five years ago,[42] the knowledge that mutations are the bearers of evolutionary change in fact proves the triumphant support, the very keystone, of neo-Darwinism, and the want of this knowledge was precisely what made Darwin's own position some-times inconsistent. How, conceptually, do continuity and the particulate character of hereditary material go together? In two ways. First, to think in terms of small building blocks, like genes, is to think in terms of parts, not wholes, or of populations of such parts, that is, of collections of partic-ulars, which again are not wholes. Moreover, as we have seen, the populations of population genetics are not even collections of whole organisms, but collections of genes or at best of gene complexes, and a gene complex, again, is a sum of independent particulars. Even though its *effect* be conceived as integrated, in its constitution it is of a mosaic rather than a comprehensive kind. It is not an organic whole, but a juxtaposition of small discrete particulars. In short, both types of concept: continuous variation or change, and particulate inheritance with its minute though discrete alterations are, precisely, devices for *not* looking at wholes or Gestalten or structures: for reducing organic structure, on the one hand, to values of continuous functions, and on the other, to collec-tions of minute particulars. Here, I believe, we really meet the ruling passion of Darwinism: in the determination not to look at structure. Structure must be explained *away*; it must be reduced to the conditions out of which it arose rather than acknowledged *as* structure in itself.

This first part of the answer is negative, and admittedly does not take us far, since particulate pattern and continuity are so obviously discordant conceptions. The second part, however, is positive: this is where popula-tion genetics steps in to reconcile the discrete and the continuous. It is a most powerful instrument for synthesis since, statistically handled, collec-tions of genes, which are particulate, do nevertheless vary continuously. Combined with Mendelian statistics – which are themselves grounded in a particulate, non-structural view of the nature of organisms – the statistical methods of population genetics allow one, therefore, to view the fluctua-tions in Mendelian statistics (which appear as the changing average of the relative frequency of one gene over its alternative in interbreeding populations) as producing a continous process of evolutionary change. Nor is that all. Statistical calculations of rates of evolution, the construc-tion of adaptive peaks and valleys, etc., which are included in the appa-ratus of population genetics, allow one to construct a highly abstract quasi-structure which appears to introduce a satisfactory – because abstract – pattern into an otherwise unintelligible flux.

What about adaptation? Again, the central role assigned this concept does not follow from the concept of continuity in itself. Leibniz saw the world as a continuous array of degrees of perception, not of adaptations. In fact, it seems odd that the very theorists who most emphatically deny any shred of purposiveness to nature, should just as emphatically declare that all significant changes in nature are adaptive in character. But the stress on adaptation falls into place once one takes the refusal to look directly at structure as fundamental. It is the shifting organism/environ-ment relation, not the form of the organism itself, that is the basic unit. Organisms, already dissolved, from the perspective of evolution, into gene complexes, are themselves constantly changing as an (equally chang-ing) environment plays upon them. Thus what changes is itself a product of two variables: average gene frequency and environment. There are no constants which would have to be assessed as patterns or achievements in themselves, only what Simpson calls the 'splendid opportunism' of life.

This is indeed a brilliant perpetuation of Darwin's vision, and its persuasive force is compelling. Schindewolf's principles are simpler. He sees typical shapes, and he sees again and again what appear to be new shapes. Therefore he assumes that living things are able to originate novel types. Mutation, he agrees, must have been the mechanism by which they originated; but the adaptive control of mutation occurs only within, not between types. The basic pattern is of change from type to type, and always, as we have seen, with the more general appearing before its specialised subdivisions. We cannot, in fact, describe specialisation except *within* a more general type: there must be some kind to become special-ised.[43] First we have types, then we acknowledge the origination of new types. Within these, there is an inner dynamic of fulfillment: orthogenesis, and at the same time radiation into specialised subtypes controlled by adaptation and selection – the last and least creative steps, leading each time to an evolutionary cul-de-sac, whether as with Lingula to arrested evolution or, as with the majority of species, to extinction.

VIII. A CHOICE OF TWO CLOSED SYSTEMS?

These are, then, two ways of looking at the evolutionary record. The disagreement between the two is not, chiefly, about verifiable or falsifiable matters of fact, but about the concepts through which these facts are to be interpreted. Each constitutes, for its proponent, a closed interpretive system *in* which he sees the facts. In such a case, the objections of one to the arguments of the other are, from the other point of view, irrelevant. Can we weigh two such systems, cognitively, at all? Or are they like an ambiguous figure which we could see one way or the other, so that we may simply take our choice? In a sense, they are: for reflecting on the evolutionary record, as biologists agree in telling us about it, we may think of it in either of these ways.[44]

Thus we may seek to lose ourselves in what Simpson describes as 'great groups of animals living their history in nature'.[45] Or we may look rather at the ascending levels of organisation into which the continuous stream of protoplasm has shaped itself. So far, then, the two theories do seem to provide alternative frameworks for understanding the data, and it seems to be purely a matter of choice which we prefer. With respect to some details within each theory, this is in fact clearly the case. For example, it is really a matter of choice whether we say that the higher categories come before the lower, or conversely, the lower before the higher. In fact, each of our two scientists, in this connection, explicitly concedes the other's position, Simpson admits that *Archaeopteryx* was already a bird,[46] and Schindewolf admits that though *Archaeopteryx* was definitely a bird, if there had never been any more birds nobody would have known it.[47] Thus, on this particular point, each system can comfortably assimilate a theorem from the other, without endangering its total structure. So far as this particular case goes, we really can look at the matter either way.

For some details, on the other hand, one point of view seems preferable, while the other fits other cases better. Looking at the two, we may find, in this case, the choice more difficult. For example, the gradual change from one extinct species of elephant to another (marked by an increase in lamellar index, i.e. in the number of enamel plates per 10 mm of tooth) is well handled by Simpson's purely phylogenetic taxonomy.[48] It is difficult to see, on the other hand, how the origin of mammals could be handled in these terms.[49] Perhaps what we need, then, is a more inclusive theory, which will assimilate adequately both sides of the ambiguity. What would be the requirements of such a theory? This is a difficult question, and one with grave philosophical implications; I can only suggest here a starting point for a possible answer.

IX. PHILOSOPHICAL REFLECTIONS

As philosophers reflecting on scientific explanation, in this case on theories of evolution, we have ourselves to stand somewhere. We have to formulate our reflections within a framework of our own. I would specify, in my own reflection on these or any other evolutionary theories, three criteria, one methodological and two ontological, in relation to which, I believe, such a theory should be assessed.

(a) *Methodological.* Each of our two palaeontologists, we found, accused the other of similar fallacies; and in particular each accused the other of arguing in retrospect. This, I think, gives us a clue to the necessary method in evolutionary theory. For this is where they are both wrong, or both right, if you prefer. Each one does indeed argue in retrospect but this in itself is not a fallacy. For what else could they do? Thucydides could not have written the *Decline and Fall of the Roman Empire*, nor could a pre-Cambrian organism, or almost-organism, have stated a theory of evolution. This is not just trivially true. The judgment of the scientist is an essential element in every scientific explanation, and especially so in a theory of evolution, since here the scientist is judging not simply the world but the world including himself and his kind, *as* the world that produced himself and his kind. The French zoologist Vandel has formulated this principle for evolutionary theory as the 'Method of Recurrence'.[50] It is nonsense, he says, to decry anthropocentrism in evolution; there is nowhere to start except where we are. This is, I believe, an essential principle for evolutionary theory, as for all historical disciplines. There is no such thing as total objectivity in historical statements; we can only look back and evaluate the evidence from where we are. All historical discourse, including evolutionary theory, is in this sense reflective. In evolutionary theory, moreover, it is in part ourselves, including our capacity for historical reflection, that we have to explain.

(b) *Ontological.* Looking back from where we stand, therefore, at the vista of evolution, what is it that we see? Here, I think, both in the light of the biological evidence (as a layman, and as some biologists, see it), and of everyday experience, we ought to admit two principles which should *both* be taken just account of in evolutionary theory. First, we should acknowledge the continuity of life, and for that matter, once embarked on an evolutionary retrospect, of life with matter. Secondly, we ought also to acknowledge the diversity of ordering principles which appear at numerous levels in this continuous process. New principles of organisation appear to have arisen out of conditions that in themselves stretch back continuously – yet they are principles of organisation which order the conditioning continuum in a genuinely new way.[51] This duality, whatever the difficulties of stating it precisely, is essential in some form to any adequate assessment of the 'achievement' of evolution.

Now with respect to this duality, I submit, our two theories are in a rather different position. Therefore, as philosophers reflecting on evolutionary theory, we can and should allow them a different weight.

The dynamic of neo-Darwinism, I said earlier, is towards the denial of structure in the sense of reducing it to its conditions. The aim of Darwinism was to shake life out of its rigidity and thus to set it free. But this liberation obeyed also other passions of the modern mind: the reduction of principles to the least possible number, the love of explanation in terms of the least common denominator, of parts rather than wholes, conditions rather than reasons. Life is all matter, *only* organised differently.[52] Even those thinkers, oddly enough, who have stressed the term 'emergence', Lloyd-Morgan and Alexander, have insisted on weighting that concept towards the reduction of principles, so that it really is hardly emergence at all.

In fact, it is from this very reductivism that Darwinism has, in large part, derived its convincing power. For what it provided, and provides, is an explanation of evolution in terms of the mechanistic tradition of explanation. A mechanistic explanation must be both *logically simple* and *automatic*. Darwinism satisfies both these demands. First, it is simple, in that it unites all the relevant phenomena under *one* hypothesis and this hypothesis is conceived in simple terms of existence and non-existence. Secondly, it is automatic, since its one hypothesis interprets organisms mechanistically, and their evolution as produced, on the one hand, by random errors (which are the sort of impersonal deviation that can happen to machines despite human efforts to prevent them), and, on the other, by the blind and automatic control of natural selection, an automatism that goes on happening by nobody's intent. It seems likely that the machine model in classical physics derived much of its authority from these qualities of logical simplicity and automatism. Certainly this has been so outside physics. Again and again, faith in our ability to build, in other fields, theories having this double character has motivated new schools of thought in psychology, social science, and philosophy. And it is this faith that supports neo-Darwinism, permitting it to carry, without stumbling, the vast epicyclical superstructure which (as Professor 'Espinasse has pointed out)[53] has been erected upon it. It is this faith also which carries the Darwinian conception past difficulties like those put forward by Schindewolf, without conceiving of them as difficulties at all. In fact, it cannot so conceive them. For to take them seriously as difficulties would be to endanger precisely the logical simplicity and automatism of the neo-Darwinian framework, the very qualities which make it, to its adherents, so *scientific* an explanation. Thus, for example, Schindewolf's view of the sudden origin of basically new types or patterns of organisation implies that the recognition of order, and of novel order, is distinct from the statistical manipulation of the conditions producing order. This, however, is to introduce a duality of logical levels: continuous and small-scale conditions, *versus* discrete and comprehensive pattern; and that means to destroy the *unitary* character of the explanation. Again, to admit orthogenesis, or spontaneous direction in evolution, would be to deny the constant co-variance of gene pool and environment, and thus to suggest as a third factor an inner dynamic in organisms, as distinct both from the non-directive control of random variation and the external steering 'mechanism' of natural selection. But such a suggestion would deviate from the belief in the *automatism* of evolution. And again, it is precisely the double automatism of gene fluctuation and natural selection that *makes* the neo-Darwinian explanation a scientific explanation in the mechanistic sense. To say, therefore, that the 'facts' suggest spontaneity in the origin and development of organic pattern is, for convinced Darwinians, not to offer scientific evidence at all, but to step outside the bounds of science. From the Darwinian point of view such 'objections', which can be formulated only in non-mechanistic language, lie for that very reason beyond the scope of science altogether, and cannot therefore be taken seriously as scientific objections. Thus Sir Ronald Fisher can say in good faith that the only rational theories of evolution are those which make its driving force consist in progressive adaptation. These are the only theories, he says, 'which make at least the most familiar facts intelligible to the reason'.[54] They are indeed the only theories which rest on the ultimate values of total logical simplicity and automatism: within the mechanistic tradition, that is to say, the only rational theories.

The question remains, however: are the criteria of mechanism (logical simplicity and automatism) adequate as tests of a theory professing to interpret living things, much less the evolution of a living thing? Simpson, like many mechanistically minded biologists, boasts of following Ockam's razor,[55] which forbids us to multiply entities beyond necessity. As they apply this principle, however, biologists appear to take it to mean that entities must not be multiplied *at all*. There must be only *one* logical level, of physico-chemical events automatically directed this way or that by some equally material relationship. Yet a one-level logic of explanation appears to be unobtainable even in physics and chemistry. In a review of von Laue's *Theorie der Supraleitung*, for example, Professor H. Fröhlich writes:

... even an attempt to derive the properties of ideal gases would find the atomic physicist (if assumed to have no knowledge of macro-physics) at a loss without the introduction of new physical concepts. He would probably start with a discussion of the motion of two and then of three weakly interacting particles, and afterwards be led to the conclusion that consideration of more particles is very complicated and unlikely to lead to any simple results. It is only after the introduction of new physical concepts which do not exist in atomic physics – such as pressure and entropy – that other simple laws of physics (the gas laws) can be found.[56]

If this situation prevails in physics, so much the more so does it obtain in biology. Professor Rosenfeld has recently suggested that biologists should admit, in the use of concepts like 'order' (of which Schindewolf's 'Bauplan' or 'type' is an example), their reliance on a logic of *complementarity*: complementarity holding between concepts of biological organisation and explanations of the physico-chemical conditions of their existence.[57] In the evolutionary context, this general relation becomes the logical duality of new comprehensive ordering principles and the antecedent particulars which were the conditions of their emergence. Such duality is, I believe, unavoidable in evolutionary theory, and should be frankly acknowledged as such.

When, on the contrary, it is not acknowledged, the result is a reductivism which, however brilliant, is basically inconsistent. Some concept of biological organisation, which implies a non-unitary logic, is indispensable to biological theory. Where, therefore, it is ostensibly dispensed with, it nevertheless creeps in again, either in the unformulated presuppositions of the biologist, or in some ingeniously contrived disguise. This kind of procedure: the pretence of understanding a subject matter in terms of fewer and narrower concepts than are in fact indispensable to its under-standing, Polanyi has called the fallacy of *pseudo-substitution*.[58] It can be found, I think, at a great many places in the neo-Darwinian literature, and at a number of crucial points in the argument of *Major Features*. I want here to point out one such instance, which is especially instructive in comparison with Schindewolf, that is, Simpson's procedure with res-pect to the concept of types. 'Type' or 'Bauplan' is, as we have seen, a fundamental conception for Schindewolf. In terms of axiomatics, it is a primitive, undefined idea, on the initial understanding of which his theory rests. For Simpson, it is a term to be avoided. In the *Meaning of Evolution* (1949), where he uses the word freely, he explains that this is popularising shorthand, and when he says 'type' he is not in fact referring to types at all.[59] In *Major Features* he is more careful; his use of the term, or of its equivalent, is, it seems to me, carefully devised so as not to *seem* to mean type although in fact it does so. He gives us first a panoramic view of a continuum of living forms varying continuously in adaptation to a con-tinuously changing environment. Out of this doubly flowing stream, he then distinguishes with some, but not entire arbitrariness, various 'types of adaptation' or 'adaptive types.[60] This is, so far, not a phylogenetic concept, nor a morphological one, and if applied strictly could not pro-duce the results Simpson wants from it. Thus, considered purely in terms of *adaptive* types whales would be fish. To prevent this, there must be some structural, some 'type' reference, implied along with the adaptive one. However, with the continuous stream of life-environment relations as our fundamental concept, we can think of adaptive types as *zones* (the two terms are from here on used as equivalent) cutting themselves out at 'ecological thresholds' in this continuum,[61] and forget, for the moment at least, the implied structural aspect of the original conception. Since these 'zones' crystallise out of neighbouring areas in the biological continuum they will in fact be close relatives, so that the whale/fish prob-lem will not arise imaginatively even though in strict logic it should have arisen. Such 'zones' are, moreover, Simpson explains, only figuratively 'zones'; they represent, he says, 'a characteristic reaction and mutual relationship between environment and organism, a way of life and not a place where life is led'. [62] But calling them zones, we can think of them almost as places, as being 'entered' and 'occupied'. Thus the need to separate out the organism as an entity, and *a fortiori* the segmentation, the 'type' aspect, is reduced and concealed.

Yet 'adaptive zone' is introduced as a *synonym* of 'adaptive type'. Although the 'type' aspect is blurred, it is nevertheless taken for granted, for the theory would not work without it. Adaptation pure and simple is an insufficient instrument for the separation of different kinds of animals or plants: as, once more, in the case of whales and fish. In practice, of course, Simpson does recognise types – general patterns of structure or function – all along. Even to count the number of enamel plates per 10 mm of an elephant's tooth means recognising some kind of organisation of jawbones and teeth. To notice the threshold at which the perissodactyl and artiodactyl limb separated out demands knowing horses and elephants from antelopes and cows. Moreover, the concept of 'adaptive zone', by providing a more elastic substitute for 'type', allows us to forget the existence of such comprehensive knowledge.

X. CONCLUSION

We should conclude, I think, that where concepts of more than one logical level are necessary to the interpretation of a set of phenomena, we ought not to pretend to be operating on one level only. In the context of evolution where we in fact acknowledge novel operational principles, we should not pretend that nothing is there but the conditions without which (admittedly) they could not operate. The sum total of necessary conditions for the coming into being of an individual, a species, a phylum or of life itself are not logically or historically identical *with* the individual, or species, or phylum, or life itself. The pretence that the whole story can be told in terms of conditions alone, that flux can generate form, that materials not yet organised can logically account for their own organisation, is a pseudo-substitution, which purports to do without what it cannot in fact eliminate.

This situation is, I believe, unavoidable in a reductive theory of evolution, a theory which attempts to treat mechanistically a subject matter radically incapable of such treatment. I suggest, therefore, that instead of Occam's razor we might adopt as a test of theories of evolution the opposite principle: that entities, or more generally perhaps aspects of reality – for principles of organisation are not entities, though they do define entities – should not be *subtracted* beyond what is honest. In the light of this principle, we should ask of any theory of evolution, does it pretend to do without concepts which in fact it does not do without?

Here Schindewolf's theory, I think we must conclude, is, philosophically, more adequate than Simpson's. The concept of type, or of organisation in some form, is indispensable, and if banished makes its way in again, or has been there, surreptitiously, all along. Secondly, the concept of novelty or originality, which again Schindewolf admits as a fundamental concept, is indispensable. Simpson's theory, being more complex and abstract, would appear at first sight to cover a wider range of phenomena; yet by the reductive nature of its abstractions it also overlooks essential aspects of the phenomena. Schindewolf's theory is simpler in its reasoning, and less unified in its conceptual structure, but, perhaps for that very reason, it remains closer to the phenomena and does more justice to their experienced complexity.

Finally let me suggest briefly the bearing of this particular study, limited though it necessarily is, on the epistemological implications of the problem of evolution. The problem of evolution is to explain how new forms of life have originated. But new forms of life embody new operational principles, and these must be recognised as such before we can so much as ask about the continuity of the conditions out of which they arose. True, they could not have arisen without those conditions, but neither are they, logically or ontologically, identical with them. This means, I suggest, that we should apply to evolutionary theory an historical form of Kant's transcendental method. As latecomers in evolution, we should ask: what are the necessary presuppositions of the history we believe we are? What are the minimal concepts through which we can assess that history as having happened, and without which we could not understand that it did happen? If, as philosophers reflecting on evolution, we wish to ask this question, we shall find in Schindewolf's theory at any rate nothing incompatible with asking it, whereas if we commit ourselves to neo-Darwinism, at least in Simpson's formulation, we shall be unable to understand, much less to undertake, such an enquiry.

NOTES

[1] I have drawn also to some extent on other writings of the same authors, notably G. G. Simpson's *Meaning of Evolution* (1949) and two addresses by Schindewolf:

'Evolution vom Standpunkt eines Paläontologen', *Schweiz. Pal. Gesell.* (1952), 374–86, and 'Evolution im Lichte der Paläontologie', *Comptes Rendus, Congrès Géol. Internat.* (1954), 93–107.

[2] It seems to me fair to call Simpson's theory neo-Darwinism if one recognises the differences introduced by the Mendelian basis of the newer type of theory. In essentials it is still Darwinism.

[3] *Origin of Species*, Chapter X, and 6th edition, Chapter VII, answer to St. George Mivart on abrupt transitions of form.

[4] See for example H. L. Hawkins, *Rep. Brit. Ass.* (Presidential Address to the Geological Section), 1936, pp. 57 *et seq.*; H. F. Osborn, *U.S. Geol. Surv. Mon.* **55** (1929), *Am. Nat.* **68** (1934), 193–235; L. F. Spath, *Biol. Rev.* **81** (1933), 418–62.

[5] I may mention, in the U.K., four scientists who have expressed themselves as, in various ways, dissatisfied with the new synthesis (although I do not suggest that they would be in argement, either with what I am saying here, or with one another): Professor H. Graham Cannon of Manchester, Professor Paul G. 'Espinasse of Hull, Professor Ronald Good of Hull, and Professor C. H. Waddington of the Institute of Animal Genetics in Edinburgh.

[6] The same point is made in a different context in Professor Cannon's recent book, *The Evolution of Living Things*, Manchester, 1958.

[7] Schindewolf calls these stages *typogenesis*, *typostasis*, and *typolysis*, and his theory as a whole *typostrophism*.

[8] *Meaning of Evolution*, p. 192; cf. *Major Features of Evolution*, p. 233.

[9] *Grundfragen der Paläontologie*, p. 404.

[10] Cf. *ibid.*, p. 273, and *Major Features of Evolution*, p. 350.

[11] *Major Features of Evolution*, 342, 370; cf. pp. 347, 350.

[12] *Grundfragen der Paläontologie*, p. 126; cf. pp. 201, 277.

[13] *Major Features of Evolution*, p. 268.

[14] *Ibid.*, p. 233.

[15] *Grundfragen der Paläontologie*, pp. 430–31.

[16] *Ibid.*, pp. 413, 430–31.

[17] A. Vandel, 'L'Évolution considérée comme Phénomène de Développement', *Bull. Soc. Zool. de France* **79** (1954), 341–56. Cf. L. Cuénot, *L'Évolution Biologique*, 1951; 'L'Anti-hasard', *Rev. Scient.* **82** (1944), 339–46.

[18] In Huxley, Hardy, Ford, *Evolution as a Process*, London, 1954, p. 3.

[19] T. Dobzhansky, *Genetics and the Origin of Species*, 3rd. edn., New York, 1951, p. 75.

[20] *Major Features of Evolution*, p. 271.

[21] *Grundfragen der Paläontologie*, pp. 425–29.

[22] *Ibid.*, p. 431.

[23] *Major Features of Evolution*, pp. 377–78.

[24] *Grundfragen der Paläontologie*, pp. 300, 455 *et seq.*, esp. p. 475.

[25] In terms of Michael Polanyi's philosophy of personal knowledge, the two are thinking on opposite sides of a *logical gap*; see M. Polanyi, *Personal Knowledge*, London, 1958, pp. 150–60. In terms of Prof. Hodges' Riddell Lectures (*Languages, Standpoints and Attitudes.* Oxford 1953), they differ *existentially*.

[26] See Polanyi, *op. cit.* Part II, 'The Tacit Component'.

[27] *Major Features of Evolution*, p. 259.

[28] 'Evol. vom Standpunkt eines Paläontologen', p. 382.

[29] *Major Features of Evolution*, p. 346.

[30] See Simpson's note, *ibid*, p. 259: 'This means the same things *descriptively* as Schindewolf ... means when he says that orthogenesis is characteristic of the typostatic phase of evolution, although I reject the theoretical implications of his statement as decisively as the rejects mine.'

[31] Based on *Major Features of Evolution*, p. 157 (Figure 17). (By permission of the author and Columbia University Press.)

[32] Selection is represented as positive (uphill) or negative (downhill), its intensity being proportional to the gradient. See *ibid.*, pp. 155–57.

[32] Selection is represented as positive (uphill) or negative (downhill), its intensity

[33] Based on *Grundfragen der Paläontologie*, p. 214 (Figure 213).

[34] *Grundfragen der Paläontologie*, pp. 280–81.

[35] *Major Features of Evolution*, pp. 391–92.

[36] Dobzhansky, *op. cit.* pp. 79–80, equates 'adaptive value' with 'differential reproduction' (i.e. with genetical selection), thus concealing from the start the fact that the two are not conceptually equivalent. In this connection it is interesting that genetical selection is so often called the only known 'mechanism' of evolution, when it is by definition not a mechanism at all, but a statistically established trend. (See e.g. *Major Features of Evolution*, pp. 144–46). Strictly speaking, the 'mechanism' resides in the mathematical skill of scientists like Fisher, Haldane, Wright and Simpson himself. It is only in the older and looser Darwinian sense that selection is genuinely a mechanism in nature. Dobzhansky's definition glosses over this difficulty; Simpson admits it at the start but tends to forget his admission.

[37] In this connection see the important essays of A. Dalcq on ontomutations, 'L'Apport de l'embryologie causale au problème de l'évolution', *Port. Acta. Biol. Vol. Jub. Goldschmidt*, Coimbra, 1949, pp. 367–400, and 'Les Ontomutations à l'origine des mammifères', *Bull. de la Soc. Zool. de France* 79 (1954), 240–55; also A. Vandel, *op. cit.*

[38] Sir Alist iir Hardy in Huxley, Hardy, Ford, *op. cit.*, pp. 122 *et seq.*

[39] This is not, of course, a mathematical continuum, but a series of minute changes conceived as functions of continuously changing particulars.

[40] T. Dobzhansky, *op. cit.*, pp. 3–4.

[41] *Major Features of Evolution*, p. 157.

[42] Sir Ronald Fisher, 'The Bearing of Genetics on Theories of Evolution', *Science Progress* 27 (1932), 273–87.

[43] Simpson's argument that the higher categories are adaptive does in fact generally involve reference to a group which separates out *within* a still wider group – as bats, or rodents, or carnivores versus ungulates, and so on. See Chapter XI, *passim*, esp. pp. 346 *et seq.*

[44] We may also, of course, think of it in other ways; I am not suggesting that these are the only two; they are the two I have been looking at in this study.

[45] *Major Features of Evolution*, p. 265.

[46] That is, it had feathers; but was 'as reptilian as avian throughout' (*ibid.*, p. 370). For a more general 'concession' on higher categories, see page 350: 'In these usual cases it is true that occupation of the zone, which in retrospect is the origin of the higher category, precedes the origin of numerous genera, species and other units that come to comprise the higher category. In *this sense*, and this only, we can agree with Wright ... that 'there seems to be a large measure of truth in the contention of Willis and Goldschmidt (also Schindewolf, G. G. S.) that evolution works down from the higher categories to the lower rather than the reverse'.

[47] *Grundfragen der Paläontologie*, p. 453.

[48] *Major Features of Evolution*, pp. 387–88.

[49] See e.g. *Grundfragen der Paläontologie*, pp. 250–51. See also the paper by Dalcq referred to in n. 37.

[50] A. Vandel, *L'Homme et l'évolution*, Paris 1949 (2nd ed., 1958). Vandel's Method of Recurrence is, in effect, an application to the problem of evolution of Polanyi's 'fiduciary programme'; see M. Polanyi, *op. cit.*, pp. 264–68, and cf. also the argument on evolution in Part IV, Chapter 13 of the same work, pp. 381–405.

[51] See Polanyi, *op. cit.*, Part IV.

[52] On the implications of Simpson's use of 'organisation', see A. Dalcq, 'Le Problème de l'évolution est-il près d'être résolu?', *Ann. Sté. Zool. Belg.* **82** (1951), 118–38, p. 125.

[53] Paul G. 'Espinasse, 'On the Logical Geography of Neo-Mendelism', *Mind (N.S.)* **65** (1956), 75–7.

[54] Sir Ronald Fisher, 'Measurement of Selective Intensity', *Proc. Roy. Soc. B* **121** (1936), 58–62.

[55] *Meaning of Evolution*, p. 139.

[56] H. Fröhlich, *Nature* **161** (1948), 37.

[57] L. Rosenfeld, 'Causalité statistique et ordre en physique et biologie', *Anal. da Academia Brasiliera de Ciências* **26** (1954), 47–50.

[58] M. Polanyi, *op. cit.*, pp. 169–70.

[59] *Meaning of Evolution*, p. 7.

[60] *Major Features of Evolution*, p. 200.

[61] *Ibid.*, pp. 201 et seq.

[62] *Major Features of Evolution*, pp. 201–20.

CHAPTER VIII

STATISTICS AND SELECTION

I

The shadow of Archdeacon Paley and the argument from design still broods over evolutionary theory. Although the inference is now not to a Contriver but to Natural Selection, the premise from which the argument starts is still the same: the phenomenon of adaptation, the remarkable fashion in which tissues, organs, organ systems of living things appear 'suited to' the functions they perform. This premise is sometimes baldly stated in the identification of animals with machines, sometimes left implicit, as with Darwin, in the direction and emphasis of evolutionary theory, but its role is crucial in either case.

What I want to talk about here, however, is not the axiom of adaptivity, as one may call it, in its whole range, but one particular form of evolutionary theory in which the pervasive importance of adaptation has the status neither of an explicit postulate nor of an implicit belief, but of a proven theorem. This is, or seems to be, the case in Sir Ronald Fisher's *Genetical Theory of Natural Selection*,[1] a work which has played a decisive part in the rise of neo-Darwinism or, as it is sometimes called, 'the synthetic theory'. It is the conceptual structure of Fisher's theory that I want to examine. Again, not all theories of evolution, not even all neo-Darwinian theories share this structure; but Fisher's argument has been so very influential that it seems worthwhile to examine it in some detail.[2] I shall take the mathematical core of the theory for granted, but want to examine the concepts carried by the mathematical formalism.

As Fisher writes, 'the rigour of the demonstration requires that the terms employed should be used strictly as defined'.[3] Yet his argument, as a demonstration of the effectiveness of natural selection throughout the evolutionary process, depends, as I hope to show, on the use of his fundamental concepts, not always 'strictly as defined', but in a number of senses, one 'strict' and several more comprehensive.

II

It is the principal thesis of all Darwinian and neo-Darwinian theories that evolution is the result of the joint action of random variation (or mutation) and environmental pressure (or natural selection). How is this basic thesis supported by Fisher's argument in the *Genetical Theory*? I am not asking whether evolution is in fact a function of random variation times natural selection, but whether or to what extent Fisher has demonstrated that it is so. The crucial text we have to consider is chapter two, on the Fundamental Theorem, although much that is said in the chapters on the evolution of dominance and on mutation and selection is also relevant.

What is to be demonstrated in chapter two, Fisher tells us at the start, is that *the rate of improvement of any species of organisms in relation to its environment is determined by its present condition*. 'Improvement in relation to environment' seems to mean 'increased adaptation'; but the term 'adaptation' is formally introduced only later, where it is defined in terms which appear to presuppose the concept of 'improvement':

An organism is regarded as adapted to a particular situation, or to the totality of situations which constitute its environment, only in so far as we can imagine an assemblage of slightly different situations, or environments, to which the animal would on the whole be less well adapted; and equally only in so far as we can imagine an assemblage of slightly different organic forms, which would be less well adapted to that environment.[4]

In other words, an organism is regarded as adapted, if one can imagine its condition as an improvement over another possible condition that would be slightly less favourable.

Thus the concept of improvement is central to the argument, and we must find out what it means. And there is also the question, in what sense the present condition of an organism 'determines' its 'improvement in relation to its environment'. Does 'determine' here mean 'cause'; if so, how, or if not, what does it mean?

III

Fisher's procedure is to establish a table of reproduction, analogous to a life table, and to specify a parameter of population increase m, expressing the 'relative rate of increase or decrease of a population when in the steady state appropriate to any such system',[5] that is, when the distribution of age groups in the population is constant. Fisher calls this the 'Malthusian parameter of population increase', and it is said to 'measure *fitness* by the objective fact of representation in future generations'.[6] Now suppose we are measuring a particular character, such as tallness, in a population. And suppose we assume, as in the light of orthodox genetics we do, that all persistent changes in organisms are produced by changes in their genetic make-up.[7] Then we may calculate the average effect on the character in question of a given gene substitution, and also the effect of this one gene substitution on m (that is, on the chance of leaving posterity). We then calculate a summation of these gene changes *and* their effects on m for all the genes in the population, and we get what Fisher calls the 'genetic variance' – in this case for example, the genetic variance of tallness. Now suppose further that what we are measuring is not tallness but fitness itself. We then find, by further calculation, that: '*The rate of increase in fitness of any organism at any time is equal to its genetic variance of fitness at that time.*'[8] This statement Fisher describes as 'the fundamental theorem of Natural Selection'.

'm', or the 'rate of progress of a species in fitness to survive', is, as Fisher says, 'a well-defined statistical attribute of a population'.[9] It is the increase (or decrease) in the chance this population has of leaving posterity at some future date. 'Progress' here means simply this statistical increase or decrease, and 'fitness' in this context would seem to be the same for all organisms: it is a statistical measurement, which has, so far, nothing to do with the particular structures or functions of particular organisms as suited to particular environments. In fact, it has, so far, nothing to do with environment at all, nor with 'improvement' in any sense other than the very restricted quantitative one: that a gene or an individual or a population may be said to be 'improved' when its chance of having descendants at some future date has risen. This is 'improvement' in the sense in which a patient has improved when he is more likely so survive.

IV

If this were the whole story, there would be no problem. It would be possible, as indeed happened far and wide in the wake of the publication of Fisher's book, to make valuable calculations, resembling actuarial tables, by which to record and, up to a point, to predict the gradual increase or decrease of certain genetic factors in Mendelian interbreeding populations. Some of the great array of experimental work in which Fisher's techniques were in fact applied is referred to in this edition.

It seems odd, however, to speak in this connection of 'improvement in relation to environment' or of 'Natural Selection'. 'Improvement in relation to environment' suggests a correlation which has so far not been considered; and similarly, 'Natural Selection' seems to describe not a purely numerical relationship, but something qualitative as well; that is, the elimination of characters less well adapted to a particular environment in favour of those slightly better adapted to that environment.

In what sense, then, is the Fundamental Theorem a theorem of *'Natural Selection'* (concerned, *a fortiori*, with 'improvement in relation to environment')? In introducing the notion of 'reproductive value', that is, the probability that a given individual or a given age will produce descendants in respect to some future date, Fisher remarks that this value is of interest, since *'the action of Natural Selection must be proportional to it'*.[10] This is indeed true. If, as Fisher points out, the reproductive value of the inhabitants of Great Britain decreased between 1911 and 1921 as a result of the First World War, then whatever selective effects *might* take place, *must* take place only upon that declining group; and therefore the action of Natural Selection would be proportional to the decline. And conversely, where there *is* Natural Selection taking place, its operation can be investigated and measured by the use of Fisher's techniques. The fact of industrial melanism, for example, has been known for many years; but Kettlewell has recently studied it with great care and precision in terms of modern population genetics.[11] Thus if normal peppered moths (*Biston petularia*) are taken by predators in industrial areas in greater numbers than the darker *carbonaria* mutants, then the increase of *carbonaria* and the decrease of light coloured individuals can be recorded as a decrease in reproductive value of the normal form (which is more likely to be eaten before reproducing – and its offspring also and so on) or an increase in the reproductive value of the *carbonaria*, that is, an increase in the probability that there will be *carbonaria* in future. In other words, once we know that selection has *in fact* taken place in a particular situation, we can record and elaborate this knowledge in the statistical terms of Fisher's theory, and within the limits of selection experiments we can apply Fisher's technique to predict their results.

This statement about reproductive value, moreover, holds equally for the fundamental theorem itself, which generalises the notion of increase in reproductive value for an active interbreeding population, and equates it with the increase in probability of survival of the particular genes forming the genetic constitution of the population in question. Thus *either* (as in the decline of the British population) the fundamental theorem expresses the limits within which Natural Selection can operate; *or* (as in the case of industrial melanism) the occurrence of Natural Selection (as known from other sources) entails the truth of the fundamental theorem.

V

But it is not at all clear that the fundamental theorem entails Natural Selection. In fact, in the example of reproductive value cited by Fisher, the effect of war, the change in reproductive value is notoriously one in which what would appear to common sense to be the better adapted are eliminated in favour of the less well-adapted: the very opposite of a natural selection effect. For natural selection, according to Fisher's own insistence here and elsewhere, expresses the tendency to *increased*, not to *decreased* adaptation.[12] More generally, we should recall also that, as Darwin himself admitted, natural selection is applicable only to such characters or functions of living things as are both heritable and adaptive (whether for good or ill) – that is, relevant to viability or the reverse in a given environment. But we might record the increase or decrease of all sorts of characters in populations without regard to their 'adaptive' value one way or another. Suppose, for example, that red hair is increasing, and hence the probability of larger numbers of future redheads is also increasing; this does not in itself tell us anything about the usefulness of red hair, unless we know from some other source that all characters that are increasing in the population must be of use to it. Darwin was well aware of this restriction on his theory, but was confident that on the whole most characters of organic beings are adaptive. And what Darwin demonstrated in the first four chapters of the *Origin* was that, given the largely adaptive nature of organisms, and variation, and inheritance, and Malthusian population increase, natural selection necessarily follows.[13]

But Fisher's statistical proof does not appear to be rich enough in its premises to permit this conclusion. In terms of the 'strictly defined' concepts it uses, the fundamental theorem is not a theorem of natural selection, but a statistical device for recording and predicting population changes. Nor is the situation altered by calling such changes 'genetical selection'. We must still distinguish between 'genetical selection', which is purely statistical, and Darwinian selection, which is environment-based and causal. They remain two distinct concepts with a common name.

VI

It is important to keep this distinction in mind when interpreting the use made by Fisher himself and by others – notably, for example, Huxley, Haldane, Simpson – in applying his formula. The fundamental theorem has been stated, Fisher says, for 'idealised populations' in which 'fortuitous fluctuations in genetic composition have been excluded'.[14] He then calculates the error due to such 'fortuitous fluctuations'. The standard error turns out to be $1/T \sqrt{W/2n}$, where T is the time of a generation, n is the number of the population, and W the rate of increase in fitness of a population measured in terms of the summation of the effect of the increase in particular genes on m. In terms of this formula, it turns out that the rate of increase in fitness becomes irregular only when it is so slight as to be of the value of $1/n$. Only then, in other words, do the random fluctuations produced in a single generation by mutation (given random mating) affect the exactness of the statistical measurement, or, as Fisher puts it, 'the regularity of the rate of progress'. And more than this: if a long enough time is allowed, even this irregularity will recede to vanishing point. For even if the value of m for different genotypes were so delicately balanced in a given generation as to show only a $1/n$ distinction, and hence a fluctuation in m and the summation of the effects of genes on m, still over a span of 10 000 generations, the deviations from regularity would be just a hundredfold less.

Now this argument is taken to show that 'very low rates of selective intensity' are effective in nature. But if we have interpreted the fundamental theorem correctly, what it shows in fact is that very slight trends in the frequency of characters in populations may be recorded if they persist over a long period of time. Whether, in nature, such trends are the result of a process that can reasonably be called 'natural selection' is another question altogether.[15]

How does Fisher deal with this question? What he argued in his 1932 paper in *Science Progress*[16] was that such trends are not, in large populations, the result of mutation alone (or even chiefly), since mutations are too infrequent and advantageous mutations still more infrequent, and therefore it must be natural selection that directs the trends which his statistics describe. This follows, however, only if mutation and natural selection are the only two possible causes of evolutionary change: and that, Fisher is arguing, is what the theory is supposed to prove, not to presuppose.

Nor is the argument of our present text (i.e. the *Genetical Theory*) more satisfactory, for it moves in an even narrower circle.[17] Since only mutation rates around the $1/n$ value are 'selectively neutral', Fisher says, random variation, i.e. mutation, will very seldom indeed be an effective factor in evolution. But what does 'selectively neutral' mean? It means precisely: with a mutation rate, that is, a rate of gene substitution, around the $1/n$ value or less. Similarly, '1 per cent selective advantage' *means* 1 per cent increase in numbers of a particular gene substitution. Any mutation rate more substantial than $1/n$ therefore automatically *becomes* 'selective advantage' and so the product of selection, not mutation; and any lesser mutation rate remains 'selectively neutral', that is, counts as mutation, not selection. All this will be a more than verbal argument only if we know from other evidence that selection, in the environment-related, Darwinian sense, is always the cause of such statistical advance. In short, genetical selection is entailed by and measures but does not entail Darwinian selection.

In terms of the concepts used in the fundamental theorem, as strictly defined, therefore, the traditional theory of natural selection, that is, improvement in relation to environment necessitated by environmental pressures, has not been touched on at all. Increase or decrease in 'fitness' or if you like 'improvement' in Fisher's strict sense, is either the measure of the effect of Darwinian selection if there happens to have been any, or the measure of whatever else may have been happening to bring about increase in certain genetic factors rather than others. Similarly, what Fisher calls 'the *progress* of a population in fitness to survive' again means strictly no more than the increased probability of leaving posterity, whatever may be the reasons. There is no justification, in terms of the concepts 'as strictly defined', for considering this as 'progress' in any other sense.

VII

Let us return now for a moment to Fisher's opening statement (at the beginning of Chapter 2) and put it alongside the fundamental theorem.

> The improvement of any organism in relation to its environment is determined by its present condition... The increase in fitness of any organism at any time is equal to its genetic variance in fitness at that time.

In the context of Fisher's argument, 'improvement', etc., clearly refers to the increased probability of leaving descendants, *for whatever reason*. And 'present condition' means present genetic variance in fitness, that is, in the probability of leaving descendants. In other words, it means the summation of such probabilities itemised in the gene pool of the relevant population.

What does it mean, finally, to say that the second of these *'determines'* the first? The fundamental theorem asserts that the rate of increase in fitness of any organism at any time is exactly equal to the genetic variance in fitness at that time. But Fisher has just said (in his proof of the fundamental theorem) that the rate of increase in fitness here means rate of increase in fitness *due to all changes in gene ratio*.[18] Now the genetic variance in any measurement in population means, according to Fisher's earlier definition[19] the summation, for all genes, of the effects on m of the excess of any one gene over its alleles. But this *is* precisely the summation of the effect on m of all changes in gene ratio. To say that the rate of increase in fitness is due to changes in gene ratio is to assert a fundamental belief of modern genetic theory. To identify the increase so caused with the 'genetic variance in fitness' is to assert an identity. How can such a statement be said to 'determine' anything? We have present rate of increase in fitness = improvement; genetic variance in fitness = present condition. Either the second determines the first in the sense that we *are* simply stating an identity; and this tells us nothing about organic phenomena except as formalising what we already know – that, as Professor 'Espinasse has put it,[20] wherever there are characters there are *some* genes that cause them. Or else the fundamental theorem is meant to direct our attention to the tables of reproduction which can be used to chart the trends in populations both at the level of individuals and at the genetic level, and to correlate trends in other measurements with trends in m. We could say, for example, that the overall increase in tallness of a population is equal to its genetic variance in tallness, because we believe that whatever trend a population is showing in respect to a given measurement has some genetic basis. To say that its overall trend in fitness is equal to its genetic variance in fitness is not to express any further 'determination' over and above this. Again, the only determination expressed here is the determination we know of from genetics, and the statistical measurements of m add nothing to this. There *may* of course be other causal connections in nature (in predator-prey situations and so on), and our statistical measurements may guide us in our analysis of, and perhaps even our search for, these. But in themselves statistics cannot specify such connections. In other words, the fundamental theorem is a guide to statistical technique which is overlaid on the causal relation of heredity and can be used as underpinning for the causal study of Darwinian selection; but in itself it asserts neither.

VIII

But how can the imposing edifice of modern neo-Darwinian theory rest on so narrow a base? In fact, it does not. It rests on the broader foundation of Darwinian thinking which is drawn into the circle of Fisher's statistical theory in virtue of the ambiguity of its central concepts.

We have only to look at Fisher's comments on the fundamental theorem to see how this works. He likens this principle to the second law of thermodynamics, as a statistical law that reigns supreme over a vast area of nature. But fitness, he says, though measured by a uniform method, 'is qualitatively different for every organism'.[21] Now this is not fitness in the sense of the mathematical chance of leaving descendants, which is a quantity, and cannot be 'qualitatively different' for each organism. On the contrary, Fisher is here referring to fitness in the sense which he goes on to amplify later in the same chapter, in the section on 'the benefit of the species': that is, fitness in the qualitative sense of an 'advantage' (of a particular kind) to the individual organism.[22] 'Fitness measured by m' will again become extremely important in the theory of the evolution of dominance (in Chapter 3) where the biologist's concern is with the probability, for certain genetic factors, of leaving a remote posterity, even though these factors may lie deeply hidden in the present population, phenotypically considered. But here (in the latter part of Chapter 2) the statistical concept of fitness has been translated into old-fashioned Darwinian fitness: in terms of our previous example, the benefit to a Manchester moth of being black and the harm to his cousin of being speckled. Of course if mostly peppered moths are eaten and black ones spared, the population will come to include more black and fewer speckled individuals. But the statistical result to the population in the future and the immediate benefit to this sooty moth on this grimy tree trunk today are not the same thing. Yet through the identity of the word 'fitness', the insistence that selection has to do with present advantage, not 'trends', is here attached to Fisher's original, stricter concept of 'fitness', which thus becomes Darwinian as well as genetical and immediate as well as long-run. So we have genetical selection for later and Darwinian selection for now fused under a single word 'fitness'.

And 'progress' is equally elastic. We have 'progress' in fitness-measured-by-m: for example, progress in the height of the population if tall individuals are on the increase (never mind why); and at the same time we have progress in an advantageous character in the sense of increased adaptation, if, say, tallness (as in the proverbial giraffe) actually, for some reason, causes its possessor to get more to eat than his shorter brothers and therefore causes, rather than simply measuring, survival.

IX

There is more to it than this, however. The advantage to the individual organism on which Fisher insists is not just the traditional one of an advantage in facing predators or the like, that is, as against our contemporaries, a lesser chance of death. It is also (and must be, in an evolutionary context) the chance of reproduction. 'It will be observed', Fisher writes,

that the principle of Natural Selection, in the form in which it has been stated in this chapter (i.e. the fundamental theorem) refers only to the variation among individuals (or co-operative communities), and to the progressive modification of structure or function only in so far as variations are *of advantage to the individual in respect to his chance of death or reproduction.* It thus affords a rational explanation of structures, reactions and instincts which can be recognised as profitable to their individual possessors. *It affords no corresponding explanation for any properties of animals or plants which, without being individually advantageous, are supposed to be of service to the species to which they belong.*[23]

Now if we omit the qualification 'or reproduction', this is a fair statement of the situation described by Darwin. Where the benefit of individuals is concerned, it is the chance of death that selection controls. True, we may also infer that, since the chance of reproduction is obviously tied to the chance of death, the future numbers of organisms possessing immediately advantageous characters, given certain environmental conditions, will increase in numbers also, if and when those conditions persist. Moths that are eaten cannot reproduce thereafter. But the *immediate advantage* to the moth in this case is not reproduction, but the omission of being eaten. Reproduction would be an advantage to the moth's descendants, and indirectly to the moth in so far as it is advantageous to it to satisfy its instincts; but in the situation of pure Darwinian selection – as in this case of the black moth on the black tree trunk – it is the character that keeps one alive that counts. Reproduction as the continuation of the species is a matter that is indirect and inferred; it is staying alive that is the immediate benefit.

Yet the 'or reproduction' alternative is essential to Fisher's *genetical* selection – and again, to the future-directed character, in particular, of the evolution of dominance: where in present-day recessives – which may some day become dominant – we are dealing, as he says himself, with the 'chance of leaving a remote posterity' by storing up characters that will some day be useful in some distant future environment. Such a chance is surely meaningless in terms of 'advantage to the individual' unless in some Biblical sense, that a man feels he is cut off from immortality if he is cut off from having living descendants. In short, we have tied together in the concept of fitness three kinds of 'selection': genetical selection, which is future-directed but only in a Pickwickian sense selection at all; Darwinian selection for now, directed to immediate advantage; and Darwinian selection for later, in reference to future advantage. The first specifies progress in the purely statistical sense of increased chance of survival of some genes rather than others; the second in the sense of short-run increase in adaptation; the third, long-run increase in adaptation.

Finally, added to these, we must notice another of Fisher's comparisons between his theorem and the Second Law. While physical systems run downhill, he says, evolution tends on the whole to produce 'progressively higher organisation of the organic world'.[24] This is progress in a new sense, and one which escapes all Darwinian considerations, though again its existence, once admitted on other grounds, might be recorded with the help of statistical methods.

Thus the larger conceptual structure of the theory is both richer and less exact than the proof of the fundamental theorem would lead us to expect. It consists of a network of concepts, in which statistical and deterministic, genetical and Darwinian meanings, short-run and long-run assessments, reinforce one another in a self-confirming circle. For if fitness means, after all, advantage to the individual, then the long-run trends of reproduction tables *must* represent real benefits in each generation, and if at the same time fitness *means* the statistical progression measured by m, then such progression must entail the whole causal context that Darwinian 'fitness' implies. It is only in this way that the 'determination' mentioned in Fisher's initial thesis can be said to make sense. And similarly, the references to internal or genetical and external or Darwinian 'selection' and the references to future (genetical or Darwinian) and present (Darwinian) 'fitnesses' confirm one another, in circular reinforcement, indefinitely.

X

This ambiguity will become clearer if we pin-point the narrower and wider senses in which the concept of 'improvement' occurs in the argument and consider their evolutionary import.

First, there is the strict statistical meaning of the Malthusian parameter: the increase in the probability that organisms possessing certain characters will leave offspring. Such an increase, though calculated for an infinitesimal period, has meaning only over time. What it asserts about organisms in time may be taken in two ways: either as a tautological statement: what has survived has survived, what is on its way to surviving is on its way to surviving; or as the retrospective appraisal of an achievement: what has survived is not what has failed to survive, but what has succeeded. But this says nothing of *why* that might have happened, either in terms of adaptive relations to environment or anything else. Tables of reproductive value record the multiplication of some characters and the disappearance of others; that is all one can say in these terms. And in terms of these tables, one may assert the probable increase of some genes rather than others.

In this sense of 'improvement' Fisher is right in saying that organisms must show 'improvement' up to the moment of extinction. In terms of the increase of some genes as against their alleles, they doubtless do. For since the sum of any gene and its alleles in any population always = 1, we can always take the increasing rather than the decreasing gene (or genes, where there are multiple alleles) and so get a picture of 'improvement'. This is really why, in the fundamental theorem, the calculated value is always positive. Not only is the time interval, dt, necessarily positive, but the changing gene ratios also can always be taken in this sense.

XI

Alongside this statistical formulation, however, consider what Fisher says about the universality of 'improvement' in his comment on the fundamental theorem. His first comparison with thermodynamics is to this effect:

The systems considered in thermodynamics are permanent; species on the contrary are liable to extinction, although biological improvement must be expected to occur up to the end of their existence.[25]

The phrase 'biological improvement' brings us back to the 'improvement in relation to environment' from which we started; it is what Fisher describes elsewhere as 'progress determined by Natural Selection', the full Darwinian improvement so emphatically underlined in the second edition. According to this conception, organisms are in fact constantly becoming better adapted to their niches in nature, as natural selection weeds out the imperfectly adapted; and therefore it follows, further, as Fisher in fact argues in some detail, that the explication of all cases of extinction must be referred to deterioration of the environment. Thus when the dinosaurs died out, for example, something must have gone wrong in their environment to bring this about. That this is actually so in nature, however, cannot be inferred from the fundamental theorem by taking 'improvement' in its strict and statistical sense.

We are dealing here, in other words, with a second concept of improvement. We are asserting that there is improvement in the sense of the appearance of adaptive relations with immediate effectiveness here and now, characters or functions which are genuinely advantageous in relation to their bearer's actual environment, and which may therefore, as compared with their absence, be called 'improvements'. These are the improvements effected by Darwinian selection, and the concepts presupposed in the assertion of such improvement are not by any means the same as those supporting the statistical concept. While statistical 'improvement' entails the conception of the organism as an aggregate of gene effects, Darwinian improvement entails in addition the conception of the organism as a machine with parts adapted to the performance of their special functions. The improvement that is effected here is that of increasing adaptation in the sense of specialisation, of fitting in better and better to a special niche in nature. The evolutionary import of such improvement is *katagenetic*, to use a term introduced by the German evolutionist Rensch. That is, it is evolution downhill, in the sense that species break up into varieties and subspecies as specialised demands are made on them. Darwin's finches are a classic example.

As against genetical 'improvement', which is meaningful only as an assessment of a trend over a lapse of time, this kind of improvement is short-run. On the other hand, it is not intelligible, as the statistical concept purports to be, in terms of gene ratios alone. Its assessment depends on the recognition of phenotypes, as wholes, not as aggregates of genes, and on the recognition of the relation of such wholes to their environment, to predators, to climate, and so on and so forth. It is neither future moths, nor gene pools, to whom it is advantageous not to be taken by a bird, but this black moth on this tree trunk today. And the evolutionary trend which establishes and maintains a phenomenon like melanism expresses the accumulation of millions of such individual escapes and individual disasters, not to genes or gene pools, but to moths, whether today or yesterday or (from some cause other than the industrial revolution) a million years ago.

XII

So much for our first pair of 'improvements'. However, as critics of natural selection theory have long been saying, there also seem to be adaptive relations which develop only slowly, and which do not appear to benefit their possessors at the beginning of their development. Such long-run adaptation appears to entail a third meaning of 'improvement'. It refers to characters or functions which will be 'better' for future phenotypes in future environments. Such improvement we may call *quasi-anagenetic*; it still concerns particular adaptive relationships and specialisations, but specialisations which accompany the emergence of new forms (rather than new sub-styles of old forms) and which develop slowly over a very long period of time. Now this seems to be the kind of situation which Fisher explicitly excludes, exiling the 'benefit of the species' approach as teleological and irrational. In terms of what he says there, one would conclude that all ultimately useful characters must have been in some direct way useful even in their minute beginnings. And it is true that this has been shown to be possible or even likely in a number of cases, for instance, primitive photoreceptors, electric organs, feathers, and so on. One can, if one likes, extrapolate such instances to all cases of clearly adaptive characters. Yet the forward reference, as distinct from immediate utility, seems essential to Fisher's own type of statistical selection, and in particular to changes in the relations of recessive and dominant alleles. Here we have recessives, either useless or even harmful, hidden away in a population over a long period of time – up to the moment of a new environmental situation which makes them advantageous and so calls forth modifiers that turn them into dominants. Only the reference to 'remote posterity' makes sense of this story. True, there is the case of sickle-cell anaemia: where we have a gene which is lethal in the homozygote, but in the heterozygote actually gives protection against malaria, and is kept going in the population by this beneficial effect. Supposing a situation in which sickling were no longer harmful, we could imagine that a character now maintained, but in a recessive state, because of the advantageous nature of the heterozygote, might become dominant and pervade the population much more completely. And we might then extrapolate this kind of process to all cases of the retention of recessives apparently for the sake of their future usefulness, but really for the sake of some other present benefit. Yet even then, the modifiers which will eventually make the now-recessives dominant must be lurking in the population ready to leap into action when the environment demands, and thus the future-directedness of the whole procedure is simply transferred to them. It seems strange to say, as Darlington does, that the genes are endowed with an automatic property of foresight,[26] yet the reference to 'remote posterity' which dominates Fisher's argument on dominance does indeed suggest some such idea.

XIII

How can Fisher and other evolutionists who follow him remain so happily unaware of this third and uncomfortably long-term 'improvement'? What happens, I think, is something like this. We have so far, three concepts of 'improvement'. *Improvement 1, statistical selection*, is the measure of *improvement 3, long-run adaptation*. But long-run adaptation is *adaptation*, and where there is adaptation there must, in the light of the Darwinian theory of natural selection, be *improvement 2, Darwinian selection* – for Darwin has proved that that is how adaptation is produced. So improvement 3, which is in fact unintelligible in terms of the dictum of improvement 2 (since in its terms all adaptation bears on immediate, not remote benefit) is nevertheless subsumed under it through the measurement of both of them by the techniques of improvement 1. Moreover, improvement 1 is expressed by a differential, which can be interpreted as a summation of gene changes *now* – short-run trends – or over as long an interval as you like – long-term trends; but at the same time, since it is one statement (the fundamental theorem) it must express one relationship, and since in Darwinian terms improvement 3 would be nonsense, this one relationship must be the situation covered by improvement 2.

The way in which 1, 2 and 3 are assimilated to one another is most evident if we place the argument about the fraction $1/n$ and the efficacy of very slight 'selective intensities' alongside the argument about immediate advantage. As Fisher himself argues, irregular variation in the rate of increase in fitness as measured by m is more apparent in a single generation than over a longer period, and therefore very small 'selective values' are sufficiently strong to establish themselves *over a long enough time*. This statistical observation is often used against those who object to natural selection theory because of the difficulty of accounting for the first beginnings of what will ultimately be useful traits. Selection can do so much *because it has so long*.[27] Yet if one speaks of trends in evolution, of orthogenesis or the like, one is told that this is nonsense because evolutionary modification always consists in the selection of the immediately useful at each step, in each generation.[28] So improvement 3, long-run adaptation, with its statistical measure in improvement 1, is used to answer one objection, while improvement 2, which can also be expressed by the same statistics, applied to short-run situations, is invoked to answer another.

XIV

Finally, we must mention a fourth meaning of improvement, the kind of improvement which is truly an advance to higher forms of life. Such improvement is explicitly mentioned only in Fisher's remark about his fundamental theorem and the tendency to 'higher organisation' (as contrasted with the Second Law and entropy). This is the only kind of improvement which represents true *anagenesis* in the sense of emergence, or the appearance of genuine novelty at a higher level of richness or complexity. But this is also a kind of improvement which neither statistical genetics nor selectionist biology can handle, since it is neither quantitative nor adaptive. It is best for selectionists to ignore it, as Darwin warned himself to do when he wrote in his copy of the *Vestiges* 'Never speak of higher or lower in evolution'. Yet the great outlines of the fossil record are there, and demand to be spoken of, especially since the fact that we can speak of them is one of the surprising results of the process they record. But evolution as macro-evolution, as the emergence of life and of higher forms of life, outruns both the concept of gene-substitution, and of improvement in relation to environment. It makes sense only as an achievement – an achievement for which statistical methods can measure the necessary, but not the sufficient conditions.

XV

One brief concluding remark. I have side-stepped here altogether the question of prediction and retrospect, of the historical nature of evolutionary explanation: a question which is very close to the philosophical difficulties raised by Fisher's theory. Evolutionary theory is essentially an assessment of the past. Fisher treats it in terms of present and future. Just how closely the philosophical confusions of this kind of argument are related to the attempt to think unhistorically about an historical subject matter, I should not at the moment venture to say.

NOTES

[1] Oxford 1930 (New York, Dover 1959).

[2] It is often said that Fisher has 'proved mathematically' the truth of neo-Darwinism. The confidence with which the 'synthetic' theory has been asserted on the basis of Fisher's argument is reflected, for example, in the contributions by Huxley, Fisher and Mayr in Huxley, Hardy, Ford, *Evolution as a Process*, London 1954; or in the conclusion of the third edition of de Beer's *Embryos and Ancestors*, Oxford 1958, or in P. M. Sheppard's *Natural Selection and Heredity*, London 1959 – to mention but a few.

[3] Sir Ronald Fisher, *The Genetical Theory of Natural Selection*, New York 1959, p. 38. I am following the Dover edition, which is in part revised, but not, with one exception, to be noted later, in ways that are philosophically relevant.

[4] *Op. cit.*, p. 41.

[5] *Op. cit.*, p. 26.

[6] *Op. cit.*, p. 37.

[7] I am speaking here (as Fisher is doing) of genic inheritance. The question of cytoplasmic inheritance introduces another dimension altogether into the evolutionary problem.

[8] *Loc. cit.* (my italics).

[9] *Op. cit.*, p. 40.

[10] *Op. cit.*, p. 27.

[11] H. B. D. Kettlewell, *Nature* 175 (1955) 943; *Heredity* 9 (1955) 323, and 10 (1956) 287; *Proc. Roy. Soc.* B 145 (1956) 297.

[12] Or if, alternatively, we say that in war it is the ('normally') mal-adapted who are *better* adapted, and vice versa, then we are using 'better adapted' to mean 'surviving' and are saying only: 'those who survive, survive'.

[13] The demonstrative character of Darwin's argument was pointed out by C. F. A. Pantin (in *History of Science*, London 1953; cf. the discussion by A. G. N. Flew, in *New Biology* 28 (1959) 25ff.

[14] Fisher, *op. cit.*, p. 38.

[15] If one has set up a selection experiment, of course they are so.

[16] R. Fisher, 'The Bearing of Genetics on Theories of Evolution', *Science Progress* 27 (1932) 273–287.

[17] *Ibid.*, pp. 131–132.

[18] *Op. cit.*, p. 37.

[19] *Op. cit.*, pp. 30–32.

[20] In an address to the British Society for the Philosophy of Science in February 1959.

[21] *Op. cit.*, p. 39.

[22] This section is added in the second edition, but what Fisher says here is implicit in the usage of the first edition also.

[23] *Op. cit.*, p. 49 (my italics).

[24] *Op. cit.*, p. 40.

[25] *Op. cit.*, p. 40. For an excellent discussion of the logical place of biological improvement in Darwinian and neo-Darwinian theory, see the article by Flew (n. 13 above).

[26] C. D. Darlington, *Evolution of Genetic Systems*, 2nd edn., Edinburgh, 1958, p. 239.

[27] See, for example, the contributions of Huxley and Fisher to the Huxley, Hardy, Ford volume, referred to in note 2, p. 171 above, or Huxley's *Evolution in Action*, 1953.

[28] *Loc. cit.*

CHAPTER IX

BIOLOGY AND TELEOLOGY

Almost thirty years ago, in his lectures on *The Idea of Nature*, Collingwood listed among the characteristics of the *modern* as distinct from the Renaissance concept of nature, the *reintroduction of teleology*. Renaissance thinkers, he said, saw nature as a machine; final causes belonged outside it, in its origin or in its use, not within it. Modern thinkers, on the other hand, he believed, were beginning to lean on a new analogy, not between nature and machines, but between nature and historical process. 'The historical conception of scientifically knowable change or process', he says, 'was applied, under the name of evolution, to the natural world'. And this application entailed, among other consequences, a return to teleological thinking about natural events. In some sort of transposition, Aristotle's recognition of goals in nature, Collingwood argued, is to be, and has been, reinstated. Discussing Aristotle, he writes:

It is widely recognized that a process of becoming is conceivable only if that which is yet unrealized is affecting the process as a goal towards which it is directed, and that mutations in species arise not through the gradual working of the laws of chance but by steps which are somehow directed towards a higher form – that is, a more efficiently and vividly alive form – of life. In this respect, if modern physics is coming closer to Plato as the great mathematician-philosopher of antiquity, modern biology is coming closer to its great biologist-philospher Aristotle.

Against the purely quantifiable, dead nature of Galileo and Newton, Collingwood is confident that his reorientation toward organic phenomena has brought about a significant reform in our whole view of the natural world. Apart from professional biologists' disputes, 'on the ground of philosophy', he says,

I think it fair to say that the conception of vital process as distinct from mechanical or chemical change has come to stay, and has revolutionized our conception of nature.

Collingwood wrote those words some time in the 'thirties – after '33–4, the year in which the course of lectures on which *The Idea of Nature* is based were first given, and in which, also, Whitehead delivered the four lectures *Nature and Life* on which Collingwood seems to be relying heavily – and justly so, for *Nature and Life* ought, together with *Process and Reality*, to have constituted a turning point in Western philosophy as marked as that initiated by the Cartesian *Meditations*.

But conceptual reform comes hard. It is startling to read, in 1964, that something like Aristotelian teleology 'is widely recognized' or that 'the conception of vital process as distinct from mechanical or chemical change has come to stay'. True, Collingwood warned us, we should not be surprised if 'many eminent biologists have not yet accepted' the revolution he describes. 'In the same way', he continues,

the anti-Aristotelian physics which I have described as the new and fertile element in sixteenth-century cosmology was rejected by many distinguished scientists of that age; not only by futile pedants, but by men who were making important contributions to the advancement of knowledge.

Scientific discoverers may indeed be and sometimes are philosophical reactionaries. Many biologists, in particular, are still held in the grip of what was called three centuries ago 'the new mechanical philosophy'. But philosophers, too, have failed by and large to accept Collingwood's 'modern' concept. By now a good handful at least of 'eminent biologists' are groping for concepts that will not, as strict mechanism demands, eliminate 'bios' from their subject, yet philosophers stand aloof from this effort: the effort to implement from the side of science the philosophical revolution which Collingwood believed already accepted a generation ago. So, for example, P. F. Strawson remarks quite by the way, in *Individuals*, that 'the category of process-things is one we neither have nor need'. But in terms of Collingwood's revolution, as we may call it, of course there are process-things, and if we do not have such a category, we do nevertheless need it, for we are ourselves among such things, so is every living species, so is the universe. Only the ghost of Newton's dead nature, cunningly masked as common sense, prevents us from admitting it.

To put finally to rest our Newtonian delusions, to renew our conception of nature as *living*, and so to see ourselves once more as living beings in a world of living beings, constitutes, it seems to me, the major task of philosophy in the twentieth century. It is far from accomplished; the rise of a new Cantabrian 'new mechanical philosophy' makes the task not only more difficult, but more urgent. As, again, Strawson's failure to notice the problem indicates, only a radically revisionist metaphysics and, even more fundamentally, a radically revisionist epistemology can do the job. I am not the person, nor are these columns the place, to do it. But I can at least ask a rather humdrum question which may help to suggest that it needs doing.

'The appearance of end', as Waddington calls it, is an ineradicable aspect of life. Teleological language, therefore, persists in biology. I do *not* mean here the language of conscious purpose; as both Aristotle and Collingwood knew, the 'ends' of nature are not plans. Organic phenomena are directive, not directed. But wherever there is reference to part-whole relations (in morphology), to means-end relations (in physiology), or to goal-governed processes (in embryology), there is teleological discourse in the Aristotle-Collingwood sense. For in each such case, something *A* is said to be for-the-sake-of, or in terms of its bearing on, something *B*, where *B* is on a higher level of organization than and acts as a norm in some sense controlling *A*. All the structures, uses, and processes so described belong to the class of what Sir Julian Huxley calls *telic* phenomena, and in investigating them the scientist uses teleological concepts.

As a first step toward Collingwood's revolution, then, let us ask: what is such teleological language doing in biology? I shall suggest a series of answers, each of which leads on to the next, and the last of which brings us to the threshold of the philosophical revolution in which, if we are not, we ought to be engaged.

The teleological reference in biological discourse may be characterized as (1) reflective, (2) regulative, (3) descriptive, (4) operational, (5) explanatory, or (6) ontological.

First, it is sometimes argued, that while teleological concepts are in fact wholly inappropriate to scientific thinking, they enter into the philosopher's *reflection* about the data of science. On this view, therefore, any reference to patterns, functions or goals in living nature must be extruded from the biologist's vocabulary. It is meta-scientific and must not enter into the speech, or the thinking, of scientists themselves.

The theoretical framework of the biologist's training is centred – implicitly and indirectly if not explicitly and directly – in the theory of evolution by Mendelian micromutations and natural selection. In terms of ultimate beliefs, this means the view that the present population of the earth's surface is derived by a repetition of a few simple mechanisms (internally, a template mechanism or something of the sort; externally, selection and other environmental pressures) from unicellular and ultimately from non-living particles of matter. This picture is associated also with an aggregative conception of the organism itself. 'Ultimately', one believes, an organism such as a mouse or a frog or a man will be identifiable in terms of material particles and their spatio-temporal relationships. 'In a sense', writes Dobzhansky, 'the development of the organism is a by-product of the processes of self-reproduction of the genes'. Recent biochemical research has given a new impetus to this atomistic conception, just as research on natural selection, against the background of the telephone-exchange theory of the behaviour of higher organisms, has given new support to the mechanistic view of causation which is associated with it. The trained biologist, in other words, works under the aegis of a guiding principle that organisms are aggregates of material particles moved by mechanical physico-chemical laws. His thinking, his research, are motivated by the search for (in Professor 'Espinasse's eloquent phrase) 'little causal thingummies'. Scientific discourse, accordingly, *qua* scientific, must be wholly non- and anti-teleological, and any speech tainted with teleology must remain wholly extraneous to science. It can be, at best, reflection about and upon it.

This ideal, however, is, to say the least, impracticable. For teleological concepts have at least a *regulative* function within the practice of biology. This was well argued by Frank Baker in a paper read to the Royal College of Science in 1934; and quoted in this volume (Chapter IV, p. 77). The scientist, Baker argued, studying the development of a fungus, for example, must understand a great deal about the orderly development of the organism before the right 'facts' can even be selected for investigation and analysis. Teleological concepts, therefore, at least *regulate* the biologist's choice of data and of problems. They are not merely metascientific.

But is that all? Can the biologist proceed to describe what it is that he is analysing without referring to structures, uses, or achievements? I think not. In other words, teleological discourse has not only a regulative, but at the least a *descriptive* function within biological research. For example, the statement that a particular spot on the petal of the cotton flower or the wing of a butterfly or the character *cv* on the wing of a fruit fly may be determined by alternative genetic loci – that is, in 'Espinasse's terms, that different little causal thingummies are causing the same effect – are assertions inferred from breeding experiments which show that the same phenotypic result is produced through breeding different strains of plant or animal. Nobody (so far) has inspected these loci directly. They are inferred from the shapes and functionings of the organisms which are held to possess them. If the scientist could not see and describe the structures, functions and developmental processes of the cotton bloom or fruit flies involved in each case, he would, like Baker's misguided chemist, be unable to begin, continue or conclude any genetical investigation whatsoever. As Professor Paul Weiss puts it (for the case of experimental embryology), 'we cannot address biophysical or biochemical questions to words', but only 'to clearly *described* phenomena'.

It may be thought, of course, that this *descriptive* function is only the starting point, and that the subsequent scientific *analysis* of organic phenomena is wholly non-teleological in character. That is perhaps the most common view of the matter. The biologist, it is conceded, may have to use two-level, teleological language to get the object of his investigation so to speak into focus, but then go on searching wholly in terms of small parts arranging themselves in causal sequences in 'mechanical' fashion. If scientists can only put their biochemical and biophysical questions to well described phenomena, the questions themselves seem to be analytical and mechanical, not teleological. Against this contention, however, we may observe that the *operations* performed by many biologists seem to entail reiterated or even continuous teleological reference. As to part-whole relations, for example, not only does classical taxonomy depend for its operations on the recognition of structures; but it has recently been argued that the new quantitative and statistical operations of numerical taxonomy also depend on the recognition of 'types'. Or again, if we consider the typical occupation of geneticists, a good deal of their time is spent, one gathers, in counting the fruitflies of wild type or of a given mutant type that result from breeding experiments: here the recognition of certain characters – *cv*, redeye, bithorax, or what you will – are essential to the actual experimental procedures; the underlying mechanical details, again, are inferred from these results. A similar situation holds in ethological experiment: where tinkering with the size and shapes of eggs or substituting red patches for actual sticklebacks, etc., are devices meant to elicit certain behaviour sequences from the subject: the behaviour patterns which display their own internal lines of organisation are still what the ethologist is working with from start to finish. Similarly in embryology: as Professor Jane Oppenheimer puts it, the experimenter can succeed only if he puts questions 'the embryo can comprehend'. The answers, in the last analysis, always come from 'the embryo alive', and the embryo answers, she says, 'at the supra-cellular level'. So in a good deal of biological work at least teleological concepts and principles are more than descriptive: they are *operational*. They control not only the way the phenomena are described but the way experimentations with them and therefore the subsequent scientific analysis of them itself proceeds. We have teleological discourse, then, firmly within the *procedures* of biological science.

It may still be asserted, however, by the die-hard anti-teleologist, that, however finalistic his techniques, his *explanations* are nevertheless wholly mechanistic, wholly in terms of 'little causal thingummies'. Certainly this has seemed to be the case with classical Mendelian genetics and with its modern heir, population genetics – for statistically treated aggregates of little causal thingummies are still particulate in nature and anti-teleological in their explanatory import. And, of course, many biological theories are of this sort. But is this true, or can it be true, of all biological explanation? The application of information theory to biology suggests a negative answer. Professor C. P. Raven of Utrecht, in a paper on 'The Formalization of Finality' [*Fol. Biotheor.* **5** (1961)] finds that, in his research, not only the problems, but their solutions are *finalistic*. *Causal* explanations of the same phenomena would be complementary to these, but could not replace them. Thus about the transmission of the DNA code he says:

On peut considérer cela de deux points de vue : d'abord comme transmission d'ordre, comme problème de la théorie des informations. C'est un problème de traduction d'un système d'ordre dans un autre système d'ordre. Mais, ce n'est pas une explication causale! L'explication causale serait de savoir comment il se peut que ces trois groupes de nucléotides attirent cet acide amine, quelles sont les forces effectives d'attraction entre ces groupements d'atomes. Les deux explications se complètent, mais ne s'excluent pas. Il faut bien distinguer; la formalisation de la finalité, c'est la rendre susceptible d'un traitement mathématique et d'une axiomatisation, mais ce n'est pas la même chose que de la transformer en causalité!

Here the *explanation* itself is teleological, in the Aristotle-Collingwood sense. It is a two-level explanation, concerned with the *particulars* comprising a *message*, in which, as in the telic phenomena it explains, the upper level controls and sets a norm for the lower. A relation of complementarity obtains between the two levels but it is a relation of *ordinal* complementarity, not the cardinal type, as in wave-particle complementarity. For the relation between a message and its particulars is asymmetric and normative. A *message* may be analysed into its *particular constituents*, but it cannot be reduced to them, let alone explained by them. The *particulars*, on the other hand, though physically caused, are *explained* by their role in the *message*. A given message can be transmitted through a variety of particulars, but in terms of its particulars alone it cannot be apprehended as a message, it cannot be distinguished from mere noise. In interpreting developmental events in cybernetical terms, therefore, the scientist is explaining a lower level (the actual particulars) through reference to a higher (the message conveyed and 'understood'). Further, such finalistic explanation is distinct in structure and explanatory power from the causal explanation which, again, is, in the ordinal sense, complementary to it. In this manner, the application of information theory to biology provides not only a 'formalization' of finality but a clear exhibition of its occurrence as an integral part of biological explanation.

There are other instances of such new teleological techniques, for example the application of von Neumann's theory of games to the theory of evolution. Or one might mention also the application of engineering principles to a problem in plant physiology by Professor W. T. Williams of Southampton [*Symp. Soc. Exp. Biol.* **15** (1961) 132]. But any one instance is sufficient to demonstrate that there are some inescapably teleological explanations of organic phenomena.

That makes five of my six possible functions of teleological language, and brings me back to Aristotle and Collingwood and the *ontological* import of teleology. There are teleological explanations. But any explanation that succeeds in explaining anything does so in virtue of our impression that we have achieved an understanding of something in the real world which we had previously found puzzling. If therefore teleological explanation is a genuine part of the knowledge of living things, it is so because living things are not only apparently but genuinely telic. It is, again, the lingering authority of the machine analogy that prevents our admitting the truth of this statement. Before we can firmly assent to the telic character of living things, including ourselves, we shall need to complete a very fundamental revision of our basic ways of thinking. We need, in other words, as I said at the start, what Strawson calls a revisionist metaphysic to bring this about. Nor will it do to say that common sense is against such speculative effort; on the one hand, it is easy to point to many areas where common sense lends us support, and on the other, even if we seem for the moment to be going against common sense, our everyday categories do sometimes need, and achieve, revision: there was a time, indeed, when the machine analogy itself was anything but common sense. As a foundation for biology it has never succeeded except as *mauvaise foi*.

CHAPTER X

BOHM'S METAPHYSICS AND BIOLOGY

PREFACE

This paper was written in response to discussions at the second of C. H. Waddington's symposia on Theoretical Biology (1967), which I had the pleasure of attending. The uniqueness of the occasion, for me, lay in the opportunity to hear David Bohm present his ontological views – in effect, a transposition into a modern coordinate system of Spinozistic metaphysics. I would by now be inclined to separate more sharply than I did then the scope and conceptual structure of evolutionary theory from the general metaphysical framework in which it may – and perhaps ought to be – imbedded.

I

A philosophical observer at a scientific conference is a kind of ethologist (or epistemethologist?) watching the conceptual behaviour of the other animals. The Second Serbelloni Symposium was outstanding not only because it brought together a group of extremely ingenious and well-trained performers, but also, and above all, because it produced a confrontation of two different conceptual patterns or, to borrow Kuhn's term, two paradigms, one orthodox and relatively restricted (and restricting) in its scope, the other heterodox and comprehensive. The result was not, as often happens in cases of deep-seated conceptual disagreement, simply the clash of two sub-groups. Rather, the first was literally comprehended – that is, described and explained – by the second, though its members were, with some exceptions, unaware that this is what had occurred. It was – I hope – a case of evolution in action: where the species doomed to extinction, innocently unconscious of its lack of 'fitness', continues hakpily to perform its traditional rites. The spectacle was instructive, but difficult to report, for two reasons. On the one hand, David Bohm in his original paper and in his 'Further Remarks' has himself indicated plainly how his 'metaphysic of process' assimilates and explains the truncated metaphysics of orthodox biology (and physics and computer science and psychology, etc.). Yet on the other hand, most of the contributors to and probably most of the readers of this volume, subscribing as they do to the still current orthodoxy, which as a matter of fact flourishes exceedingly at present, rather like the horns of the Irish elk, are unlikely to see the pertinence of Bohm's metaphysical remarks to their own methodology, and so are unlikelier still to see anything but *im*pertinence in my remarks on these remarks. The poor best I can offer in these circumstances is to try to put Bohm's speculations and, by implication, the metaphysics of the orthodox majority, into their historical context in terms of the major development of philosophical thought in the past three centuries or so.

II

It is otiose, yet necessary, to point out once more that the major trend of modern thought has been held captive by the brilliant success of the scientific revolution of the seventeenth century. The revolutions of the twentieth century have occasioned *some* fundamental rethinking of basic principles, and may yet – if Bohm's predictions are correct – have more far-reaching effects than they have, explicitly, had so far. But the chief model of 'scientific method' is still that of the Galilean-Newtonian philosophy. And there was something deeply incoherent about this philosophy from the start. Bohm indicates the source of this incoherence when he points out that the acceptance of Newtonian mechanics depends in the last analysis on an act of faith. Why should our mathematicizing be true of nature? There is no intrinsic reason, for example, why Newton's geometrical proof of Kepler's second law should demonstrate anything about what goes on in the sky. For an Augustinian, what we think, when we think clearly, and what there is in nature, both, if at different levels of perfection, express God's being. Descartes, with his sharp and simple dichotomy of cogitation (= mathematicizing) mind over against extended matter, has to invoke God *ex machina* to hold the two together. But he is still sufficiently an unquestioning Augustinian, so that, for him, the invocation works. With the secularization of thought (metaphysical as well as scientific: as Bohm quite correctly states, they are never wholly unrelated, since 'metaphysics' is just the most comprehensive range of anybody's thought, whether he knows it or not), the Cartesian dichotomy becomes unstable. Its uneasy synthesis in Kant, with nature reduced to phenomenon and mind to moral will, depends, still, on Kant's undoubting pietism for its ultimate support. That gone, it is only a short, inevitable step on the one hand to the *Nullpunktsexistenz* of the Sartrean for-itself, which frankly lives by contradiction, and on the other to the fruitless and equally self-contradictory objectivism of the contemporary philosophy of science (and of many scientists). (For the fruitlessness of the latter, see the outcome of Wittgenstein's *Tractatus*, for its self-contradiction, the argument of Russell's *Human Knowledge*; or see also the critique of E. Straus in *Vom Sinn der Sinne*, or of course Whitehead.) Of course each lingering remnant of the divided cosmology tries to account for the whole: Sartre's dialectical reason' serves up a caricature of nature; modern epiphenomenalism, a caricature of mind. Along these lines there is just nowhere further to go – and there never was; but it has taken us three hundred years, and indeed may take still longer, to find this out.

The incoherence lies not just in mind-body dualism, which has long since given way to a belief in matter-in-motion as the sole reality, but, as Bohm emphasizes, in the deep-lying divisiveness of our conceptual framework along a number of related lines. To cut off mind from nature is to cut off subject from object, so sharply that science itself (the product, after all, of subjects) becomes irrational and reality meaningless. Science becomes computation-for-the-sake-of-prediction-for-the-sake-of-computation-for-the-sake-of-prediction . , 'understanding' a merely subjective addendum, and 'truth' a dirty word, dropped in weak moments, like words with one less letter, but decently avoided for the most part in polite society. And the world so known? It used to be, and, as Bohm points out, for many biologists still is, the seemingly solid one-level nature of Democritean atomism, where faith that God made and keeps united our thoughts and their objectives gives way to the equally, if not still more, irrational faith that more complex orders *must* be explained out of, and exhausted by, those that are simplest, and ultimately out of the one 'real' order of matter in motion. Taking subject and object together we have, in Whiteheads' words, 'a mystic chant over an unintelligible universe'.

For what the subject-object dichotomy entails is a separation of order from the ordered, of meaning – which shrinks to a game with meaningless counters – from what is meant – which shrinks to an infinite aggregate of

equally meaningless data. In philosophical jargon, it entails, as the literature (and indeed, literature) richly shows, a radical division between value
and fact: in Bohm's terms, harmony becomes a little secret preference of
our own, and beauty a private vagary, rather than, as it is, the criterion of
our access to reality. In contrast, Bohm's linking of understanding,
beauty, and the timeless orders that govern emergent process may herald,
in my view, a comprehensive alternative to the self-denying ordinances of
modern thought: self-denying in that it alleges itself to be only the com-
pulsive outcome of its own neural processes. (Waddington is right, of
course, in calling Bohm's view Whiteheadian, but it may prove more
viable than Whitehead's own cosmology, since it can be developed,
I believe, without recourse to a doctrine of eternal objects: a radical
*in*coherence, I feel, in Whitehead's system.)

True, there are some other signs also of relaxation in the cramping
cosmology that still governs most scientists' minds. On the 'subject' side,
books like Hanson's *Patterns of Discovery* or Kuhn's *Nature of Scientific
Revolutions* may help. The slowly growing influence of Polanyi's *Personal
Knowledge* is a hopeful sign, although unfortunately many of those who
profess to accept its conclusion have grasped only the most superficial
theme of its complex argument. On the 'object' side, not only Bohm has
argued (and Waddington in his summary accepts this) that physics itself
has come to the end of its Democritean chapter, and so a new and richer
synthesis may be in sight. But for one thing Bohm's position on physics
still appears to most professionals as extreme heterodoxy, and on the
other hand authoritative biologists, especially in molecular biology – both
Crick and Watson, for example – still argue that a complete one-level,
particulate ontology is, if not already here, just around the corner. And
often, I believe, even when they seem to moderate their position by pointing to the richer complexities of modern physics, biologists are just concealing, to themselves and others, their real reductivism behind a screen
of Gibbs statistics, Volterra-Lotka equations and the like. There is still
a long way to go before we are out of the woods.

III

Meantime there is a further fundamental disability in the Cartesian-New-
tonian world view which was exhibited most beautifully at Bellagio, both

in its representatives and in Bohm's manner of transcending it. This is a disability also stressed by Whitehead: the incapacity to develop an adequate conception of life.

Not only Descartes' *bête-machine* but the principal thrust of Kant's transcendental analytic make it plain that in terms of the chief modern tradition this must be so. At the very start of the first Critique Kant distinguishes between acts which I perform and so have as things in themselves, but cannot know, and intuitions (*Anschauungen*) which are passively present to me, and which alone can supply content for my knowledge. The knowable, in other words, is passive; agency cannot be known. But as Whitehead argued, as Suzanne Langer argues, as Bohm argues, what we know as alive we know precisely in and through its *activity*. Bohm adduces here Piaget's account of child development; one could cite also von Weiszäcker's *Gestaltkreis* or Goldstein's concept of preferred behaviour. This is also, I believe, the fundamental (if sometimes concealed) import of Waddington's stress on epigenesis: organisms *are* not simply, they *act*. And acts are, in Bohm's terms, *creative* processes: they bring into being an order that was not. But thought in the Cartesian tradition has restricted itself, in viewing the object of knowledge, to its passivity: to what Bohm calls equilibrium and dynamic processes, exluding creative process. In the beginning was God's creation, thereafter a 'clockwork' universe, but no creative creature. This is again a consequence, if you will, of the subject-object dichotomy, of which the logical outcome on the subjective side is the Sartrean for-itself, and on the object side Crickian reductivism. Bohm argues, of course, that even in physics a concept of creative process has been found indispensable; my point here is simply that without a concept of act, that is, of creative change of order, there can be no adequate concept of life.

But, it will be objected, that is just what the conference did move to develop: we had brilliant applications of automata theory, statistical dynamics, chemical engineering, and so on, to biological problems. Of course there is no limit to such applications, and in the appropriate context they are all to the good. The question remains: what *is* the appropriate context?

The context within which the most articulate participants at the Symposium approached their problem was thoroughly functional. Almost everyone used, constantly and as self-evident, the term 'biological' as synonymous with 'functional', 'adaptive', 'conducive to survival'

– strictly, in evolutionary terms, conducive to leaving descendants – or 'produced by Natural Selection'. The best, indeed the perfect, specimens of this breed of thought were Longuet-Higgins, Gregory, and Maynard Smith; even Waddington, though not *quite* orthodox, appears in his summarizing notes as an interesting mutant of the same species.

The fundamental principles of this reigning form of biological thought are two: first, uniformitarianism extrapolated to the faith that ultimately all explanation is one-levelled in terms of least particulars. (It's all, after all, physics and chemistry – see Arbib's remarks on Longuet-Higgins.) Secondly, that the only allowed supplementation of such a monolithic materialism is the reference to adaptation personified in the concept of Natural Selection, that is, of the 'mechanism' by which the less adapted are eliminated in favour of the better adapted. Now naturally, I hasten to say, I too cross myself when speaking of Natural Selection. Yet for all my efforts I am still unable to understand quite what it really means, especially when I hear from Dobzhansky about 'selection *sub specie aeternitatis*' or from Simpson that 'whatever we see, selection sees much more', and so on. Let me try once again.

Look at Longuet-Higgins' summarizing statement: 'The secret of life is the ability of living creatures to improve their programs'. 'Improve', when challenged, he altered to 'adapt'. Now this is plainly the neo-Darwinian *credo* using the contemporary tools of population genetics and information theory. Whatever computer techniques it embodies, what it adds to the Democritean platform – the ultimate reduction of all process to its simplest level – is simply the axiom of adaptivity: the thesis that living things are adaptation machines. This thesis is then supposedly sufficient to generate evolution. What does it mean and how does it work?

Medawar recently stated the core of contemporary evolutionary theory as consisting in the two propositions: (i) that the terrestrial populations existing at a given future time will differ statistically in some degree from those existing today; and (ii) that the genetic constitution of those populations will have some connection with their changed phenotypic characters. We may add to these Waddington's emphasis on the role of the phenotype in the selective process controlling (i) and (ii). These are all perfectly harmless statements which no one, philosopher or otherwise, would want to challenge. But they tell us nothing at all about the episte-

mic relation between discourse about least particulars: gene pools or
populations of gene pools, and discourse about cells or organ systems or
organisms, nor about the ontological import of such discourse, all of
which, in Longuet-Higgins' 'Of course in the end it's all physics and
chemistry', they presuppose. Nor do they tell us how from a time when
there were no living cells or organ systems or organisms there came to be
such entities. When we try to go further here, we find a conceptual
confusion which needs to be disentangled before an adequate theory of
evolution, emergent or otherwise, can be formulated.

I have tried to analyse this confusion a number of times elsewhere
and so, of course, have numerous other people. Langer in her new book
has some good arguments; and indeed Bohm's argument on the necessity
of the transfunctional seems to me absolutely conclusive. Let me refer
here to the analysis in Chapter III of this collection. As I argued there,
adaptation is a crypto-teleological concept, but teleology even when
explicit is itself dependent on the prior evaluation of the ends evoked.
And in the case of evolutionary biology, the end is not simply survival
but the survival of – a type, a mode of living, an order of orders, in
Bohm's language, adjudged as significant in itself. That is the only
judgment that can fill in the tautology of survival-for-survival-for-survival.

IV

Such a judgment, however, you are all, or nearly all, unwilling to make,
or to admit that you make, and that despite the undeniable force of
Bohm's argument on function/transfunction. Why not?

There are temperamental reasons: the passion for model making, and
social reasons: the prestige and power granted to machine makers in our
society. But the fundamental reason lies, I believe, in the basic ambiguity
of the concept of 'mechanism' itself. The world of the seventeenth century's 'new mechanical philosophy' was, apart from its Infinite Designer,
a one-level universe, whose laws would ultimately be specifiable in terms
of its 'hard, impenetrable' least parts. Its laws are, in Bohm's phrase,
automorphic, expressing the simplest order of matter-in-motion. Take
away the Designer and you have a self-regulating system, just what, three
hundred years later, automata theorists have triumphantly learned to
produce. But a machine, as Polanyi has demonstrated (both in *Personal*

Knowledge and recently, August 1967, in *Chemical and Engineering News*) is, essentially, not a one-level, but a hierarchical system. It demands operating principles ordinally complementary to, i.e. depending on, but controlling, the laws of physics and chemistry which govern the behaviour of its parts. Because, however, we have put it together out of discrete parts which we control and can specify, we can easily neglect this two-level structure and hold that physics and chemistry alone 'produce' and 'explain' the machine. The clockwork seems to make the clock. Now of course the clock needs its clockwork. But you cannot in terms of physics and chemistry alone say anything about telling time. You cannot in terms of physics and chemistry alone distinguish any message, whether the time of day or the hereditary program of an organism, from a noise.

Why is *this* message so hard to put across? Because of the compulsion of Democritean thinking times the self-deception of 'utility'. Machine thinking, as I have argued for the case of evolutionary theory, is crypto-teleological. But when you make the end explicit, in engineering terms it is still a means; and you can keep running, like Yellow-Dog Dingo, without ever stopping to face the fact that *some* intrinsic value, some harmony, in Bohm's language, some timeless order, is the controlling principle which all the while governs your unending course. Or if you're pushed, it is still the minimal order, the maximally meaningless order, survival, or motion-for-the-sake-of-motion, that turns out, allegedly, to be your goal.

Bohm asks: Is nature more like an engineer or an artist? Admittedly, the two have much in common. The artist must control his material craftsmanly: he is also an engineer. And the great engineer also achieves beauty. Moreover, the emphasis on engineering concepts as against *mere* physics and chemistry (see for example Gregory's *The Brain as an Engineering Problem*) is in fact an advance from a one- to a two-level ontology, and that is all to the good. But the difference in ends, and therefore in the logical structure (the order of orders) of the two enterprises is what matters here. The engineer makes artifacts, without intrinsic significance or intrinsic reality. They are in essence means to the perpetuation of what is. The artist makes new realities, richer orders that never were on land or sea, dependent of course on conditions specifiable in terms of lower levels, but neither predictable nor explicable in terms of them. Such harmonies, such emergent orders, have to be apprehended, not through manipulation of means for means for going on going, but through understanding: understanding, again in Bohm's terms, as the union of observer and observed in the presence of beauty.

To the physics or engineering minded, such formulations may well appear absurd, wildly metaphysical, 'subjective', 'irrational'. Of course, so did Galileo to the good Aristotelians. I can only point, in conclusion, to convergences in other contemporary writers with Bohm's cosmology. The three principal orders he specifies in his paper are identical with those distinguished by Merleau-Ponty in *The Structure of Behaviour*, and Merleau-Ponty like Bohm points to artistic creation as the paradigm on which we ought to lean if we would understand our own way of being (the order of intelligence). And of course the analogy of nature and art (not artifaction, or invention, but creation, artistic discovery) forms the central theme of Suzanne Langer's recent work. There is no doubt, in my view, that we must acknowledge and implement philosophically Peirce's insight: that while logic is subordinate to ethics, in the sense of practical or engineering knowhow, ethics *sive* engineering is subordinate to aesthetics; our sense of beauty, of the intrinsically meaningful, dominates, whether we will or no, our grasp of what is real, of what is worth making real or allowing to perish. We all seek, in our own way, as Plato saw, to achieve immortality, to find a timeless order, through begetting on the beautiful. And only he who has seen the beautiful itself 'can breed true virtue, since he alone is in contact not with illusion but with truth'.

CHAPTER XI

DARWIN AND PHILOSOPHY

The theme of Darwin's influence on philosophy has been a recurrent one, notably in John Dewey's lecture in the semi-centennial year and again in J. H. Randall, Jr.'s defense of Dewey in the centennial year of 1959.[1] Randall was, in effect, defending Dewey against the charge of another centennial essayist, J. S. Fulton, who write:

An essay on the philosophy of evolution in the century since the publication of Darwin's *Origin of Species* can be written in two sentences. By the end of the first fifty years, everybody in the educated world took evolution for granted, but the idea was still intellectually exciting and its philosophical exploitation was entering upon its period of full maturity. By the end of the next fifty years, evolution belongs to 'common sense' almost as thoroughly as the Copernican hypothesis and other early landmarks of the scientific revolution; but the idea is no longer exciting, and evolutionary philosophy is out of fashion.[2]

Now it seems obvious, at first glance at any rate, that Randall is wrong and Fulton right. In general, we don't talk evolution in philosophy these days. Why not? Let us ask once more, what has been the destiny of Darwinian theory in connection with philosophy, both professional and popular? This entails three questions. First, what is the fundamental move of Darwinian explanation? Second, how does it fit into the history of nineteenth and twentieth century thought? Third, what is its relation to present-day problems in philosophy?

First, what Darwin did was both grand and simple. He showed how the myriad features of living things which fit them to cope with their environments, their ways of being *adapted*, could be explained, not as designed *for* this purpose, but as produced by ordinary cause-and-effect relations. Thus what *looks* purposive becomes explicable in a perfectly humdrum causal fashion, and life on earth takes its place, as the heavens had done earlier, in a plain naturalistic cosmology. There had of course been anticipations of Darwin's theory, not to mention its concurrent formulation by Wallace; there is no time to go into that question here – nor is it pertinent, for it was the *Origin* that put the theory on a new and imposing scale – backed by all the experience of a great naturalist – and, as C. F. A. Pantin has demonstrated, put it in *deductive* form.[3] Given the obvious facts of heredity and variation and the Malthusian parameter of population increase, a 'struggle for existence' (metaphorically, not literally, understood) must follow, and from this in turn must follow the survival of the fitter (not the fitt*est*), that is, natural selection as the process by which relatively superior adaptations arise and relatively inferior ones disappear. What looks purposive is explained causally, and by a logically compelling argument.

Moreover, the situation is still basically the same when you build in first genetics and then biochemistry. When, after three decades of opposition or neglect by eminent geneticists (roughly 1900 to 1930), the theory of Natural Selection was reconciled with Mendelism in what has come to be called the synthetic theory, it was genetics that was assimilated to the theory of Natural Selection as the underpinning it had previously lacked. For the measure of changing gene frequencies in interbreeding Mendelian populations *is* the measure of Natural Selection in the sense of such causal organism-environment interactions as produce heritable results (beginning of course with the internal environment within the organism, indeed, within the genome itself). The biochemical revolution, moreover, has only served to confirm once more the power of Darwinian thinking. There has been some talk of 'neutral mutations' (by adherents of the King Crowe school), but the random occurrence of single base pair substitutions in DNA is easily subordinated by orthodox evolutionists to the framework of an account of changing selective pressures. Of course mutations are random with respect to selection and so may be 'neutral' as well as 'deleterious' or 'useful'. But there are always plenty of them available, and which ones persist is dictated by the causal processes summarized as Natural Selection. However precise biochemical analysis becomes, in other words, the Darwinian explanation (of the way adaptive structures are favored and maintained) is confirmed rather than undermined, or even modified, by this increased precision.

One could find countless examples of modern, yet classically Darwinian, work, the whole literature of population genetics, for example. Take as one instance of this kind of pure Darwinian thinking, for instance, E. O. Wilson's application of ergonomic theory to the problem of the evolution of sociality in insects.[5] Darwin and others after him had considered the phenomena of altruistic behavior in the social insects stubbornly resistant to explanation in terms of natural selection. But today, with advancing knowledge of insect behavior, its interpretation in population-genetical terms, and the application of a benefit/cost calculus analogous to that of game theory, these same phenomena can not only be assimilated to a classically Darwinian framework of explication; they provide a striking instantiation of its explanatory power. At the level of the colony, the 'altruism' of the worker bee or the slave ant turns out to be the highly probable outcome of selective processes. There is still plenty of controversy in this area, but even the arguments are purely Darwinian. Nobody considers that anything *but* selection could be the controlling concept in an evolutionary debate.

One other example: on a more massive scale, and assimilating a vast range of recent data from such fields as biochemical genetics, biochemical evolution and Precambrian paleontology, Lynn Margulis, in her *Origin of Eukaryotic Cells*, has presented an impressive argument for the endosymbiotic theory of eukaryote evolution. She supports, with modifications, Whittakei's five-kingdom taxonomy, from Monera, then Protists (the earliest eukaryotes, which in this classification excludes bacteria and blue green algae), both of these originating in the Precambrian, and then the three great 'modern' kingdoms, Plants, Fungi and Animals.[6] Dr. Margulis's reasoning is indeed 'speculative', a suspect quality to most modern biologists, and revolutionary for our concepts of taxonomy and phylogenesis. But my point here is that her argument uses a very wide range of biochemical and genetic knowledge in the service of straightforward, classically Darwinian, concepts: heredity (whose origin one can now also hypothesize, as Darwin could not), variation and multiplication, resulting in the persistence of certain forms of life rather than others in appropriate environments.

So much for my first point: the structure of Darwinian explanation. Now let us look, equally sketchily, at the major trends of philosophical thought in the nineteenth and early twentieth centuries and see where Darwin's achievement was influential in shaping (or at least giving a special expression to) the philosophical tradition. (Perhaps I should say here parenthetically what I shall have to stress in the last part of my argument, that scientific discoveries, however important, do not in themselves 'solve' philosophical problems. They sometimes generate meta-problems by their own conceptual confusions or paradoxes, and they sometimes – and that is the case that chiefly interests us here – put constraints on philosophical reflection by favoring certain very general metaphysical or cosmological ideas. But back to philosophy!)

The history of philosophy in the nineteenth century and longer – shall we say till 1927 and *Sein und Zeit* on the continent of Europe and until the influence of late Wittgenstein in England in the late '30's and thereafter (we English speakers are notoriously slow in catching on, or catching up) – the history of philosophy in this period is the history of the destiny of Kant's critical system. Kant had limited our theoretical powers to appearances, to the phenomenal only. According to him, we can organize *a priori*, and therefore *know*, only the appearances of sense, inner as well as outer. Practically, however, he held, we know, in some non-theoretical sense of 'know', that we are free beings with a duty to follow autonomously a moral law. The tension between these two aspects of our nature is painful, even for Kant; but they nevertheless harmonize for him in the light of God's creation and of our fallen state.[7] The secularization of thought in the nineteenth century, therefore – which of course also set the stage for Darwin – loosened the tie between the theoretical and practical as Kant had conceived them. It left us on the one hand with a 'scientific' or better 'scientistic' tradition that was both mechanistic and phenomenalist, that is, believing that causal explanation is universal but that we can never penetrate through phenomenal cause and effect relations to the reality behind appearance. And on the other hand, *via* Fichte and Romanticism, it left us with a voluntarist tradition that stressed, in one form or another, the unique reality, however dialectically expressed, of Will or Act. These two demi-traditions culminated, one, in the philosophy of the Vienna Circle, and the other in existentialism. The world came to be seen not, as Schopenhauer would have it, as Will and Idea, but alternatively as Will or Machine. Marxists, among others, I dare say, will object to this dichotomy, for Marx supposedly reconciled science and dialectic in the shape of dialectical materialism. In reply I would have to argue that all dialectical philosophies are romantic, or at least that the attempted gap-bridging failed. But even supposing I could succeed in this perilous venture, I cannot attempt it here. For the sake of my present argument let me, with a *small* apology, permit my dichotomy to stand: either machine or will, a necessary nexus of mere phenomena, or the upsurge of reality as pure act. The question here is: where did the impact of Darwin fit into this story? And the answer is: on both sides. From, say, about 1870 to 1930, in both styles of philosophizing, different though they were, evolutionary thinking reigned supreme. Not, indeed, that Darwin was the *only* influence here: on both sides at least two other factors were important: the development of historical method and nineteenth century progressivism. The latter in its more 'scientific' version celebrated the triumphs of technology and of liberal politics; but in the Romantic tradition as well the doctrine of progress had its influence: see for instance Victor Hugo's poetic praise of progress.[8] But the acceptance of organic evolution was certainly a major factor in this development. On the science-oriented side, there was in England the popular developmental theory of Spencer, which both antedated and supported the acceptance of the *Origin*. In Germany there was the mechanistic monism of Haeckel, whose simple doctrine was certainly swallowed whole, along with Spencer *and* Darwin, by the educated middle class in America (I can attest to this from my own early recollections of popular evolutionism). Then, a little later, on the whole, but equally influential, was the Romantic version: the emergence theories of Bergson, Alexander, Whitehead and Collingwood, which were still *just* in vogue when I was a graduate student in the early thirties. In fact in 1933-34 I heard Whitehead's lectures in cosmology, which represent the grand culmination of this tradition.

Yet Fulton is right: all that is over. So I want to ask, thirdly, and this is my major question: what remains of evolutionary theory today? What is Darwin's present influence on our views of nature and of man? In general, of course, it is correct to say that evolutionary thinking has been assimilated to our common sense view of nature, and of man in relation to nature, so much so that it is, because wholly uncontroversial, philosophically uninteresting. On three of four counts, any philosophical argument on any problem has to take the outcome of the Darwinian revolution as among its unquestioned premises.

First, Darwinism (again along with other factors, historical and anthropological relativism, for example) has forced us to recognize mutability, the omnipresence of change. Not, as some people have naively argued, that change is good in itself, but simply that it *is*. The acceptance of organic evolution eliminates an Aristotelian theory of nature, dependent as such a theory must be on the existence of permanent natural kinds.

Aristotle had of course been banished from physics by the first scientific revolution, but religious sanctions combined with the plain facts of organic diversity had so far kept the new naturalism from incorporating biology. There seemed no way to bring the staggering variety of organic forms and the apparent teleology of organized beings under simple physical principles. So Kant predicted there could be no 'Newton of a blade of grass'. But Darwin not only unified living nature through transformist principles – many others had tried that too – he did it, as we have seen, by the simple but sweeping move of rendering the seeming purposiveness of organic structures and functions susceptible of causal explanation. Indeed, it was *through* that move, through the concept of natural selection as explanatory of organic change, that the *Origin* persuaded so many of the fact of evolution, precisely because it brought what had before seemed simply mysterious within the purview of scientific investigation. Yet in so doing, paradoxically, the Darwinian revolution undermined not only Aristotelian species, but the simple eternity of the Newtonian world itself. It put process before permanence, development before structure. True, this emphasis is not now, explicitly at least, current in philosophical discussion, at least in the English-speaking world. The slogan that we should 'take time seriously' has now a quaintly old-fashioned ring. But if, without acknowledging it, we *have* a metaphysic (and we always do), it is a metaphysic of mutability.

Secondly, with the permanence of Aristotelian nature we have also abandoned metaphysical necessity in favor of a recognition of fundamental contingency – and here too evolution, along with historiography as well as history itself, has played a role. Everything there is, we must acknowledge, might have been otherwise. This again is paradoxical: for the fact that we can seek a causal explanation of any natural event (and so everything is in some sense necessary) seems tied to the insight that everything is contingent, and so, in another sense, nothing is necessary. At each stage of evolution there are chances of success and risks of failure. Each new population appears as the heir of past successes and the replacement of past failures. But successes and failures are always the effects of processes that *could* have gone another way. Neither God's Providence nor the all-embracing Natura Naturans of Spinoza, let alone Hegelian dialectic, which attempts to synthesize the two, can stand against this basic principle. Again, it is not talked of in philosophy – though some discussions of determinism perhaps reflect it – but surely we have it in our bones.

The firmest lesson of Darwinism for metaphysics, however, – thirdly – is of course the lesson of our own animal nature, our demotion from super- natural support to a place in nature comparable to that of any other living thing. By now, indeed, not only the doctrine of the 'descent of man', but many lines of research in many biological fields serve to confirm this change in our view of ourselves. Indeed, so pervasive is this change that it is often misapplied, even by distinguished scientists, let alone by pure popularizers. (For a discussion of these misuses, see Leon Eisenberg's article in *Science* on 'The *Human* Nature of Human Nature'.[9]) The most intimate influence of biological research on philosophy at present, I think, comes from neurophysiology. Herbert Fingarette, for example, in presenting his theory of self-deception which supplements philosophical argument with empirical support from work on commisural patients.[10] Or more sweepingly, Richard Rorty's argument for identity theory as against functionalism in the philosophy of mind seems to rest on a general confidence in the explanatory power of physiology or physiological cybernetics.[11] But be that as it may, certainly the attempt to overcome Cartesian dualism, which still remains, alas, the major philosophic task of the waning twentieth century, found its first massive support in the Darwinian theory. No doubt about it: whatever kinds of strange fish we are, we are organic beings, not half bodies and half immortal souls.

There is a fourth point, however, on which the lesson of Darwinism is more ambiguous. In some contexts, as we have seen, evolutionary theory and the nineteenth century belief in progress were mutually reinforcing. But *is* evolution progressive? From the beginning the views of scientists have been conflicting. Darwin is said to have written in the margin of this copy of Chambers' *Vestiges*, 'Never speak of higher and lower in evolu- tion' and T. H. Huxley's friend Kingsley wrote in *The Water Babies* about a strange land where evolution went backwards – from man to amoeba, so to speak, as in terms of environmental pressures interacting with variable heritable structures, it very well might. Huxley's grandson, however, thinks otherwise, and so do many leading evolutionists today: Simpson, Dobzhansky, Stebbins, for example, not to mention others like Kimura or Thoday: but all in terms of differing criteria of 'progress'. (The various views are well summarized, as well as criticized, in George Williams' *Adaptation and Natural Selection*.)[12] For the Romantic evolu-

tionists, of course, the rise to higher levels, somehow propelled by
tendencies in life itself, was the very heart of evolutionary truth. But
what is 'progress' on a cosmic scale? Can evolution be both 'opportunistic', as in Darwin's terms, it is, and inherently progressive? Or is
'progress' what we call it when *we* look back and see the billions of years
of living history leading to – ourselves? There are, indeed, in the neobio-
sphere, innumerable populations of organisms structured in terms of a
greater number of levels of self-regulation than were the Monera of the
pre-Cambrian or even much later forms. This is clearly true of multicellular eukaryotes as against protists, again of organisms possessing a
central nervous system as against their forerunners, and so on. Stebbins
gives a table of such levels of organization, for example, in his *Basis of
Progressive Evolution*; others may wish to enumerate such levels somewhat differently.[13] There *is* progress, somehow, from blue-green algae
and bacteria to mice and monkeys, quite aside from men. But what in
such contexts does the concept of progress mean exactly, and how do we
measure what it represents? On this question, it seems to me, the message
of Darwinian thinking is much less clear. I am inclined, indeed, at the
moment, to think it safer to equate evolution with the theory of organic
change, and to reserve 'higher' and 'lower' or their analogues for the
analysis of systems once they exist, not for the causal explanation of their
origin. Living systems *are* hierarchically organized, and hierarchically
systems can be richer or poorer in their hierarchical structure. But these
systems-theoretical considerations perhaps serve chiefly to confuse when
we try to inject them into the evolutionary account itself. Once life
originates (and that's another story), we have at every period already
some self-regulating systems, mutations which permit and environmental conditions which determine the rise and spread of new such
systems. Sometimes a new level of homeostasis results, sometimes not,
but a new level of organization demands no new ground rules for the
historical account. (On this point Slobodkin's essay on 'The Strategy of
Evolution' seems to me definitive.)[14]

So far I have been talking about the influence of Darwinian thought on
what may roughly be called contemporary metaphysics. There are other
areas in philosophy, however, where we find, not so much the assimilation as unspoken presuppositions of the implications of evolutionary
theory, as the explicit application of the theory itself to philosophical

problems. Let me look briefly in conclusion at three of these areas: evolutionary ethics, functionalism in the theory of the social sciences, and evolutionary epistemology. All three, I'm afraid, are extensions of a powerful scientific theory to areas where it is misplaced, where its application serves to obscure the problem and where, therefore, instead of explaining, it explains away.

First, evolutionary ethics. Unless we go back to Dewey this is in any case an evolutionist's, not a philosopher's theory, and from a philosophical point of view hard to take seriously. What it *is* to be an ethical animal, as Waddington puts it, is not explained, as he tries to explain it, by the fact that we came into being, as animals, through mutation and natural selection, but with this strange propensity to moralize.[15] Nor is it explained by analogy with 'evolutionary progress', itself, as we have seen, a muddled concept. The problem of ethics is not how apparent norms (or ends) arose, nor how biological processes run parallel (if they do) to processes of ethical judgment. The problem of ethics is: what real norms (or ends) are: norms that are human artifacts, but real to us, as real as anything in our lives can be. Ethics, in other words, is a critical inquiry about (1) the general character of norms of conduct and goals of intentional action, and (2), and more fundamentally, about what it *is* to be an ethical norm or an end of intentional action at all. We all distinguish somehow right from wrong and good from evil. What kinds of claims are we making when we do this? Evolution does set constraints on our answers: we are not, evolution tells us, or ought not to be, appealing to supernatural revelation for our judgments of 'ought' and 'good'. But apart from depriving us of this simple if superstitious account, the story of our evolution, the explanation of how we originated from earlier anthropoids, cannot in itself clarify critically, as philosophical ethics has to do, the *nature* of moral values or of responsible choice. We are animals, but culture-dependent animals, tradition-dwelling animals, and it is inherent in culture, in the human variety of tradition, to work by self-imposed, historically received, but individually authorized, standards. What this means, is the problem of ethics, not how it all began – the problem of human evolution. Evolutionary studies, like Portmann's comparative study of early mammalian development, may indeed have an interesting bearing on this question, but in themselves they do not even put the ethical question, let alone answer it. Only if we decide, as

Hobbes did, that adaptation for survival, and absolutely nothing else, is the sole standard of rational choice, can we equate the answer to these two questions. But that is already to *have* an ethical theory, which *then* turns out to be convergent with the theory of natural selection. As an ethical theory, it needs philosophical grounds, in terms of a theory of the nature of moral standards, not of the origin of the creatures who adopt, and follow, them.

A similar confusion infects functionalism in the social sciences. Here there is an apparent homology with the evolutionary situation which leads to a parallel explanation that begs the very questions which social philosophy needs to raise. The claws of male isopods, let's say, are adapted for mating. Their origin can be causally explained in terms of Darwinian theory as making more probable for the organisms bearing the appropriate genes the production of greater numbers of organisms bearing the same genes. Survival is here the only criterion of 'success' – that's what evolution is about. Is it the same with social institutions? It might seem so. For example, Max Gluckman has described rebellion in some African societies as a device to decrease dissatisfaction and so keep the society going.[16] Thus, as with the rereading of apparent 'purposiveness' in causal terms by Darwinian biology, what looks like intentional behavior is turned back to receive a purely causal explanation (though the continuity here is cultural rather than genetic – already a substantive difference). Moreover, in these terms, again, as in every evolutionary context, survival is the sole criterion of excellence. But will this really do? Don't we want to be able to *compare* societies, to say that sometimes the superior society succumbs to its inferior? As in the case of evolutionary ethics, so here, the problem of standards, the question on what grounds we judge what makes one functioning structure (in this case social structure) better than another – the philosophical question is unseen or prejudged before it has even been raised.

What, finally, of evolutionary epistemology?[17] Isn't natural selection here a proper paradigm for the growth of knowledge? Don't those theories survive which predict successfully and those go to the wall that don't? So don't we judge claims to knowledge in terms of something analogous to the algebra of gene-frequencies: a greater probability of leaving descendants, that is, in this case, scientists who believe the theory, in the future? But remember that the algebra of gene-frequencies, like the biochemistry of DNA, is the *carrier* of Darwinian explanation, the one its formalization, the other its material; neither *is* the explanation itself. As an explanatory principle natural selection is simply shorthand for a vast nexus of cause and effect relations between organism and environment. But a causal explanation, a when-then explanation, of cognitive achievements, omits, and must omit, any account of their claims as *knowledge*. Hence Dewey's confusion in calling mathematics an experimental science because children have to learn it. Hence also, I must say with all due respect, the confusion inherent, for all its great achievements, in the very concept of 'genetic epistemology'.

That is not to say that there is not a real and significant continuity between cognitive activities and other behaviors on the part of living things. All knowledge is orientation of some kind (but even that is more an ecological than an evolutionary statement). What makes us claim, however, that some sorts of orientation in our world constitute *knowledge* is not simply their success in guiding us in our surroundings, as the plankton is oriented in the streaming ocean or the robin to the red breast of its mate. Illusions, if systematic, can also guide successfully. When we say we *know* something, we are not saying it is true simply because it will survive, but contrariwise, it will survive because it is true. We are claiming, not just that we have made a beautiful theory, a theory so elegant that it will continue to fool people for a long time to come, like advertisers crying their wares, we are saying that we have found, in some limited respect and within the canons of some discipline in which we have competence, the way things work: we are claiming, sometimes in anticipation of our theory's empirical consequences, sometimes by virtue of its very elegance, that we are in contact with reality. A scientist, Norman Campbell said, is a man who passionately believes that nature will conform to his intellectual desires.[18] This claim to truth, this gamble on being in contact with reality – of course it is a gamble and we may always be mistaken: another lesson of Darwinism if you like: fallibility follows from the acceptance of metaphysical contingency – this risky claim, then, is precisely what epistemology has critically to examine. By what standards, of accuracy, of objectivity or disinterestedness, of systematic relevance, do we judge statements to be true or false? On what grounds are such standards supported? Again, it is this critical normative discipline that is the work of philosophical reflection. To try to make a scientific theory, even, or especially, a seemingly comprehensive theory like the Darwinian, do the job for us, is to blind ourselves to the reflective task inherent in being the kind of self-questioning animals we are. Granted, we are not as simply unique, even as animals, as we used to think. Chimps, too, Jane Goodall has taught us, make and use tools; chimps too recognize, at least in some laboratory situations, their own mirror images. Each easy cutoff between men and other animals becomes more delicate, it seems, with advancing research. Possibly even language. Yet it is still true that we live more massively than other animals *in* artifacts, *in* culture, *in* language. We live therefore within our own norms for making, needing to accept those we are taught in infancy and youth, yet needing also to remake them. It is these strange unnaturally natural, or naturally unnatural, normative structures that we need to examine in every philosophical field, in epistemology the structures, for instance, that we call sciences, or theories, or empirical laws. To confuse their organization, their axiology, with the biological and psychological roots of their origin is to forget, in Eisenberg's phrase, the *human* nature of human nature, to live that unexamined life which, as Socrates told us long ago, it is not worthwhile for a man to live.

NOTES

[1] J. H. Randall, Jr., 'The Changing Impact of Darwin on Philosophy', *Journal of the History of Ideas* **22** (1961), 435–462.

[2] J. S. Fulton, 'Philosophical Adventures of the Idea of Evolution, 1859–1959', *Rice Institute Pamphlets* **46** (1959), 1.

[3] C.F.A. Pantin, 'The Origin of Species', in *The History of Science*, London 1951, p. 129ff.

[4] See e.g., T. Dobzhansky, *The Genetics of the Evolutionary Process*, New York 1970.

[5] E. O. Wilson, *The Insect Societies*, Cambridge, Mass., 1971.

[6] Lynn Margulis, *The Origin of Eukaryotic Cells*, New Haven 1970.

[7] See G. Krueger, *Philosophie und Moral in der Kantischen Kritik*, Tübingen 1931.

[8] See Victor Hugo, *Poésie*, Collection l'Intégrale, Paris 1972, Vol. 2, pp. 560–61; Vol. 3, p. 663.

[9] L. Eisenberg, 'The *Human* Nature of Human Nature', *Science* **176** (1972), 123–28.

[10] Herbert Fingarette, *Self-Deception*, London 1969.

[11] Richard Rorty, 'Functionalism, Machines and Incorrigibility', *Journal of Philosophy* **69** (1972), 203–220.

[12] George Williams, *Adaptation and Natural Selection*, Princeton 1966.

[13] G. L. Stebbins, *The Basis of Progressive Evolution*, Chapel Hill, North Carolina, 1969.

[14] Lawrence B. Slobodkin, 'The Strategy of Evolution', *Amer. Sci.* **52** (1964), 342–357.

[15] C. H. Waddington, *The Ethical Animal*, London 1960.

[16] Max Gluckman, *Custom and Conflict in Africa*, Glencoe, Ill., 1955.

[17] See for instance Sir Karl Popper's *Objective Knowledge*, Oxford 1973, p. 67.

[18] Norman Campbell, *What is Science?*, New York 1952.

CHAPTER XII

THE ETHICAL ANIMAL: A REVIEW

The Ethical Animal, by C. H. Waddington, George Allen and Unwin, London, 1960, pp. 230.

The problems of philosophy may be perennial, yet some of them press harder on some generations than on others. The fact that judgments of value differ from man to man and from culture to culture was noticed by Herodotus, and was certainly very plain to Plato. For us this diversity, especially at the intercultural level, raises the most urgent philosophical problem of our day. Can we evaluate systems of evaluation, can we grade cultures, and if so, by what right?

It is this problem to which Professor Waddington's argument is directed; if he had solved it, this book – or his earlier *Science and Ethics*, of which the present book forms in a sense a revision – would have constituted a major philosophical achievement.

What, and how successful, is Waddington's answer? 'It is the thesis of this book', he writes, 'that the framework within which one can carry on a rational discussion of different systems of ethics, and make comparisons of their various merits and demerits, is to be found in a consideration of animal and human evolution'. His solution differs, however, he declares, from that of other evolutionists. Julian Huxley's evolutionary ethics, he rightly observes, is viciously circular: by our ordinary ethical standards, we judge evolution to be good, and then we justify those same standards by reference to evolution. Simpson and Dobzhansky, on the other hand, lay too much stress, for his liking, on the factor of freedom in the sense of arbitrary choice. His solution, he argues, is non-circular, at least in any vicious sense, because it is based on a criterion over and above the source of our individual ethical judgments, a criterion which he calls 'biological wisdom'. Moreover, it is rational, since it is grounded, not in an arbitrary choice, but in 'an induction from the properties of individual ethical systems'. Biological wisdom, in other words, claims to be, not a super-ethic, but a scientific generalisation enabling us to judge between the diverse ethical systems which the ethicising animal, man, has developed in the course of his evolution.

What is this biological wisdom? Although Waddington's argument is often obscure and confusing, I think I may hazard the statement that biological wisdom is said to involve four things. Firstly, the biologically wise man recognises the fact that human evolution has its own peculiar, *sociogenetic* mechanism: in brief, the transmission of culture through authority and learning from one generation to the next. It is the function of 'ethicising' to mediate this process. Secondly, the wise man takes account of the difference between *cladogenesis* and *anagenesis* in evolution. Within a given organisational type we have, in cladogenesis, according to Waddington (after Huxley, after Rensch), a spread of diversification on a given level of evolutionary development. In anagenesis we have, still within the given type, a forward movement from a less to a better developed version of this type. Examples: *cladogenesis*: the proliferation of browsers in the equid genealogy, even during the period when the development of grazing types had begun; *anagenesis*: the advance along the line of grazing animals now perfected in the modern horse. Of such 'forward' movements, in turn, those are peculiarly significant which are 'open-ended', i.e. which lead not to evolutionary dead-ends but to possible avenues of further advance. This distinction leads the wise man, I can only say, to do one or both of two things, for the text here is elusive and ambiguous. Either (thirdly), he singles out the one line of anagenesis characteristic of evolution in general: this would be an overall direction of evolutionary advance, from type to type, rather than anagenesis as originally defined, which is restricted to progress within a given type. It would be, in effect, an anagenesis of anageneses which the biologically wise man approves and seeks to promote. What this trend is or might be, however, is again obscure. In one formulation suggested by Waddington, for example, it is 'the increasing ability to utilise for the maintenance of life more and more complex relations between environmental variables'. Here it seems to be a question of 'better' techniques for survival, the maintenance of life being the common end. In other passages, however, Waddington speaks of an overall tendency in evolution toward a 'richness', a 'depth', a 'fullness' of life: now not life itself, but in some way or other a *good* life is what evolution is said to have produced. Or, on the other hand, fourthly, one infers from some of Waddington's statements that it is a possible open-ended anagenesis, not in evolution as a whole, but simply within our own organisational type that the biologically wise man apprehends and appreciates. Thus in one passage, for instance, it is said to be the development of the super-ego, characteristically produced by our socio-genetic evolutionary system, which is to be fostered. (The dogmas of psychoanalysis, I may say in passing, are accepted by Waddington with a good deal of naiveté as established truths of science. Many psychiatrists, let alone philosophers, would be inclined to question his confident acceptance of authority here.) Elsewhere, on the other hand, he writes in more general terms: 'our argument leads to the conclusion that biological wisdom consists in the encouragement of the forward progress (anagenesis) both of the mechanism of the socio-genetic evolutionary system, and of the changes in the grade of human organisation which that system brings about'. This is a summary clearly in terms of point four; yet I suspect that it would not be possible were not the more general conception of evolutionary advance lurking somewhere offstage.

I hope this is not an unfair summary of this extraordinarily cloudy argument. Some of the difficulties of restating it may emerge from the following criticisms.

Let me say first that I heartily agree with Professor Waddington when he regrets the narrowness of much academic philosophy, and the failure of philosophers to bring relevant material from psychology and biology to bear on their reflections. Even here, however, he has misinterpreted many of the examples he cites against us. When, to take just one case, Professor Emmet tells us that the concept of 'evolutionary progress' may be one in which we are reading back into the natural world concepts derived 'from analogies with human actions and purposes', Professor Waddington comments: 'She is basing her thought on the implicit assumption that human actions and purposes could be something completely external to, and independent of the natural world'. But it seems to me quite obvious that exactly the contrary is the case. Professor Emmet is not arguing that our purposes are ever wholly independent of nature, but that nature is sometimes independent of our purposes – and 'progress' being a distinctively human and purposive concept, we have no right to impose it without further guarantee upon that nature of which we are, *qua* nature, a small and insignificant part.

'Progress', indeed, is one of the concepts which appears in just such an uncritical use in this very book. What is evolutionary 'advance' or 'improvement'? Is it purely improved techniques for survival that are meant? If so, there are two points to be made. First, in terms of survival value, in terms of a strict selectionist theory of evolution, we have no right to speak of progress as between one kind of organism and another. As both Darwin and T. H. Huxley recognised, there is, or ought to be, no distinction whatsoever between higher and lower in evolution so conceived. The insects have survived and so have we, and they have survived a good deal longer and are a good deal more numerous than we are. Secondly, however, if, as Waddington suggests, some sort of more complex organism-environment relation somehow 'objectively' marks an 'advance', and if we have in fact achieved such a relation, then nevertheless, if it is survival techniques we are judging, and if biological wisdom is to lead us to choose in the light of such a criterion between one ethicising system and another, the system which will lead to 'the maintenance of life' – any life, if only life – will be the one to choose. Now that may well be, for example, the system which practices the most efficient brainwashing: a procedure which Professor Waddington appears to deplore. Yet on 'scientific' grounds, and in terms of survival value alone, I see no foundation whatever for his deprecation of it. It is precisely in terms of our *ethics*, not of any scientific generalisation, that we deplore it. 'The maintenance of life', however, is only one of Waddington's multifarious formulations of the chief line of evolutionary advance. If, on the other hand, it is not life, but richness and depth of life in terms of which we are to make our choice, if it is an 'active, creative life' we are after, then this is indeed progress in the light of our own values, our ideals, which we are, as Professor Emmet warned us not to do, reading back into nature through the accordion-like use of that expansive concept 'evolutionary progress'. This, again, is no scientific generalisation, but the statement of an ethical appraisal.

Again, a similar ambiguity affects Waddington's use of 'anagenesis': as I have already pointed out, this concept is introduced as applying within a single type of organisation, but is somehow, without warning, extended to refer to an advance *between* types, an advance in evolution as a whole. Yet how can we, who are *within* one type, judge 'objectively' such overall advance?

Yet again, for example, we are told that we can see more in evolution, for our purposes, if we look at it 'organismically' rather than purely mechanistically. Evolution in itself includes no values, no 'harmonies' or 'wholes', and 'organicism' is simply read into the evolutionary story by the all-too-human assessor of it for the purpose of assuaging his human needs. In that case, however, the levels of organisation Waddington continues to talk about have no *scientific* meaning, but are pure 'as-ifs' used to soothe our ethical feelings, and references, for instance, to 'the clear-cut hierarchy of organic forms recognised since Aristotle' give us no information whatsoever about nature, let alone a 'scientific' yardstick by which to judge between the merits of divergent forms of life or systems of authority.

I could go on listing such ambiguities and inconsistencies, all of which stem in one way or another from a failure to distinguish between philosophical (in particular moral) problems and causal analysis, between the desir*able* and the desired, between ought and is. One passage will perhaps illustrate this fundamental confusion at least for philosophical readers. In discussing criticisms of his earlier book, Waddington quotes the following statement by Professor D. D. Raphael:

Waddington thinks the answer to my question: 'What am I to do and for what reasons?' is the same as the answer to the question 'What will I do and for what causes?' This is why he thinks that a causal account of how ethical judgements have come to be what they are can supply a criterion or rational ground for the ethical judgements we should make. But since the two questions and the kinds of answer they seek are of different logical types, Waddington's argument for using the direction of evolution as the direction for ethical judgements rests on a logical confusion.

In answer, Waddington now replies, 'Here I take it that 'reasons' are verbal arguments deducible from some theory which accounts for certain phenomena in terms of the causal properties of the components of the system in which the phenomena occur'. It seems to me superfluous to comment on this extraordinary statement. But, Waddington continues, this is not the main point: more important is that Raphael's second assertion is still in the first person, whereas his 'wisdom' refers to some 'larger entity, such as the human species as a whole, or even the living world as a whole'. He reformulates Raphael's statement, therefore, to conform to his own belief:

Waddington thinks the answer to the question 'What would it be wise for me to do and for what reasons?' can be deduced from the answer to the question: 'What has the world at large been doing during its history and from what causes?' This is why he thinks that a causal account of how individual ethical judgements have come to be what they are can supply a criterion or rational ground for the judgement between different ethical beliefs which it would be wise for us to make. These are two questions and the kinds of answer they seek are of different logical types. Waddington's argument for using the direction of evolution as the criterion for judgement between ethical beliefs rests on acceptance of this.

Without denying that the whole of history may have a bearing on what it would be wise for me to do, I may, I hope, simply assent wholeheartedly to Raphael's criticism, which is clearly valid against this book as against the earlier one, and suggest, in conclusion, a further revision:

Waddington thinks the answer to the question 'What would it be wise for me to do and for what reasons?' can be deduced from the answer to the question 'What have living beings been doing since the Cambrian and from what causes?' ... since the two questions and the kinds of answer they seek are of different logical types, and therefore, though each has relevance for the other, neither is deducible from the other, the one referring for its answer primarily to my ultimate ideal of what life ought to be, the other primarily to a factual account of what life is, Waddington's argument for using the direction of evolution as the direction for ethical judgements rests on a logical confusion.

Criteria of objectivity, of relevance, of accuracy, are indeed necessary to ascertain what goes on; but what happens cannot, apart from our evaluative assessment of it, *create* criteria in its turn. If *oughts* can be deduced from *is*'s, it is only from the *is* of an *evaluation* (like the utilitarians' 'all men value pleasure', etc.) that they can be derived. Only by smuggling 'progress', 'richness', 'criterion', 'rational' into nature, can we pretend to draw such qualities out of it again. This is a fair, a necessary procedure if we know what we are doing, but a snare and a delusion if we think that the practice of 'normal scientific methods' and not the articulation of our deepest hopes and most personal aspirations is what we are about.

CHAPTER XIII

EXPLANATION AND EVOLUTION

I want to discuss some problems about explanation in biology. The leading question about biological explanation has been the question whether, and if so how, it differs from explanation in other fields. Many claims and counterclaims are made by biologists as well as philosophers in this connection. I want to examine in particular the claim of Francisco Ayala that biology is irreducible to physics and chemistry because, and only because, of the teleological structure of evolutionary theory.[1] But to reflect on Ayala's claim I must have some notion of what makes a theory in general explanatory and also some notion of what makes an explanation teleological. While I can't claim to have a pat answer on either of these puzzles (especially the former!), let me make some preparatory remarks about both before looking at the alleged teleology of evolutionary theory in relation to its explanatory force.

First, then, there is the general question: what makes an explanation explanatory? There was a time when philosophers of science thought this problem tractable, or even solved. One had only to separate cleanly the 'context of discovery' from the 'context of justification' and then some version of the 'covering law hypothesis' or some account of the 'hypothetico-deductive method' would do the job. Controversies about confirmation theory, or about verification *versus* falsification, took place within this purified context. What the subject had been purged of, however, of course, was any relation to science. Now, thanks to the historians of science and to some philosophers coming to philosophy of science from the practice of the sciences, it has become plain to almost all concerned that science without 'the context of discovery' is indeed Hamlet without the Prince of Denmark. We have to understand science, and scientific explanation, within the context of scientific activity – within history –, not just in terms of a formal reconstruction of its dead bones.

So much is clear. But how is this more concrete, and therefore less precise, account of scientific explanation to be formulated? There are a number of theories in the market, so to speak, but no one of them so far has become current coin. Michael Polanyi, who has certainly come as close as any one to stating a full-fledged philosophy of science, has never developed a theory of explanation as such. I asked him long ago about this – for the problem of explanation seemed, and still seems, to me central to philosophy of science – but he professed to be without interest in it. Perhaps an account of the factors in discovery is supposed *a fortiori* to include, or to be identical with, an account of explanation. I don't know. But in the present context I think we *can* bring some of the principles of Polanyi's account to bear on our problem.[2] Two in particular. First, it is important that Polanyi has substituted for the stock criterion of 'predictability' a criterion of 'unpredictability'. What the traditional concept of 'fruitfulness' of theories really meant, he argues, is that a great discovery – notably Copernicus's or Kepler's – is fraught with more consequences, systematically interrelated consequences – than its discoverer could have foreseen, and that his confidence in its truth consists in his confidence that this would prove to be the case. This is of course at first a hope, a hope, in Polanyi's words, that one is in contact with reality, and as such its content is specifiable only with hindsight. That is why, afterward, it looks as if the whole thing had been predictable: just as any free action, which is always a gamble (and the affirmation of a theory is of course an action) always looks in retrospect thoroughly determined.

What more than this little can we say about theories and their explanatory power? We may mine something from a second theme of *Personal Knowledge*: the analysis of 'scientific value'. Not every statement of fact and not every mathematical formula, Polanyi reminds us, gets accepted as scientific. Which ones make it? There are many considerations that may determine the admission of a concept or a statement or a formula to the scientific fold. In general Polanyi distinguishes three classes of criteria: accuracy, systematic relevance, and intrinsic interest.

The first, accuracy, seems obvious. It is expressed in the well-known thesis that only the quantifiable is scientific (which in other formulations goes back at least to Galileo, or perhaps to Democritus). But in most seasons accuracy is not enough; to get into science, statements, whether of alleged 'fact' or of conceptual, usually mathematical, relations, must have bearing on some field of science, on some range of subject matter or theory which has received, or by the present discovery is to receive, some overall conceptual ordering. This is the criterion that was misunderstood in the interpretation of explanatory power as 'deductive'. In that form it has its classic statement in Hume's *Treatise*: the hypothesis is best from which the greatest number of observable consequences can be deduced. It gets a more 'historical' or 'psychological' coloring in Born's account of discovery as the finding of a gestalt which no one before had seen. It should be noted also that the ordering concept or principle that lends coherence need not be wholly contained within the acknowledged universe of discourse of the science in question or even of science in general. Metaphysical concepts or maxims, if they mesh well with thought or practice in some area of scientific inquiry, may also operate in determining the systematic relevance of a theory (though I think it is correct to say they never operate alone).

Polanyi's third ingredient of scientific value, the intrinsic interest of the subject matter, though it will concern us briefly later, does not appear so directly related to the problem of explanation. Intrinsic interest compensates, in Polanyi's view, for the lesser accuracy of explanations in some areas. Thus we will take a less precise theory about living things or human actions, he suggests, than about crystals or electrons, because the former are in themselves of such overriding interest to us. But on the one hand I am not at all sure that this is true any longer of biology or perhaps even of social science, and on the other it is in any case the interest of the subject itself (the explanandum) that is in question here, not the intrinsic interest of the theory invoked to explain it. But I hope that in general the concept of scientific value and its ingredients, as well as the reconstructed concept of fruitfulness, will assist us in examining the case before us.

My second preliminary question concerns teleological explanation. Where if anywhere do teleological explanations occur in science? I have tried to answer this question elsewhere but now think I was in part mistaken and so would like to try again.[3] First let's see where teleological questions might occur in scientific investigation and then ask to what extent they are asked in biology and to what extent they get teleological answers.

Questions get asked, to begin with, not about isolated phenomena but about phenomena in context: about systems or parts of systems. (Tycho's Nova, one supposes, was interesting not so much in itself as because it cast doubt on an accepted context.) What sorts of questions, then, does one ask, in scientific inquiries, about systems of phenomena or parts of such systems? The *first* question would seem to be one of identification or classification: what is it? For Aristotle, indeed, that was the controlling question, but in the context of scientific explanation as we have come to understand it, this kind of question is soon replaced by or assimilated to other kinds of questions. Thus one may ask of a system or partial system, *secondly*, what does it do, or how does it work? Or a *third* kind of question: what is it made of? Or *fourth*: how did it come to be, or, how did it arise? Or *fifth*: what is it for? (Five is often preliminary to two, but they're not the same question; that's important) Or *sixth*: what good is it to us, how does it, or can it, serve our needs? Which of these questions are teleological? Five and six are clearly so. Six, which we may call a question of subjective teleology, is the kind of inquiry that seventeenth century thinkers condemned – see the Appendix to Part I of Spinoza's *Ethics* – believing that with it they were banning all teleological inquiry from the study of nature. Question five, however, the question of instrumental teleology (or teleonomy if you prefer that term) remains as a legitimate question, not, indeed, as for Aristotle, for whole systems, but for parts of systems, where the part is taken to have some function in the operation of the whole. The only kinds of whole systems for which we still ask this type of question, I believe, are artifacts. But artifacts are parts of human culture; when we ask what a machine is for we put it into the larger context of human action. But we don't any longer put natural objects taken as wholes into such larger contexts, neither into the Aristotelian context of the eternally recurring reproduction of the kind in the individual, nor into the Judaeo-Christian context of divine agency. We ask what an elephant's trunk is for, relative to the elephant, but not what elephants are for – unless, be it understood, we are asking about elephants as parts of an ecosystem, and then we are taking them as functioning parts in a larger whole. Questions of subjective teleology, then, are forbidden but questions of instrumental teleology are perfectly good questions to ask about parts of systems taken as instrumental in the functioning of larger wholes.

Question four, however, how did this system or subsystem arise, may also be put in teleological form. We may seek either (4a) the predetermining causes of the genesis of the system under consideration, from past to present, or (4b) the reasons for its origin in some end to which it has been directed, from present to past. The answer to the first kind of historical question would be causal: it would ask us to specify the necessary and sufficient conditions, which being given, the result inevitably followed. The other sort of question would be teleological: it would demand that we seek an endpoint of the process which, so long as nothing overwhelmingly intervened, tended to come about by some means or other. In the first case the temporal antecedent would univocally necessitate the consequent; in the second case, of historical or developmental teleology, the temporal consequent would tend to induce its antecedents, although there would be no strict determination of just what those antecedents would be. Given a set of causes, a unique effect follows, but given an end, any one of a number of means may arise to bring it about. Thus there seems to be in historical teleology a many-one relation of means to end that is lacking in the causal case. Cause and effect relations are inflexible; means-end relations are not so.[4]

Now since evolutionary theory answers a how-it-arose question, it is plainly here, if anywhere, that its teleological questions, and correspondingly its teleological answers, would be located. Its teleology would be, of course, not subjective, and not instrumental either, but developmental. But before we finally get down to asking if this is in fact the case, let us try to sort out the place, if any, of teleological answers in biological theory, apart from evolution.

Are there teleological explanations in biology, in the study of currently existing individuals or ecosystems as distinct from the succession of populations that constitute the evolutionary past? That is: are there explanations corresponding to questions of type 4b or 5, questions of developmental or instrumental teleology? There are certainly what Huxley calls telic *phenomena*, either *developmental*, that is, finite series of events which usually tend to a given endpoint, the maturation of a limb of a sense-organ, the development of linguistic competence or of the power of locomotion, and so on; or *instrumental*, organs which act in a functional way in a larger system: the hand for grasping, the larynx for speaking, mitochondria for respiration and so on. But are the *explanations* of such phenomena teleological? That biologists use teleological language is clear; but what are they doing with it? First, such language is heuristic; it helps identify the system to be studied. Secondly, it is descriptive: we have to describe the system we are studying and to describe it without means-end language, whether developmental or functional, would be tedious to the point of absurdity. Perhaps, thirdly, teleological principles may also serve as so-called 'regulative' maxims. The investigation of the Golgi apparatus, for instance, was guided by a puzzle about its possible function. More recently, the interpretation of electromicrographs has certainly occasioned new physiological or functional hypotheses as well as simple descriptions. All the way back to Leuwenhoek and his little animals, indeed, theorizing about function has interacted with description of what was there to be seen. But I'm not sure to what extent this kind of consideration survives into the scientific *explanation* of the phenomena. The answer to the question, what is it for? operates in the initial stages, and thus again heuristically, but when it gets going investigation seems to be directed to questions 2, 3 or 4a: how does the system work, what's it made of, how, causally, did it get that way? True, the hand is seen to be a hand, not just by looking at its shape, but by identifying what it's used for; it is described in terms of its prehensile capacities, and so on, and one may perhaps be guided in the study of its development by the know-ledge of what this organ will eventually have to do. But this is all on the road to explanation, not explanation itself. Apart from evolutionary theory, is there any context in which biological *explanation* is genuinely teleological, either instrumental or developmental: that is, where an explanans entails reference to the evocation of a means by its end or to the end-point of a process as evoking the steps to its achievement?

Let's look briefly at two very famous cases of biological explanation and see where teleology enters into them: first, Harvey's explanation of the action of the heart and second the explanation of heredity in terms of the genetic code.

Harvey discovered the circulation of the blood and declared the heart to be a pump-like muscle whose contraction pushes the blood around the body out through the arteries and back through the veins. Descartes agreed that the blood circulates but held the heart to be an expanding, furnace-like chamber which rarefies the blood by its heat and so blows it out through the arteries, whence it returns, condensed, through the veins. They agreed as to what the heart was for – to promote the circu-lation – but disagreed on their explanation, that is, on the mechanism by which the goal was achieved. In one case: a pump, contracting and so pushing; in the other, a balloon-like furnace, expanding because of the rarefaction of its contents and so expelling that contents as the result of the rarefying process. The explanation is in each case mechanical, in terms of an operating principle (an answer to question 2). It is the process we are studying that has a 'goal' or performs a 'function' which we must identify and describe in order to know what we are investigating. But the explanation, whether Harvey's correct hypothesis or Descartes's erroneous one, is in terms of 'how', not of 'what for'. There is an agreed end, the circulation. The explanation specifies the means. So we have a mechanical explanation of a telic phenomenon; but the explanans itself does not specify a goal; it specifies a mechanism. The end provides the explanandum, not the explanans.

Now it's true, of course, that Harvey's recognition of what the valves in the veins were *for* formed an important step in his explanation of the heart's action. I once saw a lovely illustration in a text, I think, by Fabricius, where he had drawn a vein and a branch of verbena one above the other: he thought that since these sets of bumps looked alike they were probably doing the same thing. Obviously, Harvey's advance over his teacher consisted partly in the advance of teleonomic over purely descriptive thinking. But Harvey's insight into the function of the valves hardly stands alone: it gets pretty smoothly fitted into an explanation of how the whole circulatory system works, that is, into an answer to our question two. (Also one: the heart is a muscle!)

What about my second example, the explanation of heredity in terms of the DNA code? A teleological perspective may be said to enter here too into the early stage of the inquiry. Crick's discovery was not based on direct inspection of the sequence of bases; as in experimental embryology, the underlying relations were inferred from the behavior of the organism. Thus a telic phenomenon, the production by the organism of a normal enzyme or an abnormal protein or its failure to produce any protein at all provided (1) heuristically, the identification of the problem; (2) the description of the explanandum – but the explanans? No: the explanation consisted in the hypothesis of the non-overlapping triplet code, the hypothesis that the order of the four kinds of bases in the DNA chain determines the sequence of amino acids in protein. This explanation specifies a pattern, a particular arrangement, which tells us what the form of the system is, how it works (answering question 2), and its actions can then be seen, if you like, in response to question 4a, in causal terms, as providing the necessary and sufficient conditions for the development which follows. Crick himself of course thinks the explanation wholly causal – as he thinks all scientific explanation is. Or rather, he thinks all scientific explanation answers our questions 3 and 4a combined: first the materials are discovered and the discovery of the effective arrangements simply tells us: given certain conditions A–E, effects F–N follow. There is just this one kind of explanation and the code is a triumphant case of it.

But the situation here is a little more complicated and I think some of the confusions about teleology may be alleviated if we stop to look at it a little more closely.

It has been argued, convincingly, I think, that Crick's reductivist view of his own achievement is inadequate. Any code, it is argued, is a very improbable arrangment, one which could not be predicted solely on the ground of the laws of physics and chemistry in terms of which we have to understand the operation of macromolecules such as DNA. Even if some arrangements are chemically more likely than others, the persistence of just *one* particular arrangement, and no other, is not explicable in chemical terms alone. The reference to a code, in other words, introduces an explanation which, though it does not contradict the laws of physics and chemistry, is not on the same explanatory level as they are. Even if such an explanation is causal, in the classical sense of when-then causality, it does not appeal to the same causal series as in the chemical case. Maybe physico-chemical explanation is not strictly causal either; that's another question, to which I'll return in a moment; but even supposing it is: supposing it simply tells us that given conditions A–E, conditions F–N necessarily follow, the reference to a code, which picks out a unique and persisting order not specifiable on the grounds of the lower-level laws alone, must be in some minimal sense at least of a different kind. Even if it too is causal, it represents a new, superimposed causal series. The explanation through reference to a code, in other words, is necessarily a *hierarchical* explanation: it looks at a physico-chemical system, which necessarily follows the laws of physics and chemistry, not in the terms of those laws alone, but in terms of additional constraints on the operation of the system. Only if the system is arranged, and with a fair degree of stability, in this particular way, will it be a system of this kind. By disorganizing the code, you kill the organism. Nothing happens to the laws of physics and chemistry nor to the molecules whose behavior is explained in terms of those laws; all that happens is that you don't

any longer have an organism to study. So the code-explanation is at a
level additional to the level of explanation of physics and chemistry.

Now this hierarchical character of biological explanation has, I
believe, been amply justified. But my question here is this: is the upper
level, the pump level or the code level in our examples, teleological in
character? I really cannot any longer see that it is. Either it is a causal
explanation on a different level: when this order, then ordinary development, when not, then not. Or it is what one might call a formal explanation answering our question 2; it tells us what kind of system this is in
terms of the order (rhythm of muscular contraction or triplet code) of the
elements that makes it possible for the system to function and thus to
exist as a system of this kind. The answer to the question what it's for
sets the stage for the inquiry – and in the code case so does the answer
to question three, what's it made of – but the explanation responds to
questions 2 and 4a, not to 5 or 4b. We could not investigate heredity and
development if we didn't first identify the normal endpoint of reproduction and growth – a new individual or population of individuals – but we
explain that development to a normal endpoint either causally over time,
through reference to a necessitated sequence of events, or formally *at* a
time, through reference to an ordering principle or arrangement of elements which constrains the operation of those elements in a certain way.
The analysis into the elements (answering question three) is of course also
an important part of such explanations: indeed the answers to questions
2, 3, 4a can all be complementary to one another; but the what-for question (instrumental teleology) has been long left behind. Nor does developmental teleology (4b) come into it: the endpoint of the process, the growth
of a new organism is certainly not being said to induce the means to its
own production.

I should add, also, referring back to my parenthetical query about the
causal character of physico-chemical explanations, that there too one
can explain either e.g., through reference to the laws of thermodynamics,
and therefore formally, or by applying these laws to the account of a
sequence of necessary and sufficient conditions and their necessitated
results, that is, causally. Even the hierarchical character of the explanation of living systems, therefore, is not unique. There is, as I have also
argued elsewhere, no pure 'causal', when-then explanation. Causality is
always dependent on *some* rationale, some laws to which a series of events

is seen to conform. Otherwise it would be pure witchcraft. My point here, however, is that even in the conspicuously hierarchical explanations of the behavior of living systems, teleology enters only into the heuristic, descriptive and perhaps regulative forerunners of explanation, not into explanation itself.

What – at long last – about evolutionary theory? Let us examine briefly the conceptual structure of the dominant, so-called synthetic theory, and then ask whether or how it is teleological and why some of its adherents think so, considering on our way the grounds for its explanatory power.

Evolutionary theory studies a global sequence of ecosystems over time and purports to tell us how – by what 'mechanisms' – they come to succeed one another. According to the prevailing view, the processes which have produced the changing organic scene are: mutation, which furnishes the materials of change; linkage, crossing-over and recombination, which have to do with arrangement or re-arrangement of the genetic material, isolation, and natural selection. Critics of the theory have sometimes argued that its reliance on mutation – random changes in the genetic material – invalidate it, since this makes the vast complexity of living forms so infinitely improbable. How could the eye originate by chance, people have asked recurrently since 1859. The theory's adherents reply, however: (1) The process is long and gradual; in any one phylogenetic chain only one small step need be taken in any one small sequence of generations; and (2) the pool of variations available to any population is so immense that no striking 'chances' are needed to provide the material for new development. Indeed, I have heard two great evolutionists, Dobzhansky and Stebbins, agree that they could forego mutation for half a million years and still get all the variants they need for evolution. Natural selection working on the existent gene pool would suffice. Of the other factors, linkage, recombination, crossing-over, simply reshuffle the material already available; 'isolation', though interpreted in a number of different ways, is in general taken as a necessary condition for selection, or in particular for speciation, rather than as itself a 'causative' factor. So the theory becomes in effect, as it is often called, a theory of natural selection. To inquire into its conceptual structure, therefore, is in effect to inquire into the concept of Natural Selection. What does this concept mean and what is its explanatory force?

When one looks at its use by biologists one finds that the concept

'Natural Selection' has three kinds of meaning which slide into and reinforce one another. It works like an accordion. First, natural selection in its modern version is sometimes said to be a formula for 'differential reproduction' (call this NS_I): it expresses the probability of survival of certain genes in a population at some future time in terms of the ratio of the frequency of one or more genes to that of their alleles in a given population at the present time. This, like any algebraic formula, however, it has often been pointed out, is tautological. It conveys no information about the world. It appears explanatory because it is mathematical. Darwin's theory of selection was mainly qualitative; Fisher, Haldane and Wright have given the theory a more exalted scientific status by giving it quantified formulation. But in this form it is only a refined tool, applicable when one knows, or believes, on other grounds, that selection in something like Darwin's sense is at work: that is, when one knows, or believes, that, genetic capacity permitting, environmental changes are inducing corresponding changes in heritable organic patterns of morphology, physiology or behavior. In this second sense, which we may call NS_{II}, however, natural selection is a concept of efficient causality, mediating the belief that all characters of living beings are functions of the demands of environment (starting with internal genetic environment). No change, no stable or significant change, synthetic theorists hold, is *ever* produced by some 'drive' or 'tendency' of the organism itself, but only by environmental pressure. This thesis too is beginning to receive quantitative formulation in the work of ecological evolutionists, and so it too takes on the aura of accuracy. But its explanatory power rests primarily on its systematic coherence. It grasps all that has ever happened to living things on this earth or ever will happen under one single explanatory principle – a principle, the force of circumstance, that draws support also from other sources: it clearly holds also for our ordinary experience in our own lives; it permits, and reinforces, as Shaw eloquently argued, our liberation from the grim hand of Almighty Providence – a metaphysical but no less powerful support; and it can be applied, as I have already emphasized, with the elegance and rigor of mathematical calculation. Further, this principle in its modern form, unlike Darwin's theory or Lamarck's, unites in detail the unifying principles of several disciplines: paleontology, embryology, ecology, genetics, biochemistry. It shows us how phenotypic change, the fossil record, has been mediated by genetic change. That wretched giraffe stretching its neck was all very well, but nobody could imagine *how* its babies had their necks lengthened too. But the synthetic theory, synthesizing Mendel and Darwin, enables us to envisage not only what happened, but how. A vast range of when-then processes governed by a single uniform set of laws mathematically expressible and fitting smoothly into our overall view of nature: what explanation could be better? True, orthodox philosophers of science have sometimes lamented the non-predictive character of the theory (it looks chiefly back, not forward) and indeed its non-falsifiable character: for anything that happens confirms it: whatever survives survives, and if organisms are conceived by definition as aggregates of responses to environmental pressures, whatever survives survives because of environmental change. But on the one hand, it has become clear that neither predictability nor falsifiability is in any case an adequate criterion for explanatory power (just try to find a true *experimentum crucis*!). And on the other hand, as synthetic theorists have been loud in proclaiming, although every evolutionary theory is retrodictive, this is the only evolutionary theory that has ever suggested any (successful) experiments and further it has indeed suggested many and beautiful experiments, which serve to solidify its proponents' faith in its fruitfulness: fruitfulness in the Polanyian sense that its adherents believe themselves to be in contact with reality, with a reality which will yet show itself in systematically related but inexhaustible ways – both in population genetical studies in the field and in studies of single mutations by biochemical methods in the laboratory.

This claim too is borne out by recent controversy. For one of the hallmarks of a 'fruitful' theory is, not the ease with which it is falsified, but the vigor and coherence with which it assimilates apparent counter-examples to its own framework of explanation. There was a small counter-revolution, for example, led by King, Crowe and others, emphasizing the randomness with which biochemical alleles must occur: a circumstance which seems at first sight to remove them from the control of selection. But against this objection it is easy for orthodox evolutionists to show that statistically random variants are at once and all-powerfully subjected (1) to the controlling curves of the evolutionary algebra and (2) to the varying needs of varying organisms in varying environments. Dobzhansky deals briefly with this controversy in his *Genetics of the Evolutionary Process* and Mayr in his *Populations, Species and Evolu-*

tion;[5] it was also handled most persuasively, for example, in the discussion at a recent meeting of the Statistical Society in Berkeley.[6] The 'anti-Darwinians', it was argued there, neglect the complexity of living forms and living environments of which the synthetic theorists – who are biologists, not just biochemists – take due account. Thus if for example there are several types of fibrinopeptide with different rates of clotting, these occur 'randomly' with no immediate relation to the needs of the organism: they seem to occur as so-called 'neutral' mutations independent of selection. But the biologist, as distinct from the biochemist, can easily show that this is just another case of variation made use of by the selective principles that govern changing gene ratios. For there are animals whose wounds will have, depending on their predators, different shapes and sizes, and depending on their habitats, will need to close up more or less quickly. They may be gouged by long horns, scratched by little thorns, they may live in icy antiseptic climates or in putrid tropics. Each environment, biological and climatic, will demand a different response or possibility of response. In other words, if you look at the when-then relations, at the necessary conditions for such and such a sort of life, you will find your 'random' alternatives articulated into a variegated yet uniformly ordered living history where they promptly fit into their proper places with no recalcitrance at all.

So far then we have two meanings of natural selection: an algebraic formula, 'scientific' because mathematical, but like any formula empty unless applied in some context; secondly, the concept of a set of environment-organism interactions exhibiting deterministic cause-and-effect connections of the ordinary when-then sort – a concept which can be extrapolated, if one finds it convincing, from 'micro-evolution' to the whole history of life. And the set of interactions so understood, moreover, can be neatly and precisely studied through application of the formula for differential reproduction that the mathematical concept provides. Now so far these are explanations like any other: a mathematical law, like the second law of thermodynamics, to which Fisher likened his fundamental theorem, applied to sequences of events which are studied, and held to be determined, by its means. What is unique to biology here?

Supposedly the fact that while this explanation, like Newtonian theory or any pre-quantum-mechanical theory in physics, is indeed both mathematical and deterministic (or mechanical), it is also teleological. Not only do the necessary conditions suffice to produce the result, the result, we are told, induces its conditions: this is the third meaning of natural selection (NS_{III} for short), where the endpoint, survival, is said to be understood as the goal of what leads up to it.

Is this (1) logically possible? (2) empirically correct? The answer, it seems to me, is 'No' on both counts. Causal and teleological explanations, in answer to historical questions, are reversed explanations of temporal succession. Where we find a set of necessary and sufficient conditions at time t_0 that necessitate a set of effects at an immediately succeeding time t_1, we have a causal explanation. When we find a condition at time t_1, which, we suppose, has produced its antecedents at time t_0, we have a teleological explanation. Now we could conceivably have correlated, but contrary, explanations of the same phenomenon: like Epicurus's multiple causes. But a *true* explanation explains how things are in the real world – our belief that this is so, we have seen, is part of its explanatory power. We have the impression, through it, of being in contact with reality, of understanding how something really works. Can the same process be at one and at the same time *really* determined from t_0 to t_1 and induced, with the less rigorous determinism characteristic of teleology, from t_1 to t_0? What-it-does and what-it's-made-of accounts can be complementary because they answer different questions. So can what-it-does and how-it-arose accounts for the same reason. But deterministic and teleological answers to the *same* question – how this system came to be – cannot both hold good at once. The explanation of human action may be an exception. Kant at least held that for human action both explanations, causal and teleological (or moral), are valid – because we cannot know (theoretically) the real moral beings that, practically, we feel ourselves to be. But scientific explanation cannot rest content with this uneasy duality. Indeed, in the case of evolutionary theory, the force of the explanation depends, not only on the systematic coherence and scope of the theory, but on the prospect of determination through uniform principles which it invokes. In this respect evolutionary theory is thoroughly Kantian science, and thoroughly indifferent to any morality, Kantian or otherwise. Indeed, that thorough determinism has been the great asset of 'uniformitarianism' ever since Darwin adopted it from Lyell. Given conditions at the origin of life, it appears, and the laws in operation in nature now, of reproduction, heredity, and variation, as well as the physico-chemical laws governing the operation of nature as a whole, we can envisage the way in which, step by small step, all the changes from Lingula to mice and men could have arisen – and indeed must have arisen. True, this is not a logical 'must' but a contingent one: given the circumstances at each stage, and within the limitations of chance variation in the systems then existent, the next step was inevitable. As we can envisage the process with Kettlewell's peppered moths, so with every organism-environment interaction. There is no more to it. But if the process is thus determinate, at each step in the succession, if it follows rigorously from each t_0 to each t_1, where is the teleology? Do Darwinian evolutionists really believe that the end, the organisms existent at t_1, have *produced* the means to their survival? Even if we emphasize the role of behavior in evolution, as Waddington keeps reminding us to do, the one-toed equid running from his predator and so on, we can hardly hold that the speed of a future Derby winner induced his ancestor to run faster than the ones who didn't get away. Thinking in terms of natural selection ($NS_{I \text{ and } II}$), of the algebra of gene frequencies applied to organism-environment interactions in Mendelian populations, we find a powerful explanation of change (or of course of stability where environments are unchanging) in organic populations, an explanation which necessitates while yet admitting ultimate contingency. Everything that happened might have happened otherwise, but given the immediately preceding circumstances at each stage, it had to happen like this. But surely the sequence in question can't at the same time have been, not necessitated, but induced from back to front. Again, where is the teleology?

The proponent of evolutionary teleology has two answers, both, in my view, lame ones. First, he says, given the extremely complex causality of mutations, which we can never exhaustively know, we have in evolutionary theory only a specification of necessary, not of sufficient conditions. So, he may say, we have here determination in the abstract but teleology in fact. But there are two troubles with this answer. First: admittedly, mutations are 'random' in that their occurrence is unrelated to the 'needs' of the organism in which they occur. Nearly all mutations are deleterious and many are lethal. So in emphasizing this real want of sufficient conditions we would again be putting chance back at the foundation of evolution, and we have the old familiar objection about the eye turning up again. No, no, the evolutionist cries, natural selection dominates chance at every throw – but then secondly, we do have on principle necessary *and* sufficient conditions; the insufficiency is simply an expression of our ignorance, not of a real breakdown of natural selection (that is, of regular deterministic organism-environment interactions) as a classically causal 'agency'. So the appeal to a lack of sufficient conditions is either a surrender of natural selection to chance or a 'refuge to ignorance'. In neither case does it support any reference to teleology.

The other answer is at least as weak. The processes of organism-environment interaction summarized, or algebraically formulated, as natural selection are said to be teleological because they produce functioning and therefore teleologically organized systems. But again there are glaring deficiencies in this defense. First, historical and instrumental teleology are not identical; much confusion results from confusing them. Secondly, explanations of functions are basically formal explanations – in terms of operating principles: answers to question 2, not to 5, let alone 4b. The end of a function, what its activities achieve for an organism, is a guide to its study, but not part of the explanation of its operation. And what's more, functional explanations (even if they were teleological) refer to parts or subsystems of organisms, not to the organisms themselves (or to populations of organisms). It is only in a Pickwickian mood that one can call a hen an egg's way to produce another egg, as evolutionary theory, teleologically conceived, has got to do. Finally, the process that produces a functioning system, or even a properly teleological system, say a government or a person, does not necessarily produce it *for its sake*. Look at the history of the British constitution, or at the origin of most human individuals: planned parenthood is still something pretty new and globally still pretty rare.

Anyhow, the great glory of Darwin was supposed to be that he showed how all the marvellous adaptations of the natural world could have been produced without purpose. And though, indeed, teleology in nature is not necessarily the same as plan or purpose, in Darwin's day, as for many evolutionists still, the only intelligible alternative to purpose was natural necessity: not an end toward which things move, but an impetus that pushes them wherever there is room for them to go. Evolution, we are told, is 'opportunistic'. There is no favored direction of development, no 'orthogenesis', things go whatever way circumstances impel them to go. But unless we personify Natural Selection in the shape of a Wall Street operator (as some Darwinians perhaps have done and as many have accused them of doing), this kind of metaphor is either meaningless or describes the complex branching of evolutionary lines consequent on its pure circumstance-determined, mechanically causal character. Evolution is more like a football-playing slot machine (where the coins feed themselves in and nobody hopes to win) than it is like an end-directed process.

Still adherents of the synthetic theory (some of them!) keep saying it is teleological. What, in empirical terms, are they saying? There is, they say, a specifiable end to the processes of evolution, always and at every stage, namely *survival*. But is this an end in the sense in which teleology, instrumental *or* developmental, would specify an end: an activity for whose sake means are organized, or a condition that tends to bring about the steps leading to its occurrence? Historically, in a sense, of course, yes, because obviously if you calculate the probabilities of survival of a certain gene or set of genes at a future time you are looking not only from now to the future, but from the future back to now, and saying, this is what the probability will have produced. But there is no causal efficacy in this calculation. If I am likely as a smoker to die of lung cancer the causality is in the cigarettes and in my lungs and trachea, not in the statistics. Moreover, as has been repeatedly pointed out, the alleged teleology in this explanation is as tautological as the formula that carries it: we are saying, what survives, survives. Granted, the tautology is not a trivial one. You may fill it in with endless elegant and accurate calculations of how events have gone, are going, will go in the interactions of organisms with their environments. But then we have NS_{II} again, classical causal determinism. What has happened to NS_{III}, to teleology? In the case of individual development we clearly have telic phenomena: this frogspawn tends, other things being equal, to produce first tadpoles and then frogs. But we cannot specify anything which natural selection *tends* to produce except the empty concept: whatever survives. It tends to produce whatever it does produce. Heuristically, indeed, we use hindsight in approaching the problem of evolution: we have to start from where we are. As we say, here are frogs, how did they get that way from fertilized frogs' eggs, so we say: here are present-day flora and fauna, how did they get that way from the condition on the earth 500 million years ago? But even in describing the process, let alone explaining it, we start from the Cambrian or pre-Cambrian and work forward. We find no 'normal' endpoint which the earlier steps in any substantive sense were *for*, but only a record, e.g. that echinoderms left descendants, dinosaurs left none, and so on. Only the hollow concept 'survival' gives the thinnest possible appearance of an end to be mediated by its antecedent steps.

This is easily apparent, for example, if you look at the Darwinian concept of the 'fit'. Does not evolution work toward preserving the healthier, more viable varieties in a population? More viable indeed, but what does that mean? Not healthier, but more likely to leave descendants. So who are the fitter in wartime? The lame, the halt and the blind – for they survive to reproduce. That is in Darwinian terms what 'fitness' means and nothing else: what survives in the circumstances, not the healthier, the better, the more worth striving for or the more striven for. Nothing is striven for in evolution, consciously or otherwise. Things happen as circumstances compel them to happen. Naturally what survives is viable, otherwise it would not have survived. To say that it did survive is only to say that it did so, not that it tended to do so. Thus the alleged teleology of NS_{III} collapses back into NS_{II} mediated by NS_I, determinate series of organism-environment interactions calculated in terms of gene ratios. At most: these survived as a result of those conditions; at the least: what survived survived. The accordion that was expanded from I–II–III is pressed back in from III–II–I.

Yet, – again – distinguished evolutionists recurrently claim a teleological structure for their theory. Why? First, I think, because of the heuristic component: because in putting the question they have to start from now, not then. Secondly, and more importantly, because Darwinism originated in the context of the traditional view of organisms as adaptation machines. Paley, we are told, was one of the few writers Darwin enjoyed reading in his enforced submission to the Cambridge curriculum. And Paley's conception of living beings as marvels of means-end devices persisted in Darwin's thinking. It was not Paley's watch he threw into the ocean, but only the watchmaker. He retained what one may call the axiom of adaptivity, the view that everything interesting about an organism must be a *means*. This metaphysic of life still haunts post-Darwinian thinking. But the principle of adaptivity, of all principles, cannot be self-sufficient. For means, or uses, must be uses to some one for some end: to the organism for the sake of its survival, or in evolutionary terms, to genes for the sake of their survival in the gene pool of a future Mendelian interbreeding population. This end, as we have seen, however, is empty. When we look back, the goal seems, because of the temporal direction of our investigation and because of our interest in the subject matter (an important ingredient here) to be: those organisms now existing, Drosophila melanogaster, Homo sapiens, Mus norvegicus or what you will. But our explanation shows these endpoints, whatever they happen to be, to be the end result of a set of curious chances impelled entirely *a tergo* by genetic, ecological and physical laws. Our 'teleological' perspective has been a jumping-off point for explanation, but it enters not at all either into the description of the phenomena or into the explanation itself.

This is *not*, if you please, to deny the power of orthodox evolutionary theory properly understood. In terms of the criteria we have been using, fruitfulness, that is, the intimation of contact with reality, accuracy and systematic scope, NS_{II} and NS_I taken together are, as we have seen, powerful explanatory concepts indeed. Intrinsic interest enters too in the fascination of the subject matter, not only the concern for our own origins, but the fascination of seeing the vast and intricate variety of living forms within a unified framework which nevertheless permits respect for that infinite vaiiety. But there is nevertheless conceptual confusion associated with the theory, it must be insisted, either when NS_I – which is a mathematical tool of explanation, not explanatory in itself – is said to tell us why things in nature in fact happen as they do – or (my chief point in the present context) when NS_{III} – which is either heuristic (here we are, let's look back) or empty (what survives survives) is said to carry the theory, which in fact resides in NS_{II}, implemented by the algebra of NS_I. And if this explanation is unique, it is so, again, only because it concerns a unique subject-matter, the question how the forms of living things arose, not because it uses kinds of explanatory concepts alien to other parts of science.

So the claim I proposed to examine, that evolutionary theory is unique because teleological, fails. But, some may object, are there not other forms of evolutionary theory, theories of emergence in particular, which are truly teleological? I can deal with this suggestion here only very briefly. Such theories if they fix the emergence of man as the endpoint of evolution run too emphatically counter to the panorama of organic phenomena to get off the ground as scientific theories. Or, if they claim, like Vandel, following Bergson, for example, two high points of evolution, generic intelligence in insects and individual intelligence in man, these serve, again, only as heuristic devices, and as inadequate heuristic devices. We must look back from *all* there is, not just from man and insects. But so far at least no emergence theorist has in fact produced a theory at all: no algebra (no accuracy), no systematic relevance enforced by connection among scientific disciplines or implemented by detailed experimental inquiries such as systematic coherence must lend itself to to become part of science. Nor, I venture to predict, will such a theory emerge. For the subject matter of evolution, however fascinating as subject matter, does not lend itself to teleological explanation. Even functional explanation, we have seen, though it responds to questions expressible in teleological language is, as explanation, more formal than teleological in its import. It shows how a system, or subsystem, works, not how, as an end, it evokes its own means. But evolutionary explanation cannot even be functional, because there is no single organized system we are looking at, whose functions we could be examining. It is, in short, ordinary causal explanation of a well-developed and subtle sort, fitting smoothly into a naturalistic view of nature, and mediated in its details by the tools of mathematics. Its naturalism gives it metaphysical support, but emergence theories have *only* metaphysics to support them – a very different case indeed.

What then of the claim that evolutionary theory rescues the uniqueness of biology? I can only hint in conclusion at the direction for an answer. My own hunch is that if we are to find, if not uniqueness, at least something approaching uniqueness in biological explanation, we shall find it in the concept of organization (or perhaps information?) and its mediation in the study of the ordering principles of hierarchically organized systems. When evolutionists have presented confused, and confusing, apologiae for their theory in terms of an attempt to ground NS_{II} on an alleged teleological selection (NS_{III}) it has been from the subordination of the concept of organization to that of adaptation that their confusion has sprung. It *is* convenient, and interesting, to ask about adaptation (or adaptedness) with respect to organisms, because organisms are open systems which have indeed to be studied in terms of their relations to their environments. But to study adaptations in the sense of functions enabling organisms to carry on the business of living is to study certain styles of organization, certain styles as distinct from others. It is organization that is the overarching and fundamental concept for the study of living things, the framework for answering the question what does it do, what operating principles make it work? This question is indeed complementary either to the question what it's made of, or the question, how it arose. But it is, and should be, the ruling question. Even this approach, however, I have suggested, is not logically unique to biology. Every scientific study of any area imposes formal laws on a certain range of phenomena. Form and causality are therefore always hierarchical components for any explanation. To paraphrase Kant, we could say that ordering principles (laws) without phenomena are empty, phenomena without ordering principles are meaningless. We have to have both together. But biological systems are more massively organized than those of non-living nature, more strikingly improbable, and of infinitely greater interest to us as living things, so that their orderly structure more clearly demands the kind of dual, or plural, explanation of form and causality, or if-then and when-then factors. It may be that the difference is only one of interest. I'm not sure; but I am confident that it is here, and not in teleology, let alone in the evanescent teleology of evolutionary explanation, that the core problem of biological explanation resides.

But perhaps an adequate approach to that problem too must await further light on the puzzles of scientific explanation itself.[7]

NOTES

[1] Francisco Ayala, 'Biology as an Autonomous Science', *Amer. Sci.* **56** (1968) 207–221.

[2] Michael Polanyi, *Personal Knowledge*, University of Chicago Press, Chicago, 1958; *The Tacit Dimension*, Doubleday, New York, p. 196.

[3] Marjorie Grene, *The Knower and the Known*, Univ. of Calif. Press, 1974, Ch. 9; cf. Chapter IX of this volume.

[4] I am grateful to Professor Lorenz Krüger of the University of Bielefeld for his criticism of an earlier version of my argument at this point.

[5] Theodosius Dobzhansky, *The Genetics of the Evolutionary Process*, Columbia University Press, New York, 1970, pp. 261–66. Ernst Mayr, *Populations, Species and Evolution*, Harvard University Press, Cambridge, Mass., 1970, pp. 126–127.

[6] Cf. G. L. Stebbins and R. C. Lewontin, 'Comparative Evolution at the Levels of Molecules, Organisms and Populations', *Proceedings 6th Berkeley Symposium on Mathematics, Statistics and Probability* **5** (1971), 23–42.

[7] Professor Lynn Margulis of Boston University and Professor Stephen Jay Gould of Harvard University have kindly read and criticized parts of my MS in an earlier version.

CHAPTER XIV

ON THE NATURE OF NATURAL NECESSITY

The trouble with philosophers, or one trouble with philosophers, or at least with one kind of philosopher, is that the questions that interest us are the questions that don't have answers, or don't have straightforward answers. Not only are we unable to lay out straightforwardly in a given case our materials and methods, but it's just when we don't have results to report either, that we want to share our perplexities with others. Some philosophers believe, indeed, that it's only the insoluble problems that count as philosophical at all. All this is to say that it's rather a puzzle than a clear solution to a puzzle that I want to discuss here.

First, though, let me say briefly what I am taking as an *un*problematic starting point. For some years I have been concerned on and off with questions about the conceptual structure of evolutionary theory. It seems to me by now that there is good reason to affirm that the theory of evolution – the currently accepted theory, in effect, the theory of natural selection – is a causal-deterministic theory like any other classical scientific theory. Its explanatory power resides in the specification of environmental changes necessary and sufficient to bring about corresponding changes in the gene-frequencies that support the heritable structures of interbreeding populations. Just as, for example, it was shown that the presence of the microorganism *Plasmodium malariae*, transmitted by the Anopheles mosquito into the human blood, causes the disease malaria, so one can specify, or hopes one can specify, systematic environmental changes (including, of course, internal environmental changes) which cause the altered gene frequencies which in turn effect changes in the populations that carry these genes. There is no teleology here, but just plain necessitating conditions.

So far, so good. But what's the problem? I don't want to open up here the question of what is meant by a 'causal' explanation as such;[1] but supposing – as my simple-looking example suggests – that there are explanations which satisfactorily specify the necessary and sufficient conditions for an event or process, and that evolutionary theory consists in such an explanation, I want to ask two further questions. First, when we say that, given certain changes in internal and external environment, certain changes in relative gene frequency necessarily follow, what kind of necessity are we talking about here? And secondly, given that somehow (yet to be discussed) selection necessitates, does such necessitation leave any room at any special points in the evolutionary process for something of the nature of 'choice', or of 'degrees of freedom' (in a logical or mathematical, not a human, sense of both those terms)?

First, however, a preliminary to my first question (the question of evolutionary necessitation). It has sometimes been alleged against the theory of natural selection that it does not in fact offer an adequate causal explanation because it admits that mutations, the materials for evolutionary novelty, happen at random. Ever since Darwin's time, opponents of his theory, or of its modern heir, neo-Darwinism (or, as it is often called, the synthetic theory) have protested that its basic explanation is not a causal one at all, but an explanation, ultimately, by reference to mere chance. For mutations, selectionists admit, are random, and mutations are the whole material available for natural selection, so to speak, to work on. So really, these critics allege, Darwinian evolutionists are saying that everything there is in the living world just happened, by a set of curious chances, to turn out the way it did. Can the vertebrate eye, can the human brain, they ask, be the result of a sequence of mere accidents? Mutations are mistakes in replication. Can the marvellously coadapted structures and functions of organic beings result from a series – an almost infinite series – of happy accidents? Now, it is true, I believe, that part of the explanatory force of evolutionary theory rests in its emphasis on contingency. Everything there is *might* have been different: this belief in the historicity of nature is one of the metaphysical strengths of the Darwinian creed. And it is precisely the question what that contingency means which will constitute the first part of the problem I want to pose. But before we can think about evolutionary contingency to any purpose, we must discard the red herring of 'mutationism'. Darwin stated, as one of the first principles of his theory, that organisms differ slightly from one another; insofar as such differences are heritable, they are now known to be founded on differences in the nature of the genes. When such a difference is considered in relation to its origin, it is called a mutation; insofar as it simply exists as an alternative to some other structure at the same locus, it is called an allele. And it is because there are such alternatives, because mutations have happened and do happen, because alleles, or multiple alleles, exist, that evolution can occur at all. But evolutionary theory does not use the concept of chance to explain how evolution happens; nor is the 'chance' occurrence of one allele or another in any given case what it is trying to explain. In other words, chance is neither the explanatory principle of the theory, the explanans, nor what it is trying to explain, its explanandum. What it explains is changes in the relative gene frequencies of succeeding populations, and nothing else. The theory depends on 'chance', as we shall see, insofar as its subject matter is susceptible only of stochastic formulation; but again, 'chance' is not *what* it explains nor what it explains *by*.

Admittedly, evolutionists speak of mutations as occurring 'randomly', and of the segregation of the genes at meiosis as 'random'. What do these admissions mean? There are two parts to the answer.

First, the statement that mutations are 'random' means only that their occurrence is not a function of the needs of the organism in which they occur. The gene producing sickle cell anaemia, for example, originates, like any such inborn error, as a mistake in the replication of the normal gene. That this condition, when heterozygotic, is associated with an increased resistance to malaria accounts for its continued frequency in African populations: a selective advantage has offset a selective disadvantage. But the origin of the mutant gene was not a function of its possible utility or non-utility to its bearer. That does not mean, however, that a biochemical explanation of its occurrence could not be found; indeed, this is precisely one of the cases where the precise chemical basis of the mutation has been discovered: the substitution of one base (U for A) in the DNA and RNA triplets codes for a different amino acid (Val instead of Glu) at a single position in the beta chain of hemoglobin and thus causes the individual hymozygotic for this peculiar condition to contract the fatal disease.[2] Now such an explanation once discovered is plainly causal and deterministic in the ordinary way. And there is no implication whatsoever in the structure of evolutionary theory that forbids similar explanations for the origin of any or all mutations. Even if every mutation that ever happened were given a good hard biochemical explanation mutation would still be random relative to the needs of the organism; that is all, in this context, that randomness means. Similarly, 'randomness' of segregation at meiosis means simply irrelevance to biological needs.

There is another sense, secondly, in which 'chance' enters into the structure of evolutionary theory, but, again, not in contradiction to, though perhaps, as we shall see, in limitation of, the deterministic nature of its explanatory force. As Darwin already observed, 'selection' needs a very large number of slight variants to work on – in modern terms, it needs the availability of a very large pool of slightly differing genes for environmental pressures, internal and external, to 'select'. Except in very small populations in very specialized environments, contemporary evolutionists believe, such a pool of variants is always available. There is thus a stochastic base, so to speak, underlying or inside the theory. But what the theory does is to show how, from one generation to the next, some of this material comes to occur more frequently than its alternatives – to occur more frequently because of ordinary cause and effect relations between environmental conditions and the organisms carrying the totality of the alleles in question. The situation is similar to that of any investigation which uses statistical methods to prove a causal theory, as in Darwin's experiments about cross-fertilization and growth, or in current investigations about the causal relation between smoking and lung cancer. That one uses stochastic material – gene frequencies – to support a causal theory – reasons for changes in gene frequencies – does not make the theory itself other than deterministic. The statement that the continued occurrence of sickling is a function of its effect on the resistance of the subject to malaria is based on statistical evidence; but it is a perfectly straightforward deterministically causal statement.

The theory of natural selection, then, insofar as it explains changes in populations on this planet, explains in a classically causal fashion; it does not ascribe population change to chance. Indeed, 'natural selection' is simply shorthand for the fact that from given environmental conditions (including, of course, internal environment) certain changes in population distribution necessarily follow.

But what does 'necessary' mean here? That is our first major question, and one to which I can only suggest some very tentative and unsatisfactory answers. The problem, really, a problem that has troubled critics of Darwinism from the beginning, is, that, if the theory doesn't (as I hope I've shown) make the whole history of life depend on the irrationality of mere chances, on the contrary, it apparently forces us to interpret the whole history of life in a thoroughly deterministic manner. And that seems equally embarrassing. Once life began – a phenomenon Darwin did not attempt to explain – the changes in population frequency and structure seem to follow everywhere and forever from the initial conditions at the very beginning of the story. True, natural selection only goes step by step, from one generation to the next, but if (with minor exceptions for cases like 'genetic drift') there is necessity at each step, then there seems to be necessity throughout. Dinosaurs *had* to perish, man *had* to originate: was all this fore-ordained in the pre-pre-Cambrian when the first life began? Evolution, of course, has only to do with differential reproduction; thus evolutionary theory as such has presumably no bearing, except in a very indirect and limited way, on responsible human choice and so on the issue of free will versus determinism in the moralist's or metaphysician's sense. Yet the evolutionary vision does seem to suggest a universal cause-and-effect relation in the history of populations on the earth which is readily associated with a deterministic view in general. If we accept selection theory as a causally deterministic theory, do we have to accept determinism as a philosophical position? One way out would be to argue with Miss Anscombe that one may always ask a causal question, and find a cause-and-effect answer of some sort, in any, or even in each, given case, without embracing universal determinism.[3] But I don't think her argument, interesting though it is – and convincing for the problem of volitional causality – will help us at this juncture. For the convincing power of Darwinism since its inception has been precisely its totalizing deterministic force. Instead of leaving before us the bewildering and inexplicable diversity of biological phenomena, Darwin showed us how a set of natural laws, the same that hold good today and held good in the beginning, could account for the origin of the whole range of species populating the earth today. This is either a very general sort of determinism or nothing.

   At the same time, on the other hand, the necessity Darwinism envisages is not logical necessity. Indeed, as we have already acknowledged, the admission of fundamental contingency was one of the persuasive factors in evolution to those it did persuade. Whatever happened *might* have happened differently. That basic principle lies deep, I believe, in the conviction of most educated adults in our culture, and it is partly to Darwin that we owe this insight (as I believe it to be). True, there is logical necessity built into Darwin's theory, insofar as the structure of his argument (in the first four chapters of the *Origin*) is deductive. But its premises are empirical. Given heredity, slight variations, and population increase, natural selection necessarily follows. But these conditions must be given as matters of fact and the consequences they necessitate are therefore as empirical as the premises from which they follow. Such necessity is sometimes called physical rather than logical. What does that mean? To Leibniz, it meant that given God's choice of the best of all possible worlds, everything in this one necessarily follows: hardly a convincing explanation for Darwin or the heirs of Darwin, since it is just the rigid control of God's Providence from which he is said to have emancipated us. Perhaps, indeed, we would say with Leibniz that of all the logical possibilities only certain ones became actual and in this sense physical necessity is somehow less necessary than logical. Yet when we attempt to apply this distinction to our present context, it doesn't seem to help. Evolutionary theory says: given the initial conditions, genetic and environmental, at any given time t_0, the consequences in altered gene frequencies at the corresponding time t_1 necessarily follow. And then our 'physical' necessity seems to turn back into plain old Laplacean determinism. So what's happened to the contingency of evolution? We have an explanation whose convincing power rests at one and the same time on causal necessity and on a lack of necessity, on historical contingency. This paradox is not confined to Darwinism: all the way back to Democritus, it has seemed that 'naturalistic' accounts of nature may be taken alternatively, or at one and the same time, as reducing everything to a necessary sequence of events and to a sequence of contingencies. The point is, I think, that Darwinian, like Democritean, explanation abolishes purpose from nature. It substitutes causes for reasons (God's reasons) and so appears to produce what in modern parlance is called 'accident': something that happened by necessity, indeed, but not through foresight or purpose. Compared to the Providence of a Platonic world-mind or a Christian Deity, such a progression of events may appear to have been all necessitated or all to have happened 'accidentally'. That's perhaps, by the way, part of the reason why critics of Darwinism insist it assigns 'chance' as the cause of population change. As destroyers of cosmic teleology, chance and necessity, though in themselves contrary, appear very much alike. But that doesn't help us much with our present question. Evolutionary theory

claims the possibility of specifying at each step of the development of life
the necessary and sufficient conditions for the changes in relative gene
frequency on which such development depends. Changes in populations
are caused (indirectly through changes in relative gene frequencies) by
changes in internal and external environment. To cause, it is held, is to
necessitate – where in this picture is there *real* contingency, something
different from Laplacean determinism?

Here I think the best we can do is to rely, in the first place, on the stochastic base of the theory. There is a very large number of variants
available for 'selection', so to speak, to 'choose' from. Out of a vast number of possibilities, those increase in frequency which, given environmental conditions, are somewhat more likely to survive than other alternatives. It is important in this context to recognize that what Darwin
– and Spencer – called the survival of the fittest is really only the survival
of the fitter – of those genes whose possession enables their bearers to
produce slightly more, or slightly more viable, progeny than their alleles
permit their bearers to produce. Some great evolutionists – like Muller
in his Harvey lecture, for example[4] – have stressed the precision of
genetic adaptation. This may indeed be the case with organisms delicately adapted for life in a very specialized environment, like the liver
fluke or the lyre bird or the Columbia river salmon and so on. But
extreme specialization also entails genetic rigidity and hence, with changing environments, extinction. Novelty in evolution – what Rensch calls
anagenesis[5] – demands as its necessary condition a large pool of genetic
variations which *can* become more frequent in one direction or another,
depending on circumstances; and the looser the fit of adaptedness, so to
speak, the more likely an explosion (relatively speaking) in a new direction. It may be objected, then, as some of the Muller school have in fact
objected, that it needs a supply of so-called 'neutral mutations' to get
evolution going – and so 'selection' is not the prime, or only, mover in
evolution – and, indeed, we are back with 'chance' as our explanatory
principle. But, once more, the stochastic base of the theory simply provides part of the initial conditions which it specifies. The explanandum is
change in relative gene frequencies; the explanans is the statement of
cause and effect relations between environmental conditions and such
changing gene frequencies. That the materials 'selection' 'acts' on are
such that (a) they occur irrelevantly to reproductive advantage and (b)

there are a great many of them, is not part of the theory itself, nor, as I have already stressed, does it matter to the explanatory force of the theory how the origin of this large pool of biochemical alternatives happened to arise. What the stochastic base of the theory – in the sense that it posits the existence of a wide range of variations in the heritable structure of organisms – what this stochastic base *does* provide is a certain flexibility, a certain range of possibilities available to the historical development of life. The theory does not explain *by* chance; what it explains, deterministically, is the change in relative gene frequencies, the materials for which, however, must be available in very large numbers, and hence can be treated in terms of stochastic laws, if the changes which the theory (deterministically) explains are to occur.

Moreover, it is here, I think, that the first 'degree of freedom', so to speak, in the history of life occurs. Supposing, for example, Lynn Margulis is right about the origin of the higher plants: a primitive protist swallowed a blue-green alga without metabolizing it.[6] Such a symbiont, being able to photosynthesize, had access to nutrient sources and metabolic processes which its near relations lacked, and so multiplied more than they. This mightn't have happened, and then there wouldn't have been flowering plants, or most of the metazoa either; but once on its way the history of the flowering plants is well enough explained in terms of the classical, causal principles of organism-environment interaction controlling changing gene frequencies, in other words, in terms of the theory of natural selection.

It is this sort of flexibility of alternative developments, I think, that we mean by historical contingency. The logic of the situation is like that of Aristotle's famous sea battle: it will or won't happen tomorrow. No one can say today. That is not to deny the law of excluded middle, but simply to assert that the future is too rich with possibilities to be predictable in advance. This much play of possibility there is at the root of evolutionary theory, even if not in the theory itself. Again, the theory itself says that, given a set of alternatives, the one that has a slight advantage over the others will increase, and that is a purely deterministic statement.

Such an interpretation, it seems to me, fits in well also with the evolutionary account of such phenomena as convergent evolution, as in the cephalopod and vertebrate eye, for example, or parallel evolution, such as that of the artiodactyl versus perissodactyl limb. The paleontologist Schindewolf, despised by orthodox evolutionists for his 'typological' approach, insisted that since running fast on two toes is just as efficient as running fast on one, natural selection by itself could not explain the development of both methods.[7] Now I still think Schindewolf was more candid than some evolutionists in admitting that there are a multiplicity of ways in which, so to speak, nature accomplishes the same object – and that the biologist has to recognize this diversity of 'types' in order to know what he's talking about at all. As Canguilhem has convincingly argued,[8] the very subject-matter of biology, if not its explanations, demands a reference to standards, types or norms, and those who allege that they can do without them have to take refuge in pseudo-substitutive language, to pretend to do without them and introduce them surreptitiously. But given the vast variety of genetic material and the concomitant variety of environments, it is not surprising – and is explicable in perfectly good Darwinian, and causal, terms – that some vertebrates came to run fast on two toes and others on one. Even where adaptedness has come to be precise, there is enough free play of circumstance in the starting points of evolutionary histories for such precision to have developed, now in one pattern, now in another.

It is, if I understand him correctly, this sort of flexibility that Manfred Eigen is referring to in his account, not of organic evolution, but of the origin of life itself.[9] He accepts the extrapolation of Darwinian principles to the whole history of life, including its very first prebiotic stages, but suggests that the sufficient conditions for this first step in evolutionary history will never be made entirely explicit, not simply because of our ignorance, much less because of the 'chance' origin of mutations, but because the variety of alternatives available leaves open a wider range of initial conditions than a deterministic theory like the Darwinian can ever fill in completely. This is the first 'degree of freedom' at the very start of biological history – and that's why from the start it *is* the account of a history, not a timeless set of deductions from the premises of an axiomatic system. This initial degree of freedom is, minimally, what is denoted, I suggest, by the maxim 'it might have happened otherwise'.

With this rather feeble statement I have done the best I could for my first question: how is physical – or historical – necessity different from logical? And I have also moved to the second part of my reflections, by specifying the first place at which a limit on universal determinism is to be read into our view of nature's history. However, before we proceed to ask where else along the course of life's history we may postulate further degrees of freedom, I must pause to look at the bearing of my remarks on the status of the neo-Darwinian or synthetic theory itself. Once more, they do *not* entail the 'mutationist' theory that evolution happened 'just by chance'. In so far as selection operates it operates causally, 'favoring' one set of alternatives over another. (Such talk is metaphorical, but I hope its meaning is clear.) Nor, on the other hand, am I suggesting that because there was a very large range of possibilities at the beginning of evolutionary history, some of which proved more advantageous than others and so persisted, 'selection' is some kind of creative agency brooding over nature for its good. Selection, Sir Ronald Fisher argued in his Eddington lecture, is a truly creative process.[10] Such a remark, I submit with all due respect, is nonsense. 'Selection' is shorthand for a set of causal relations. Lewontin indeed has gone so far as to say it is not an explanation at all, but simply a statement of certain empirical facts.[11] It certainly isn't a 'mechanism' like a heart or an internal combustion engine, let alone an agent capable of imagining and bringing into being novel creations. To personify it in this fashion is to compound the confusion already inherent, as I have argued elsewhere,[12] in evolutionary teleology or in the notion of evolutionary progress. Natural selection explains – or, if you prefer Lewontin's emphasis, describes the necessary and sufficient conditions for systematic changes in relative gene frequencies, and that is all it does. Whatever else there is to study in the organization of living things – and there is plenty – does not fall within its scope.

Given then, our first degree of freedom, the stochastic character of the material available at the outset for evolutionary development, is the record of evolutionary development wholly deterministic from there on? I think not. To begin with, as we noticed earlier, natural selection explains causally only from step to step, and the redundancy of available material noted by Eigen at the origin of life may, perhaps does, recur at the nodal points of any new departure. This is not to say that natural selection specifies only necessary, not sufficient, conditions. It presumes that both are specifiable for whatever it does explain; but there is, so to speak, a good deal of noise in the background of the information it conveys.

A second degree of freedom, moreover, may be postulated, very much later indeed in the history of life, with the new patterns characteristic of animal behavior. A biology teacher describing instructions for classroom experiments once complained to me: 'They say "put an earthworm in a jar", but the earthworm may not like it in the jar.' I'm not suggesting, of course, that the earthworm's behavior might directly influence the course of evolution; as we have all been told repeatedly for the past forty-five years, evolution is something that happens to populations, not individuals, and not as the direct effect of what individuals may 'try' to do. At the same time, however, as Waddington has stressed,[13] though evolution is a record of changing gene pools, it has to work through the phenotypes elements of whose genotypes may or may not survive into the gene pool of the next or future generations. And once one gets to animals, how they behave is going to make a difference to the question which ones get to reproduce and how many offspring they have. Now it may be argued, and is indeed argued, by such writers as R. L. Trivers, for example,[14] that such behavior may be calculated in terms of benefit-loss theory so as to be wholly assimilated to the classical principles of natural selection in terms of what is called 'inclusive fitness' (the Darwinian fitness of parent × offspring or the joint fitness of siblings, and so on). Thus, he argues, parent-offspring conflict, for example, can be calculated in terms of its effect on the probability of the production of further offspring by the parent as well as on the probability of future reproduction by the offspring. This calculus subjects much behavior which previously seemed extraneous to the scope of selection theory to its deterministic principles. This has been done by E. O. Wilson and others, for instance, for the social insects.[15] But such a sweeping application as Trivers and his supporters want to make of this kind of calculus seems, to say the least, a little hasty. On the one hand the theory entails talk of 'choice' of sex by parent or offspring as early as meiosis and other such anthropomorphic locutions which to a literal-minded philosopher are simply unintelligible; and on the other hand it assimilates much too quickly the culture-dwelling nature of the human animal to purely biological considerations. What I want to suggest here is something different, which would provide a limit to 'inclusive fitness' calculations as well as to the calculation of alternative alleles at a molecular level: just the flexibility of goal-directed behavior in animals whose behavior one can reasonably describe in these terms. Larry Wright, in a paper on

'Explanation and Teleology', has proposed the following formula for such behavior: '*S* does *B* for the sake of *G* means: (i) *B* tends to bring about *G*, and (ii) *B* occurs because (i.e., is brought about by the fact that) it tends to bring about *G*.'[16] Wright calls the 'because' here 'etiological' in order to avoid the usual contrast between causal (i.e., deterministically causal) and teleological. In other words, he argues for a more flexible conception of causality than the traditional when-then, necessary and sufficient conditions type. I would be happy to accept this reform in terminology, but since I have been taking evolutionary theory to be a classically causal, that is, deterministic, theory and trying to exhibit both its explanatory force and its limitations in these terms, I cannot stop now to translate my own usage to accord with his. All I want to do here is first to acknowledge, with Wright, that there are behaviors which are intelligible and explicable in teleological terms. The rabbit ran into the burrow to get away from the fox; the rat pressed the lever to get at the bait, and so on. These are, as Wright convincingly argues, empirically defensible, non-circular, and non-anthropomorphic (or not perniciously anthropomorphic) locutions. And secondly, I want to suggest that we have here a class of behaviors which again provide a base of alternatives for 'selection' to act on, not directly, of course, but via the effect on relative gene frequencies of the goal-directed behavior of the organisms the probable survival or non-survival of whose genes are in question in a given case and at a given juncture of evolutionary change. This does not mean, indeed, that evolution at this point becomes a telic phenomenon or evolutionary theory a teleological theory. On the contrary. It means simply that the necessary conditions for a given evolutionary change – say, the development of tree-dwelling primates – include the varying behaviors of the members of certain populations, some of whom, in this case, climbed better and got away – or perhaps got the fruit on a higher branch, and so lived to reproduce or to produce more viable offspring, and so on. Since the behavior of such animals is goal-directed, its explanation may include teleological concepts; the change in relative gene frequencies which determines the changing character of this population's descendants, however, is explained in the same classically causal terms as any other evolutionary change. But the alternatives which provide the material on which selection works this time include alternative goal-directed behaviors as well as alternative genes. Thus if the very large numbers of variants available in the gene pool place an element of chance, and thus of contingency, at the basis of our picture of evolution, though not into the theory of natural selection itself, we may suggest that, analogously, we have with this second nodal point in evolution – in other words, in animal behavior – a second 'degree of freedom' of a different nature from the first.

We have so far, then, two limits on evolutionary explanation, first, in the stochastic base of the theory, and second, in the alternatives of goal-directed animal behavior. The third limit I want to suggest, finally, is the hardest to deal with: the point at which not just behavior, but responsible, that is, human behavior, enters into interaction with the organism-environment interactions summarized as 'natural selection'. Here, obviously, we have a complicated feedback situation which is extremely difficult to sort out. It is certainly not enough to contrast, as evolutionists have often done, 'biological' with 'cultural' evolution. It is true, of course, that the handing on of a verbal or behavioral tradition from human parents to their offspring, and reciprocally, the offspring's rebellion against tradition, and the communication of a modified tradition to its peers and its (and their) offspring, occurs much more rapidly and is implemented by very different kinds of occurrences from the development, say, of an evolutionary phenomenon like industrial melanism, or, on a larger scale, the adaptive radiation of modern mammals to sea and air as well as to innumerable specialized ecological niches on the land. It is true, too, that the attempt to assimilate human history, or parts of human history, to a purely 'biological', let alone an evolutionary, framework, is a tricky and often disastrous game. And of course, conversely, to try to interpret man without any reference to his biological nature, including his evolution as the kind of animal he is, is to produce an arid philosophy of mind. But the simple contrast between the two kinds of evolution is also over-abstract and misleading. It follows, in effect, from a contrast between nature and artifact which produces an over-simple view of both, a view which has proven, both theoretically and practically, catastrophically mistaken.[17] Man is an animal dwelling *in* culture: his very anatomical and physiological development demand the artifacts of culture to permit his maturation as a human person.[18] At the same time, through his interaction with nature, and through his delusive conception of himself as separate mind in control over unthinking matter, he has transformed nature itself until it threatens in return to obliterate him, unless he first obliterates himself. And equally obviously the effect of responsible (and that includes irresponsible) human action on nature has had everywhere dramatic effects on the direction natural selection has taken. Darwin took as his model for the concept of 'natural selection' the work of the great British sheep and cattle breeders. They chose with forethought: Bakewell built the Leicester and Bates the Shorthorn for the demands of a market. 'Nature', on the other hand, Darwin argued, 'chooses', so to speak, perforce. Given certain circumstances, certain slightly altered circumstances follow. But since the rise of man, and with him of deliberative action (such an origin of action, said Aristotle, *is* a man), *our* choices have been among the initial conditions permitting the course natural selection may take. As in the case of the eighteenth century livestock breeders, and equally in the case of the enterprising businessmen who hired the lumberjacks who made our dust bowl, so everywhere human choices have channelled organism-environment interactions both for our own species and for countless other species as well. It is only here that freedom in the sense of responsible action enters – not into evolutionary theory – but into its initial conditions. And here again, the alternatives open to human persons and societies present a range of possibilities which selection itself works from, but does not control – though it may of course take its revenge, since we, too, like all populations, are subject (once more, metaphorically speaking) to its power.

NOTES

[1] The problematical presuppositions of the modern concept of 'causal' explanation are touched on in the first essay in this volume; and I agree both with Professor Anscombe in *Causality and Determination*, Cambridge University Press, Cambridge, 1971 and Professor Larry Wright in 'Explanation and Teleology' *Phil. of Sci.* **39** (1972), 204–218, both referred to briefly below, that the concept of 'cause' needs thorough philosophical reexamination. For my present purpose, however, in the context of reflection on the structure and explanatory force of evolutionary theory, I think I must rest content with using the term in its classical 'when-then', and even deterministic, sense.

[2] See T. Dobzhansky, *The Genetics of the Evolutionary Process*, Columbia University Press, New York, 1970, pp. 48–9.

[3] See reference in note 1.

[4] H. J. Muller, 'Evidence of the Precision of Genetic Adaptation', *The Harvey Lectures*, Thomas, Springfield, Ill., 1948, pp. 165–229.

[5] Bernhard Rensch, *Evolution above the Species Level*, Columbia University Press, New York, 1959, p. 281ff.

[6] See reference in Ch. XIII above.

[7] See references in Ch. VII above.

[8] G. Canguilhem, *La Connaissance de la Vie* (2nd ed.),Vrin, Paris, 1965 and *Le Normal et le Pathologique*, PUF, Paris, 1966.

[9] M. Eigen, 'Self-Organization of Matter and the Evolution of Biological Macromolecules', *Die Naturwissenschaften* **58** (1971), 465–523.

[10] R. A. Fisher, *Creative Aspects of Natural Law*, Cambridge University Press, Cambridge, 1950. Cf. Dobzhansky, *op. cit.*, p. 430.

[11] R. C. Lewontin, 'The Bases of Conflict in Biological Explanation', *J. Hist. Biol.* **2** (1969), 35–45, p. 41.

[12] Chs. 12 and 13 above.

[13] Cf. e.g. C. H. Waddington, 'The Paradigm of the Evolutionary Process,' in *Population Biology and Evolution* (ed. by R. C. Lewontin), Syracuse University Press, Syracuse, 1968, 37–45, p. 38–9.

[14] See e.g. R. L. Trivers, 'Parent-Offspring Conflict', *Amer. Zool.*, (in press).

[15] See reference in Ch. 11.

[16] *Op. cit.* (note 1), p. 211. Wright's formulation is a modification of that proposed by Charles Taylor in his *Explanation of Behaviour*, Routlege and Kegan Paul, London, 1964.

[17] See e.g. my 'Hobbes and the Modern Mind: An Introduction', in *The Anatomy of Knowledge* (ed. by M. Grene), Routledge and Kegan Paul, London, 1969, pp. 1–28.

[18] See Chs. XV and XIX below.

CHAPTER XV

ON SOME DISTINCTIONS BETWEEN MEN
AND BRUTES

Everyone appears to agree nowadays on the need for a revaluation of human values; but some think we must for the first time set man into his proper place in the whole of animal nature, while others are certain we need to reaffirm his special place outside the sphere of the merely natural. Whether the problem is one of abstract metaphysics or practical pedagogy, this issue is involved.

Perhaps a re-examination of some of the historical alternatives may, if not answer, at least help to clarify the question. Put at its crudest, the current opposition seems to be something like this: on the one hand, it is said that, like other philosophic theories of the past, the traditional cleavage between man and nature is the result of overabstraction so gross that the principal facts of the case are overlooked or actually denied. The considerable, in fact, central, role of impulse, habit, and the irrational generally is denied in favor of a reason which actually men are most infrequently seen to exercise in its purity even in the abstruse fields of logic or mathematics, let alone in the tangled and pressing concerns of their daily existence. Of course, every theory of human nature includes its account of feeling and emotion, but the objection in the main is that traditional theories treat feeling as something bad to be suppressed rather than as the real driving force it actually is: a force morally indifferent in itself but operating for good or ill under a variety of circumstances. On the other side such biological treatment of human nature is said to eliminate any standard of good or evil by which to discriminate one use of emotion from another. If men are not distinguished from brutes by some unique character which itself supplies a standard of value, then, it is thought, no end is left for human endeavor, nothing remains but the Hobbesian sequence of appetites and aversions in which no need or appetite can be criticized as bad or harmful. As a result of this lack, every excess of cruelty and sadism is sanctioned; and, in the resulting war of all against all, the good and decent elements in human nature – which under ordinary circumstances can be observed to exist equally with the bad and brutal ones – are swamped by the more violent impulses thus dangerously unleashed. The practical result is the stasis of Corcyra or Europe under the Nazis; the theoretical result is the ideal tyrant of Thrasymachus or the leader-principle of the S.S. man. Both sides see evil practical consequences from the alternative view: the one in the suppression of fruitful and progressive impulses, the other in the loosing of brutal drives which by a natural process suppress the better ones. But, quite apart from the moot question of the effect of any philosophic theory on anybody's practice, the principal charge philosophically seems in both cases to be one of inadequacy: because men are described as different from other animals or because they are described as like them, the facts of human nature as common experience reveals them are falsely reported.

The charge of inadequacy is justified on both sides, I think. In historical terms: the Cartesian separation of men and brutes is wrong – but not every distinction between men and brutes is essentially Cartesian (though personally I suspect that every honest Christian one is so, unless, like the Kantian, it renounces its own metaphysical foundation); the Hobbesian identification of the desires of men and brutes is wrong, but not every theory that makes men similar to other animals is Hobbesian.

First, as to the Cartesian distinction. Certainly the theory of 'brutism' is, judged by the criterion of adequacy, one of the most fantastic theories about anything anywhere in the history of Western thought. To think of the animal body as a machine with, in man, a completely incorporeal yet communicating entity mysteriously attached raises all the insoluble and unnecessary problems of interaction that everyone has been pointing out for three centuries. As a result, human psychology is either absurdly or merely conventionally described – the latter, e.g., in Descartes's pallid recapitulations of Senecan morals to the Princess Elizabeth, a subject which obviously interested him not at all. For orthodox principles of morality could be automatically deduced from his metaphysics and physics, and there was no problem about it – since, in fact, a decent Stoic-Christian morality can easily be fitted into any simple dualistic account of reason and the passions. But if human nature is superficially dealt with in the Cartesian account, the unfortunate remainder of animal creaton receives infinitely more flagrant mistreatment. The experience of the owner of two *tournebroches* who visited Port Royal can be duplicated by anyone who has ever had a dog, even a fairly stupid one (see S. Alexander's posthumously published essay, *The Mind of My Dog*), or for that matter by anyone who has had any acquaintance with any of the higher mammals. On the one hand, there seems no reason to hold that the very similar facial expressions, for example, which are thought to denote certain feelings in human beings do not express any feeling in other animals. And, on the other hand, the dissimilarity in the thought-processes of other animals suggested by the absence of speech is amply offset by striking similarities in other behavior patterns associated with processes of inference – see the example of the *tournebroche* or, if you want more up-to-date material, Köhler's apes. It is not too sweeping a statement, I think, to say that no theory which makes human thought and feeling differ *toto coelo* from the thought and feeling of other creatures can ever be accepted by anyone who has had any ordinary experience of animals at all; there are simply too many facts the theory has to ignore or at least fantastically to reinterpret to make them fit.

Of course, attacking poor Descartes for his brutism is beating a very dead horse. Descartes's philosophy is not one of the eternally recurrent Weltanschauungen but a peculiar synthesis suited to a very special time and place: a synthesis which outside the atmosphere of seventeenth-century French theological circles – or, better, seventeenth-century French Augustinian theological circles – does not synthesize at all. But there are two reasons why more organic theorists of human nature can still relevantly attack a theory as obviously dead as Cartesian dualism. The influence of that view in modern thought runs so wide and deep that even as staunch a phenomenalist as J. S. Mill can unquestioningly differentiate the permanent possibility of feeling from the permanent possibility of sensation, though nothing in the phenomena as he views them could justify such a distinction. Even more important, it can be fairly maintained, I think, that the Cartesian theory of human reason and animal nature represents the core of the Christian tradition and that the Jansenists were quite correct in so accepting it. Certainly, the disagreement between Descartes and the Thomists on the relation of thought to sensation and of the *knowledge* of mind to that of body (as represented in *Objections* I, II, and VI) was a very genuine disagreement. But stitching an Aristotelian theory of knowledge into the interstices of a Christian world view does not in the least alter the fundamental necessity for any orthodox Christian position: that *animus* must be radically different from *anima*, that the single respect in which man is made in the image of his maker must be entirely separate and distinct in him from everything bodily. In some medieval accounts of the human soul[1] one feels that the transition from the finest animal spirits to the mind itself is so gradual that one can imagine *animus* as the rarest and subtlest portion of the *anima*; that is what Descartes himself suggests when he runs over the things he had 'imagined' himself to be.[2] But though such an identification may make things easier for the lay imagination and furnish charming matter for the artist, theologically and philosophically it is surely most dangerous; and no serious Christian theory would dare admit it. In short, a Christian world view demands an Augustinian-Jansenist-Cartesian conception of the relation of mind to body and human to animal nature, and the importation of pseudoscientific phraseology from the Philosopher about the way of knowing mind and nature can serve only to obscure, not to eliminate, that necessity.

Even the apparently noncommittal view of Locke on human reason is still Cartesian and depends explicitly, moreover, on a divine sanction. If we still believe in the Jeffersonian revision of Locke's law of nature, we have unfortunately no logical or rather metaphysical right to do so unless we take the Cartesian world and its all-wise maker with it – or unless we have an alternative account of man and nature from which the right to life, liberty, and the pursuit of happiness is equally deducible.

But if Hobbesian psychology is really the exclusive alternative to the Cartesian-Christian conception, our philosophical as well as practical situation is pretty desperate. For if Cartesian brutism does less than justice to the common facts of animal life in general, so does the so-called selfish system to our common experience of human feelings. Hobbes himself admits the test of his philosophy is in our own hearts[3] – and, while we may find there the fears and expectations he describes, we find too much that for their sake we are asked to explain away. In his definitions of the passions Hobbes gives cursory treatment to such feelings as 'kindness', for instance; but they have, of course, no motive force, not even, as in Hume, for mere moral judgment, let alone as sources of action. Some kindly feelings, for instance, gratitude, Hobbes even describes in such a way as palpably to contradict the facts, or at least to omit important and relevant parts of them:

To have received from one, to whom we think ourselves equal, greater benefits than there is hope to requite, disposeth to counterfeit love; but really secret hatred.... For benefits oblige; and obligation is thraldom; and unrequitable obligation, perpetual thraldom; which is to one's equal, hateful. But to have received benefits from one, whom we acknowledge for superior, inclines to love; because the obligation is no new depression; and cheerful acceptation (which men call *Gratitude*) is such an honor done to the obliger, as is taken generally for retribution.[4]

Or in treatment of family relations, for example, Hobbes, of course, takes no account of any natural ties of affection but indicates simply the obligation to obedience on the part of the child toward whoever nourishes it:

For it ought to obey him by whom it is preserved; because preservation of life being the end, for which one man becomes subject to another, every man is supposed to promise obedience, to him, in whose power it is to save, or destroy him.[5]

In *Behemoth*, when he presents himself with the problem of a man ordered by the sovereign to execute his own father, there is indeed some suggestion in the way his interlocutor puts the question that such a man might feel some hesitation in obeying. But the basic question, whether there is such a thing as filial affection, is hedged; and the answer is made that, after all, the order is unlikely anyhow, and if it is given it must be obeyed only if decreed as a law rather than as a special order having reference to a particular person.[6] To be sure, Hobbes can, if he likes, explain away the kindlier passions to his own satisfaction; but if each man's introspection is really the test of the correctness of the theory, then there is strong evidence against it somewhere in the experience of most of us. Granted, with Hume, that the gentler passions are weaker than Hobbes's basic fears and appetites; they are there nevertheless in some measure, as each of us can verify. And, what is more, no system of education could induce them if there were not some spark in us for such habituation to work on. The same argument – that used by Hume in the *Inquiry* against egoism – is valid against Hobbes's political psychology. Perhaps many of us act like Hobbesian men all the time, and all of us most of the time; but many of us admire actions Hobbes would condemn as foolish, and a few even practice them. And, though Hobbes would, of course, account for such admiration or practice as the consequence of hearing or reading seditious doctrine, such miseducation would not have taken effect were there not something in men's natures to respond to it.

Nevertheless, Hobbes's inadequacy is not the result of an equation of men and brutes or a failure to distinguish human reason as a unique directive element distinct from animal passion. In the first place, Hobbes does distinguish men from other animals, and much more shrewdly than the Cartesians. Strauss seems to think the basic Hobbesian distinction is contained in the observation that man alone among animals is vain or proud; and the reference to pride as the basic human passion, he thinks, invalidates Hobbes's mechanical account as a consistent system.[7] If, however, one takes the Hobbesian philosophy of motion at its face value, interpreting the relation between Hobbes's philosophy of nature and human nature, for example, as Tönnies does, one may take as fundamental the distinction of the *Leviathan*: brutes and men reason, with prudence, from effects to causes; men also, in science, from causes to effects. The latter type of ratiocination, starting from definitions, depends on the invention of language, which makes man if not different in kind from brutes at least an infinitely cleverer brute in the techniques of satisfying his animal wants. This distinction, unlike the Cartesian, may be amply confirmed by observation of animals as well as of men.

But not only does Hobbes himself make a clear distinction (if not two distinctions) between men and other animals; the inadequacies of his system can be largely eliminated even in a philosophy which stresses the similaiities of various species, namely, in Hume's moral philosophy. The examples Hume uses to show such likeness, especially in the chapters on pride and humility and love and hatred in animals,[8] are admittedly very weak; for he seems to have confined his observations principally to barnyard fowl, who are surely among the vertebrates least endowed with thought and feeling. But his illustrations can be bettered without altering the basic analogy which is fundamental to his whole system of knowledge and morals. It is an important part of the evidence for his theory of mind that it explains a wider range of phenomena, namely, animal and human, than any other. In fact, it is, Hume says, only another hypothesis accounting equally well for animal as well as human thought and feeling that could, by the rules of empirical evidence, be said to have equal probability with his own. Yet Hume, stressing the likeness of men and brutes, takes full cognizance of the benevolent aspect of human nature, giving man's limited generosity place beside his more self-centered feelings. Sympathy is not, to be sure, by any means the primary motive force for most human action; but it sometimes functions as motive in the direct form of kindness or benevolence and serves in its weakened (because generalized) form as the source of moral judgment.

What is mistaken in Hobbes, one concludes, is not the equation of men and other animals but the reduction of both to mere motions. One is no better off with men and brutes made mere machines than with the Cartesian world of machines and incorporeal minds to observe them. One is no better off with a nature dead through and through than with a dead nature and a live reason mysteriously functioning in it. The selfish system is the logical result of a thoroughgoing mechanism in which motions to and from, that is, appetites and aversions, in the individual body are the only possible passions. On the other hand, Hume's use of any and all given feelings as data, with the flexible range of imaginative association operating on them, allows the admission of much wider data and more inclusive treatment of them and, therefore, a much more adequate account of the variety and range of human character and feeling.

Perhaps, then, by the simple inclusion in our data of the gentler as well as the more violent feelings, we many produce a description of human nature that will give a sufficient basis for moral judgment without recourse to a supernaturally implanted reason. As a matter of fact, Hume's ethics has descriptively, I think, a high degree of adequacy; moreover, despite the current rejection of Hume's atomic psychology, the pragmatic philosophy seems to me in the main an attempt to restore Hume's position – with a good deal less precision than Hume achieved. But, despite its competency to describe more varied phenomena of human nature than most ethical systems, Hume's ethics has two serious drawbacks. First, a question may be raised about the adequacy of its metaphysical basis – and again the same doubt would apply to its contemporary descendants in the philosophies of pragmatism, since, despite the rejection of Hume's atomic theory of mind, his fundamental position is maintained: i.e., the reality of universals is denied. But, of course, if Peirce's critique of nominalism or Plato's refutation of Protagoras in the *Theaetetus* should be correct, such a philosophic basis would be mistaken, no matter how adequate the conclusions presumably deduced from it. That objection, however, goes far beyond the range of this discussion. More immediately relevant is the objection raised by Hume himself: in a letter to Hutcheson he suggests that, should the universal existence of the moral sense fail to be confirmed in experience, his and Hutcheson's ethics would collapse.[9] Explicitly, of course, pragmatism stresses the relativity of value-judgments rather than the simple uniformity of human feeling Hume relies on. But Hume certainly in his emphasis on custom and habit takes equal account of the variation in such judgments; and, on the other side, modern pragmatism as an ethics still implicitly demands a simple faith not so unlike the eighteenth-century one: a faith that really everyone is at heart an awfully nice fellow. Neither Hume's system nor modern pragmatism gives us any defense when we meet with individuals or groups who are definitely not in the least nice fellows and whose philosophy of human nature is not at all nice either. In the polite circles of eighteenth-century Edinburgh and Paris or in the bigger-and-better heyday of our roaring twenties the mere description of human good nature may have looked very charming; but in Plato's generation, for instance, it was clearly not enough, and it is just as clearly not enough in this one.

Yet a return to the Cartesian distinction with its absurdity and artificiality is no satisfactory escape. Are there other traditional ways out to be inspected? Plato himself, faced with a not dissimilar moral situation, could distinguish something unique in human nature as a source of moral standards and at the same time see a continuity in the whole range animal nature. In the metaphor of the many-headed beast, the lion and the man, for example,[10] the likeness and difference are equally pointed. So are they (with greater emphasis on the continuity of animal life) in Diotima's account of the urge for immortality.[11] Or, in the myth of the *Phaedrus*, for instance, the soul throughout the universe has a uniform function, but human souls have a unique insight that puts them high in what is nevertheless presumably a continuous hierarchy:

For the soul that has not seen the truth will not come into this (i.e. human) shape. For a man must understand what is spoken according to form, bringing together what comes from many perceptions into one by means of reasoning.[12]

Moreover, in several of his accounts of human psychology, Plato indicates that the passions which men presumably share with beasts are something to be controlled, not extinguished – and that, as the theory of the θυμοειδὲς suggests, reason itself never operates without the motive force of passion accompanying it. So again, as in Hume, we may recognize the continuity of animal nature and acknowledge the importance of feeling in men – but with the recognition of the unique element in this most godlike of animals, the element which should control the brutal aspects of man's nature however seldom it may, in fact, effectively do so. And, certainly, Plato, unlike Hume, does not fail to supply a sufficient metaphysic to support his moral system. But again there are at least two major difficulties in the way of taking Plato's psychology as the solution ot our problem. For one thing the metaphysical foundation of Plato's ethic eliminates any but a strictly intellectual standard of morality; and an ethic which allows no virtue distinct from intelligence is simply too uncomfortable for use by most ordinary mortals. One can so define wisdom as to eliminate really clever bad people; but to eliminate good, kind, decent stupid or at least simpleminded, unintellectual people is, I should think, rather too much for most of us. But the principal difficulty in Plato's account of men and animals is in the metaphysic itself on which it rests: despite the enormous spread and influence of something called Platonism, I suppose there has never really been a Platonist and never can be. From the unlucky Dionysius on, every supposed follower of Plato has been infinitely less elusive, more stock and settled and therefore dead than the creator of the dialogues himself. Some few people have the gift of illuminating Platonic metaphysics (most have not) – but it is not a doctrine to be taken over and imposed as a solution on our particular problems. We may or may not find indefinite depths of suggestion and illumination in the dialogues – see, for instance, Diotima's discussion mentioned above beside Hobbes's bungling treatment of the family – but there is no Platonic system in the framework of which we can neatly put man into his place in nature and feel satisfied that our moral standard is provided and our problem solved.

The question remains: How can we retain a sense of man's uniqueness sufficient to provide a standard of value that can hold in the face of the relativity of human feelings and judgments, yet without resorting to the implausible and barren conception of a Christian-Cartesian soul split off from all natural kinskip? One more suggestion may be found in Kantian morals – where man is acknowledged to be an animal (in fact, from the cognitive point of view he is nothing else), yet at the same time when looked at morally something radically different. In a review of an Italian work on evolution[13] Kant declared himself in sympathy with the conception that man has developed from and is biologically akin to four-footed beasts; but he uses the evidence in a peculiar way. Moscati had suggested that certain disorders of the female reproductive system result from an upright posture obviously lately and awkwardly assumed. Kant declares this confirms his view that, while we are animals, we are botched, bad animals – and our purpose in life is therefore something different from the satisfaction of our animal needs. So Kant could accept, I should think, a thoroughly physiological interpretation of human behavior and still find that, though all this be true, there is something more. The ethical situation remains what it is, at a tangent to the natural or biological explanation, complete and self-sufficient though the latter must always appear. Kant's teleological language in talking of human action (what Nature intends with us, etc.) is puzzling and his whole puritanical account of man's character most lamentable and one-sided: what less loveable creature than the misanthropic philanthropist he so admires? Nevertheless, the general conception of men's complete animal and yet non-animal nature is, if one may take it out of context, most suggestive – as is the second and least abstract formulation of the moral law: 'Treat every human being always as an end and never as a means'.

But, of course, it must be said that, despite the ostensible division of morality from metaphysics or theology, Kant did have a supernatural faith, and a most austere one, to sustain his sense of duty; and without that support most of us, I suspect, have difficulty in discovering in its Kantian purity the moral law within on the presence of which the whole system admittedly rests. So we come back to the question suggested by our glance at Hume and Plato: Can an ethical system really stand on its own without an adequate metaphysic; or what acceptable metaphysic can justify the ethical standard we wish to support? That is a problem far beyond the scope of this discussion, but one on which it looks as if the question in hand must ultimately depend. Short of its treatment, one can merely point out some of the factors that make our present perplexities about human values something a little better than a Hobson's choice between the Hobbesian machine-man and the Cartesian-Christian separate and immortal mind.

NOTES

[1] See, e.g., Hugh of St. Victor, *De medicina animae*.
[2] *Meditation* ii.

[3] *Leviathan* (Everyman ed.), Introduction.

[4] *Ibid.*, chap. xi, p. 50.

[5] *Ibid.*, Chap. xx, p. 105.

[6] *English Works* (ed. by Molesworth), VI, 227.

[7] Leo Strauss, *The Political Philosophy of Hobbes*, Oxford 1936, chap. ii, *passim*; cf., e.g., *De cive*, chap. v, art. 5: 'Nam primo, inter homines certamen est honoris et dignitatis; inter bestias non est: unde *odium et invidia*, ex quibus nascitur seditio et bellum, inter homines est, inter illas non est'.

[8] *Treatise*, Book II, Part I, sec. 12, and Part II, sec. 12.

[9] *Letters* (ed. by Grieg), Oxford, 1932, I, 40, No. 16.

[10] *Republic* 588B.

[11] *Symposium* 207A f.

[12] *Phaedrus* 249B.

[13] *Werke* (ed. by Cassirer), IV, 437.

CHAPTER XVI

THE CHARACTERS OF LIVING THINGS

I: *The Biological Philosophy of Adolf Portmann*

I

The dramatic advance of biological research in the past few decades has been proceeding on two very different fronts. The triumphs of bio-chemistry, in its detailed study of the nature and regulation of metabolic processes; the startling advances of molecular genetics, in particular DNA 'code-cracking'; the revelation, through the electron microscope, of a whole new world of complex organization at the minutest level: in short, those fields of research designated by the package title 'molecular biology': these form one such ever-widening front of advancing knowledge. A little less conspicuously in the headlines, but as ingenious in its methods and certainly as significant in its philosophical implications, is the almost equally new and equally rapidly growing science of animal behavior, or ethology, which studies, not the minute component parts of animals, but their action patterns, whether as individuals or in groups.

The difference in the research procedures of these two groups of scientists is obvious. One set works chiefly analytically. What the visitor sees in their laboratories is blackboards full of calculations, expensive electrical and electronic equipment, carefully isolated preparations of the appropriate tissues or micro-organisms or metabolic substances; but anything that *looks* like a plant or an animal is conspicuous by its absence. True, molecular geneticists still perform breeding experiments, but usually with bacteriophage or other borderline organisms invisible to the naked eye. Even in electron microscopy, where techniques of looking, of 'pure observation', are undoubtedly crucial, the structures 'seen' are far indeed from our ordinary field of vision.

Ethologists, however, must *watch* devotedly hour by hour and week by week, animals, living animals, in laboratory conditions, in zoos, or, best of all, in the wild, managing or submitting to their environments in a very great variety of species-specific ways. They do, indeed, perform experiments of great ingenuity and sophistication, interfering with the environment of their subjects in such a way as to infer from altered or constant behavior the fundamental action patterns which certain situations call forth. They spend much time, for example, trying to discover whether a given action pattern is 'innate' or 'learned'. In every case, it is not tissue cultures, proteins, or genes, but whole, individual animals or groups of animals they are observing. However abstract and elaborate their theoretical explanations of such behavior, they never escape this base: it is *what animals do* that they are talking about, and this is a very different subject matter from that of their molecule-oriented colleagues.

True, some very eminent ethologists have sometimes insisted that, so far as their program goes, they hope eventually to translate their statements about the behavior of animals, of great tits or sticklebacks or chimpanzees, into statements solely about muscle contractions or glandular secretions, and these would in turn be smoothly translatable into the terms first of biochemistry and ultimately of the science of sciences – as they consider it – mathematical physics. Whether this 'ideal' is realizable we shall have to consider in terms of the position taken by the scientist we shall be meeting in this chapter. For the moment, however, it is sufficient to notice that what, say, biochemists and ethologists *now* spend their time doing are two very different sorts of things. Their activities differ, in particular, in the directness of their relation to the ordinary *perceptible* world of living things that surrounds us and of which we form a part. True, it is a far cry from the amateur 'bird watcher', for instance, to Lorenz's research as director of his Max Planck Institute; but the ethologist's skill is an extension and extrapolation of the primal fascination described by Lorenz himself as 'that mysterious charm that the beauty of living creatures works on some of us.'[1]

Take as typical of the ethologist's activity an example from the work of G. P. Baerends on digger wasps, quoted by Portmann in his book *Animals as Social Beings*. Of a particular female under observation, Baerends reports:

On 24th July, 1940, 000 completes the first stage of Nest 61; on 25th she opens Nest 68 (unknown to me till then) and brings food here later. When I am putting in a plaster nest here, I find the young larva with its first caterpillar. It seems to me not quite healthy, and I find that the wasp, after her next unladen visit, has stopped looking after the nest. On 26th she inspects Nest 61, but the larva has not yet emerged (A). On 28th she digs Nest 84; when she arrives there with her first caterpillar, I have a caterpillar in there for her. She throws out the strange caterpillar, puts her own in, and lays an egg on it (B). On 30th she again inspects Nest 61; I have previously replaced its contents with a cocoon. She closes the nest, and I do not see her there again (C). Now she begins to dig Nest 356, but does not supply it with food till 2nd August, because the weather in the two days between is very bad. On that day she also visits Nest 84, but the larva here has not yet emerged. So she brings no food here (D), but builds a new nest, which, however, is destroyed by me in trying to replace it with a plaster nest. On 3rd August she again visits Nest 84, into which I have previously brought six caterpillars. She then closes the nest (probably for good but this I cannot say for certain) and stops bringing food here (E). On 4th she works first on the first stage of Nest 340, and finishes this stage; later she comes to inspect Nest 365 and then takes a caterpillar there as food. On 5th she digs Nest 423, brings the first caterpillar there and lays an egg on it. Then she inspects Nest 356 in its third stage. Before she brings a caterpillar there, I replace the larva with a caterpillar and egg; she brings in her caterpillar, without letting herself be disturbed by the larva's absence (F). After an hour, however, she makes an inspection here. There being no larva in the nest, she stops bringing food here (G). The contents of the nest afterwards rot from the rain. On 6th she opens Nests 340 and 423; the eggs have not yet been hatched, and she stops bringing food here (H and I), but digs a new nest 440. From 7th to 14th the weather is very bad, and it is not till 14th that she starts on the third stage of Nest 423. On 16th I take the larva out of the nest, and she makes an inspecting visit, after which she closes it for good, without there being a larva in the nest (J). Then she begins on the second stage of Nest 440 – and I give up my observations of her.

Although, therefore, this wasp has been disturbed in many ways, she has always reacted 'logically' both to the nest's natural content and to its content as altered by me [Figure 1].[2]

What is the ethologist watching? He is observing the behavior of an individual living being. Plainly, that is not altogether the same thing as watching a sunrise or taking pointer readings. Nor indeed is it the same as observing in detail the pattern of a butterfly's wing or a peacock's tail. The latter example can in itself illustrate what I mean. To describe in detail the pattern of the peacock's tail is not the same activity, for example, as to watch the use of his fan by the peacock in courtship display. Even apart from the study of the physical nature and distribution of colors in the feathers, one can distinguish further the morphologist's account of the appearance of the pattern itself from the ethologist's study of its functioning in the life story of its possessor. The ethologist is watching the animal in a manner analogous to the way in which he might watch another human being. He is taking account of an individual as in some way a center of activities. This is *not*, be it said at the outset, to predicate of the individual animal a 'consciousness' like ours; but it *is* to acknowledge the existence of a center of drives, perceptions, successes,

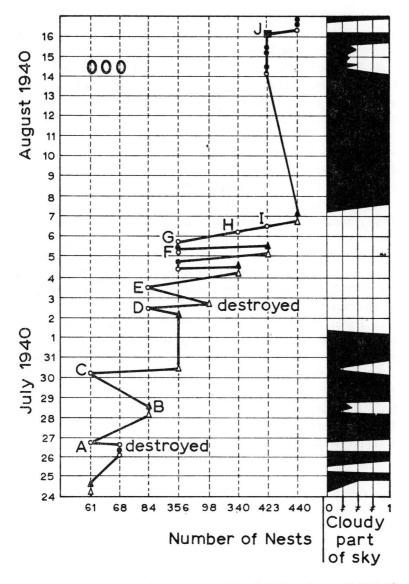

Fig. 1.   Diagram of activity of digger wasp 000 from July 24 to August 16, 1940. The black on the right shows the cloudy parts of the sky: the increased activity on cloudless days (July 26 and August 14 and 15) is striking. Such a diagram gives some idea how much the wasp's success in rearing depends on the weather.

and failures of which, although we have no direct access to it, we do take account in our dealings with the individual in question. And this acknowledgement clearly marks off the ethologist's from the morphologist's study. Both these types of research, however, the morphological *and* the ethological – and that is the main point here – are directed to macroscopic, visible features of the organic world, rather than to molecular structures accessible only through highly indirect and analytical techniques.

I have been drawing out to some length a very obvious distinction between the procedures characteristic of different branches of biological research. This distinction, however, with which Portmann opens his own *New Paths in Biology*, may serve us as a key to his conception of his subject and in particular to his account of the essential criteria of living things.[3] Portmann is far from minimizing the importance for biology, and for man's technical applications of biology, of the advances in the molecular or, for that matter, in the population–genetical field. Nor is he in fact, professionally, an ethologist, but a student of comparative morphology and comparative development. But he does oppose, for good reasons, which I shall shortly examine, the tendency to *reduce* biology to its physico-chemical base. Instead, he suggests a starting point for a new and richer philosophy of living things – a philosophy which will permit respect for all facets of biologists' activity, including in particular those which bear directly on plants and animals as they strike our human senses. It is from this general contrast between the imperceptible and the perceptible, which he has emphasized in many of his writings, that we can best approach his thought.

II

Let us return for a moment to the second of the two types of research we have mentioned, that is, to the ethologist and the morphologist taken together. We may describe them as students of *perceptible organic form*. For if the ethologist studies behavior, it is still perceptible form in the spatiotemporal sense that he is investigating, for example, in the sequence of the digger wasp's nesting activities quoted above. This whole class of perceptible patterns in space and time Portmann calls *authentic phenomena*, as distinct from all the rest of the subject matter of all the sciences, including that of molecular biology, which he refers to collectively as *inauthentic phenomena*. This distinction is both more important and more revolutionary than it appears at first sight, and it will need, I think, a brief historical excursion to see what is entailed in it.

The term 'phenomena' has a venerable history. Astronomical theory was traditionally concerned to 'save the phenomena', that is, to find a formulation in terms of the mathematics of the circle which would be consistent with the movements of the heavenly bodies as seen. Positivist philosophers of science, such as Duhem, have interpreted the whole of scientific activity in these terms. Theories, from this point of view, are constructs, in themselves empty of meaning, intended to guide the observer from one set of 'phenomena' to another. Yet in its transformation through the intellectual revolution of the seventeenth century, the reference of the term 'phenomena' itself has been strangely altered. The modern positivist's 'phenomena' are not what appear to the ordinary stargazer, let alone to the ordinary natural historian, bent curiously over leaf or chrysalis. For a philosopher like Duhem, it is the 'phenomena' accessible to the ingenuity of the mathematical physicist that count, and these are, in general, data obtainable only in highly contrived experimental situations. This change, like so much in the scientific revolution, can be documented in the writings of Galileo, on the one hand in his dictum that 'nature is written in the mathematical language' and on the other in his exclusion of color, sound, taste, and smell from natural realities, that is, his insistence on what Locke was to call the distinction between primary and secondary qualities. Both these innovations, which are, indeed, closely linked to one another, are announced in Galileo's polemical work *The Assayer*, and it is worth looking at both passages in some detail to see how his program and that of the then new mechanical philosophy are related to Portmann's distinction.

The first passage runs:

Philosophy is written in this great book, the universe, which stands continually open to our gaze. But the book cannot be understood unless one first learns to comprehend the language and read the letters in which it is composed. It is written in the language of mathematics, and its characters are triangles, circles, and other geometric figures without which it is humanly impossible to understand a single word of it; without these, one wanders about in a dark labyrinth.[4]

So authoritative is the place of mathematical physics in our conception of scientific knowledge that we take this pronouncement as the enunciation, trail-blazing in its time, of what is now a truism. Applied mathematics is the paradigm case of science, science the paradigm case of knowledge; *of course* someday all we know or can know will be statable in strict mathematical form. But let us think again and ask, first, what Galileo's statement really means, and second, whether it is true.

The language of nature, Galileo tells us, is one that we must learn. Certainly; all languages must be learned. But this is a language, he seems to be suggesting, which by no means everybody knows. It is not like one's mother tongue, assimilated in infancy by any normal human child; for certainly he is convinced that at least the wretched Sarsi, against whom his polemic is directed, has never learned it. The language of nature, then, is in some sense a foreign language. And indeed, for most of us, the language of mathematics, which *is* for Galileo nature's language, has to be learned in school, not at home; it is a secondary and artificial acquisition. It belongs, in other words, not to the life world, but to the secondary, painfully constructed world of learning. But how do we learn such a secondary, a foreign language? Either its alphabet is written like our own, and we have to learn the meanings of the words; or there is a foreign alphabet, a different kind of character, to be learned before we can get as far as trying to understand the words. The latter situation holds for nature as Galileo sees it: not even the letters of its language form part of our ordinary environment, for, he insists, we have to learn them too. And again, this is, of course, true of the language of mathematics: we have to familiarize ourselves with its formalisms before we can make use of them in order to understand what they have to teach. To the ordinary person, then, the language of nature bears to the perceptible surface of the things around him a relation rather like that which, for a native English speaker, a page of Chinese or Arabic bears to a page of ordinary English prose. Until he is trained to do so, he cannot make out so much as its constituent elements, let alone their meaning. In short, the universe that 'lies open to our gaze' is, if Galileo is correct, a volume written in a secret code, which only the trained cryptographer can interpret. The rest of us can only 'wander in a dark labyrinth'.

Is this true? Are all of us blind to nature who have not learned to decipher the applied mathematician's code? Is there no mother tongue of nature that we have learned in early childhood and on the ground of which we acquire the second, formal speech of mathematics?

Two aspects of our conceptual situation obstruct an honest answer to these questions. One is the product of that very revolution in which Galileo was engaged. The 'naturalistic' world view which it engendered has been often described; but what is seldom acknowledged in such expositions is the sediment of ordinary, prescientific human experience out of which this, like any world view, developed and which it cannot wholly replace. In other words, the 'objective' world of science, and especially of applied mathematics, must be distinguished from the 'life world', the world shared by all human beings, in and out of which it develops.[5] 'Nature' within the main tradition of modern science, in other words, is an *objectified* nature, and this is Galilean nature. If, moreover, Galilean nature was a construct of the seventeenth-century revolution, so is it a construct, within the life world, in the life history of each of us. We have each of us a dwelling place on this earth, a biological environment, within which, and transcending which, we assimilate, in the process of education, the intellectual framework of our culture. The distinction between the *environment* (*Umwelt*) of animals and the *world* of man (*Welt*) can be transposed into a distinction between the primary and secondary worlds of man himself. As human beings we dwell primarily, and immediately, in the life world; as participants in modern Western culture we acquire, as we mature, the intellectual framework of modern objective thought, within that primary frame. But – and this is the point in our present context – as we acquire our cultural heritage, we come to dwell in it also. We assimilate it to our persons and identify it on the one hand with our primary world and on the other with reality itself. So nature comes to *mean* to us Galilean nature, and the existence of the primary life world is ignored.

Second, most of us live and have always lived in an urban, industrial environment. Man is an earth-dwelling animal among other animals; urban man lives densely jammed among other human beings, indeed, but his nonhuman environment is chiefly one of artifacts. He is cut off by roads and pavements from earth itself, by smog and steam from the sky, by electricity from hewing wood and drawing water, by processed and packaged foods from man's age-old wrestling with the elements in heaven and earth to gain his sustenance. He may make parks and gardens for his own amusement, climb rocks for sport, or relax at the seaside in the sun. But the abstract language of mathematics may well be, for him, the only language of nature he has ever learned. Yet he sleeps and wakes; he has been born and will die. And he is in the first instance part of a biological unit, the family – and we have all been told how difficult it is for him to mature successfully into full humanity if he has been deprived of this normal biological beginning. Even in his technocratic world, moreover, he needs to re-create for himself analogues of nature, not only islands of living nature itself in parks and zoos but colored surfaces and artificial rhythms to fill in the unnatural emptiness of life seemingly founded in artifact and abstraction: the demands of the life world persist. But in an industrial environment these needs are disguised and neglected, and the sophisticated city-dweller can easily identify the objective world of science and technology with 'nature' itself. He is a Galilean to the point of denying the existence of nature in the primary sense.

If, however, we can, by acknowledging them, hold in abeyance these disabilities, we must admit that, taken at its face value, Galileo's pronouncement is false. Of course nature includes the geometrical shapes which are his favored mathematical 'characters' but much more too: colors and sounds and tastes and smells. This is, indeed, just what Galileo denies in the other crucial passage I want to examine here. Swayed by the authority of physics and its mythology, we have too sheepishly followed him in his denial and again have overlooked the full implications of his statement. In the passage in which he initiates the distinction between primary and secondary qualities (or, strictly speaking, revives the ancient Democritean distinction), he puts his thesis as follows:

Now I say that whenever I conceive any material or corporeal substance, I immediately feel the need to think of it as bounded, and as having this or that shape; as being large or small in relation to other things, and in some specific place at any given time; as being in motion or at rest; as touching or not touching some other body; and as being one in number, or few, or many. From these conditions I cannot separate such a substance by any stretch of my imagination. But that it must be white or red, bitter or sweet, noisy or silent, and of sweet or foul odor, my mind does not feel compelled to bring in as necessary accompaniments. Without the senses as our guides, reason or imagination unaided would probably never arrive at qualities like these. Hence I think that tastes, odor, colors, and so on are no more than mere names so far as the object in which we place them is concerned, and that they reside only in the consciousness. Hence *if the living creatures were removed,* all these qualities would be wiped away and annihilated. But since we have imposed upon them special names, distinct from those of the other and real qualities mentioned previously, we wish to believe that they really exist as actually different from those.[6]

How true, we say. This is the story of Eddington's two tables: the hard, brown, smooth, coffee-stained or ink-stained object in my study which is my table and the buzzing congeries of billions of subatomic particles which is the physicist's table, the *real* table. But think of the condition on which the whole construction depends: 'if the living creatures were removed'. This great book of nature, in other words, would be a nature deprived of life. And *a fortiori* it would be a nature without ourselves. But surely it is our world, the nature we live in, and the nature that we are, that we are trying to read. Yet the letters of Galilean nature exclude the language of life, including the life of man. And this is quite literally so: for modern naturalism has, and can have, no adequate concepts for the interpretation of organic phenomena nor for the disciplines that deal with that most strange though most familiar animal, man.

Yet we do have a world of color and taste and smell and sound which is not just part of our individual subjectivity but part of our biological environment, genetically determined as much as our more 'mechanical' powers are determined. To this, of course, the Galilean answers: but all these perceptions are 'determined' by physics and chemistry: sound waves, light waves, and the chemistry of taste and smell. And it is true, of course, that *conditions* for all our perceptive powers, and those of all animals, are given by physicochemical laws. But in its qualitative nature, perception has laws of its own, which do not indeed contradict the laws of inorganic nature, but are nevertheless not identical with them.

Portmann presents as an example of this fact the human 'world of color'. Color perception has its own laws which are laws of 'the living creature', and this is both our nature and the nature of the world as the human living creature's world. It was this aspect of color vision that was the primary theme of Goethe's *Farbenlehre*, not in contradiction, but in supplementation, of Newton's more artificial experiments. Goethe's 'color circle' presented a sequence of *seen* hues with their own intrinsic order – an order that was simply not discoverable in terms of the physical theory of 'pure' colors. Physicists have recently begun to pay attention to Goethe's results; moreover, the work of E. P. Land on color vision, though controversial, has raised dramatically the question of laws of visual perception distinct from the physicist's classical trichromatic theory. To put it briefly: Land has shown that the superposed projections of two black and white pictures, one through a red filter, can produce, for the spectator, a picture filled, not, as one would expect, with a uniform pink, but with a variety of colors, including greens and blues, strikingly similar to the colors of the original subject. Whether or not this result 'disproves' the classical theory, it certainly suggests – and very powerfully – that the laws of vision are more subtle and complex than the theories of Young, Maxwell, and Helmholtz would have led one to suspect.[7]

Such laws, of course, by no means invalidate the physicist's investigations of isolated colors nor the classical laws of color mixing where the physicist's special conditions prevail. No one is suggesting that they do. But the point is that there are natural laws characterizing many levels and dimensions of reality. The laws of more complex entities, such as organisms, do not indeed conflict with the laws of physics and chemistry, but neither are they 'determined' by them. And to investigate adequately the regularities of such more complex existences, the scientist needs to apply his analytical techniques to the *natural* context: he has to take account of the living world of his subject. It is on the reality and significance of this richer context that Portmann is insisting.

It is, then, the whole range of perceived and perceptible phenomena, the nature which would not be at all were the living creature, with his sense organs, not existent or not at least possible, that Portmann calls *authentic phenomena*. What for Galileo is authentic because independent of living things, and existent out of all relation to them, is for Portmann *inauthentic*, just because of its want of such relation. It is not genuinely appearance because it does not and cannot appear to anyone. And conversely, what for Galileo is a 'mere name', given reality only by the living creature's presence, and therefore inauthentic, is for Portmann authentic, just because it does, or can, properly be said to *appear*. His nature is nature with living creatures in it, not, by a contrary-to-fact condition, reasoned away, and its qualities, rather than being 'mere names', call on us to name them.

True, this is not a simple reversal: movements are also 'authentic' for Portmann insofar as they appear to our perceptions, and they are, of course, truly part of nature for Galileo. Yet even the motions apparent in full natural reality are not what the physicist studies: from Galileo's inclined-plane experiment on, he has studied progressively more abstract relations of ideal motion, contrived through ingenious experiment, remote from the movements of everyday life. Rivers flowing, plants growing, fishes swimming, birds flying, our own changes of attitude, our speech: all these, not the physicists' pointer readings, are authentic phenomena. Nature for Portmann is in the first instance nature seen and heard, not nature reasoned out in the physicist's dream of a mathematically ordered world.

Authentic phenomena, then, are the perceptible surface of the things around us: rocks, clouds, stars, and above all the pullulating variety of plant and animal life as seen and heard. Now, most modern biologists, it must be admitted, would make short work of this conception. They would agree in fact with Galileo: such surfaces will appear mere names once science has 'explained' them, and such explanation, they insist, is partly imminent, partly already achieved. Granted, such biologists argue, that colors, sounds, smells, tastes, do indeed enter, in important *functional* ways, into the life histories of plants and animals. They are confident, however, on the one hand, that such matters can be explained as effects of underlying macromolecular causes. And on the other hand they insist that all such macroscopic functions with all the perceiver-perceived relations they entail are also to be explained by mechanical causes, in particular by the so-called 'mechanism' of natural selection. Bees are attracted by colored petals; the colors of flowers are therefore 'explained' by pollination through the flight of bees. Thrushes take more pink and brown snails on the green leaves in summer, more yellow snails on the brown earth in winter; color polymorphism in the common land snail is thus 'explained' by the resulting alternations in 'selection pressure'. And, of course, both the colors of the snail shells and the vision of their predators, the thrushes, are themselves explained in physicochemical terms. It all boils down in the long run to matter in motion in the classical Galilean form.

To this purely functional interpretation of authentic phenomena Portmann makes two objections. First, he does not agree that all the great multiplicity of situations in which animals *appear* to other animals can be interpreted *exhaustively* in terms of natural selection. Selection can indeed explain the change of such phenomena, but not their origin. Thus color polymorphism can, of course, be maintained by change of selective pressure; but the snails and their colors must pre-exist for selection to work on them. Nor is mutation, the major complementary principle of neo-Darwinian explanation, adequate to account for the origin of complex patterns. Not only are nearly all mutations deleterious but they too are, again, minute deviations from complex genetic formulae already in existence. To extrapolate these two well-established principles of micro-evolution to the whole range of living phenomena in their whole develop-ment is, Portmann believes, a procedure unjustified by the present state of our knowledge.

Second, the range and complexity of authentic phenomena, Portmann argues, far exceed those functional needs which alone natural selection can explain. To explain the elaborate pattern of the peacock's tail, for example, by natural selection in the orthodox fashion meant attributing to the female a kind of artistic connoisseurship in her preference for the more clearly elaborated and 'harmonious' effect. Recent experiments with these and allied species, however, have suggested two more plausible but restricted explanations for the functional aspect of the peacock's fan. First, the female displays from an early age a marked response to small kernel-shaped objects; this is plainly a drive connected with feeding behavior and would account for the selective advantage of clearly developed eyes in the fan. Second, the fan and the whole courtship dance seem to furnish an over-all index of the liveliness and individual superiority of a given male. Now, both these explanations again are selective; in neither context, however, need we invoke the anthropomorphic conception of a developed artistic taste on the part of the female. But what then of the whole intricacy of the pattern, which is, relative to both these ends, superfluous? We cannot explain them, Portmann insists, by reference to any automatic external control, but must acknowledge that they have a value of their own:

We know of many patterns that exist without any possibility of direct natural selection and the special formal properties of which can in no way be attributed to selection. The morphologist is led to the assumption that the genetic endowment of an organism includes special systems of factors whose role is directed to the development of complex patterns. Such genetic factors are directed to these achievements just as other genetic processes are aimed at the construction of a brain, a kindey or a heart.[8]

This is not, be it noted, to deny the importance of selection in general and of the selective control of patterns in particular. Indeed, Portmann him-self has published a textbook on protective coloration.[9] What he is denying is simply that the whole intricate detail of living pattern is to be explained *solely* in selective terms. Sometimes and in some respects it is so and can be shown experimentally to be so. But not therefore always and in all respects. Selection explains the maintenance of some patterns – as in stick insects, for example – but not the maintenance in all their detail of all patterns, nor, indeed, the origin of any. To claim that it *must* do so is to make an unwarranted extrapolation supported in its comprehensive import, not by experiment, but by the demands of an oversimple methodology and an overdogmatic metaphysic.

Further emphasizing and expanding this point, and more directly contradicting the neo-Darwinian reduction of all organic phenomena to their functions, Portmann divides authentic phenomena further into two classes: *addressed* and *unaddressed*. By *unaddressed phenomena* he means such appearances of living things as do not have their *raison d'être* in being directed to the sense organs of other animals of the same or other kinds. The superfluity of detail in the peacock's fan, over and above selective utility, would be such a phenomenon. So is the black pattern on the wing of the mother-of-pearl butterfly, where size and background color have been shown to be the necessary conditions for mating behavior, but where the lively black pattern can vary or even be absent without effect. 'The pattern as such', Portmann comments, 'however "optically" it may affect us, is in the details of its form, even in its existence or non-existence, functionally insignificant... Yet it is there'.[10] Such phenomena are unaddressed in the sense that they do not have their meaning in an act of vision. Even more plainly 'unaddressed', moreover, are the patterns of animals without vision, where a 'technical', selective explanation of shapes and colors is on principle impossible. Portmann describes the dramatic variations and arrangements of color in sea snails, which the marine biologist observes, and which to him, again, are optically effective, but which can have no optical significance for their eyeless owners, nor, presumably, in such elaborate detail as they display, for the inhabitants of their normal environment. Even if they serve to warn off predators, *any* bright color would suffice to meet this need; the whole detailed pattern, and the variety of patterns, far transcend the selective demand.[11] Yet there they are: the oranges and blues and purples and golds in glowing regular array. From a Galilean point of view, this riot of color and immense diversity of constant form must be in some so far wholly unintelligible way purely coincidental to the mechanisms of living, while these in their turn must be but epiphenomena of the working of inorganic laws. Yet that 'the living creature' does display to our perception colors, sounds, smells, and, at higher levels of life, responds to these displays in others is a massive fact of the living world around us, a puzzling, even a mysterious, fact, but still a fact. We have no right, in the name of science, to deny the fact just because our favored kind of explanation, a Galilean explanation, could not allow it to exist.

On the contrary, the appearance of living things, the way they show themselves in surface pattern, in rhythms of movement or voice, constitutes for Portmann one of the two basic characters of living things as such. This fundamental character of life he calls *Selbstdarstellung*, life's showing of itself on the surface. The concept is difficult to render in a single English word; with some hesitation, I shall translate it as 'display', emphasizing, however, that its meaning is wider than the usual biological concept. Ordinarily display means an active posturing of an animal, as in courtship display and the like. As a rendering of *Selbstdarstellung*, however, I am using 'display' in a sense similar perhaps to its use in merchandising. It includes active display, but also all the passive show of animal shapes and patterns. These, Portmann has demonstrated, have their own laws, clearly genetically determined with as much complexity and regularity as any more 'functional' characters. All animals but the very simplest have an outer layer, a skin, marked in characteristic, often symmetrical and highly complex patterns, in contrast to the less regular arrangement of their internal organs. Even in very simple transparent animals, Portmann points out, the internal organs are often knotted up in the corner as if not to interfere with the symmetry of the overall body design. For such species, moreover, as for the sea snails, visible patterns cannot (except partially and indirectly) serve a 'useful' purpose, because they have no eyes; yet pattern is nevertheless constant, intricate, and universal in these cases as it is also in all higher forms of life. Why? No one can say; but it is a basic character of living things that it is so.

By hardheaded evolutionists, Portmann will doubtless be accused of 'teleology', 'vitalism', and 'mysticism'. What of these charges? Note, first, that it is the neo-Darwinians whose thought, in this context, is overteleological. They must have a 'function' for everything, because otherwise selection could not control it. Descriptive laws, simply accepting a massive body of phenomena, though not 'explaining' them, they cannot stomach, if these phenomena plainly resist their favored form of teleology, that is, 'improvement' through the external control of natural selection. Portmann, on the contrary, is not invoking any mysterious 'purpose' for display; he is simply acknowledging its existence, admitting openly a massive reality to which Galilean thinkers must close their minds, as well as their eyes. The patterns of animals, feathers, scales, and so on, are indeed, he insists, in principle 'explicable' in terms of their foundation in a genetic basis, as all persistent characters are. And they may someday be explicable not only causally but in the sense that we may find some new – other than functional – perspective out of which to understand them. For the present, however, we should honestly admit the existence of organic display as a given, an ultimate, fact of the organic world. We have no adequate theory to explain the fact; only an adumbration of causal laws, in genetics, which could account for its persistence and modification, though not for its ultimate origin.

Is this 'mysticism'? In a sense: openness to unexplained phenomena is openness to mystery, and Portmann likes to stress the mysterious complexity of life in a way which intensely irritates mechanistically minded biologists. But why is it 'unscientific' to admit the unexplainted, to turn afresh to the phenomena? To face and acknowledge mysteries may be, not the rejection of science, but the first step to discovery. Besides, are not scientists supposed to pride themselves on 'open-mindedness'? Yet it is scarcely 'open-minded' to deny the very existence of the whole range of non-Galilean phenomena which constitute most of the living world and therefore of our own lives as living beings.

Nor, finally, is the acknowledgment of display as a basic character of living things a renewal of 'vitalism'. It does, indeed, involve resistance to reductivism: to the view that life is 'nothing but' statistical variations in the gene pools of populations. But it has nothing to do with any addition of a mysterious 'life force' over and above the obvious appearance of the things themselves.

III

It is from Galileo's rejection of secondary qualities that I have tried to approach Portmann's concept of authentic phenomena and of display as a basic character of living things. 'Display', however, is not simply color, sound, smell, as such, but the exhibition to the senses of animals of perceptible forms and patterns characteristic of other animals, whether of their own or other species. It is display *of* – what? Living things are not mere surfaces, nor are they, as used to be said, simply 'sacks full of functions'. Just as their superficies, their appearances to one another, form a significant, indeed essential, aspect of their nature, so does what very broadly speaking one can call their 'inner life'. A second essential character of living things, in other words, inseparably allied to but contrasted with display, consists in the fact that organisms are *centers* of perceptions, drives, and actions. This is most plainly true of ourselves and of mammals close to ourselves, as we feel, in the scale of living things. We *know* that we are conscious and that our awareness of our own identity somehow matters through all the variations and vicissitudes of our actions upon and passivities to our natural and cultural environments. We know, moreover, that our conscious thought is only a narrow center of a much broader field of unconscious processes. The totality of our 'sentient' life far outruns its self-conscious, wide-awake core: it ranges from the narrow circle of focal awareness through a continuum of gradations all the way to the wholly unconscious. Indeed, many of the transactions through which we both master and submit to our environment, and through which we make it a human world, are on principle entirely out of range of consciousness; yet they are nevertheless *our* achievements. Such, for example, is the achievement of color vision.

We do indeed *see* the colors and know that we see them, but we cannot 'see' or in any direct way apprehend what our nervous system has done to achieve this end. Taking consciousness, then, as a narrow center of a much wider range of inner dealings with the outer world, we may extend this extension to include another phenomenon which, again, in our experience of the world, we immediately and indubitably confront, namely, the fact that other animals, especially higher mammals, appear and behave, in a way analogous to our own, as 'centers of doing and letting be'. Stimulus-response theory, for all its vaunted 'objectivism', cannot explain away the massive confrontation with animals as living centers: the experience, for example, of Rilke watching a panther pacing his cage:

Der weiche Gang geschmeidig starker Schritte,
    der sich im allerkleinsten Kreise dreht,
ist wie ein Tanz von Kraft um eine Mitte,
    in der betaubt ein grosser Wille steht.

Circling, revolving, lithe prowling strength soft-footed,
  The narrow ring he treads is like a dance,
A dance of power with at the center rooted
  A great will stupefied, a will in trance.[12]

Rilke's poem begins with the 'look' of the panther (*sein Blick*), all but destroyed by the bars of his captivity. And it is, indeed, on those occasions when we meet the eyes of an animal with our own – just as we encounter our own kind 'face to face' – that we acknowledge most directly the centered depth of animal life. A dog, unnaturally bound to human life, directs to his master a mute appeal that seems at first sight to make him, like a Hegelian slave, a personality only in relation to another, fuller personality. Yet, turned back into itself by the impassible barrier that language erects between ourselves and other creatures, his look displays at the same time a structured resonance of mood, of character, sometimes almost of something analogous to wisdom.

But the most eloquent animal countenances belong to our closer kin, the apes. The quality of this life, so familiar to us, yet so foreign, Portmann illustrates with the picture of a male gorilla from the Basel zoo, or of a gorilla mother and child. How rich a center of experience meets us here: the muteness of the gaze serves but to deepen the reverberation of the mood that confronts us. Whatever our *theories* of animal behavior or animal evolution, we must acknowledge quite simply and factually the presence here of a center in which the living being's dealings with its environment are drawn together and from which they radiate.

As we move further from our own obvious near kin, the life we confront seems more alien to us, but at a few removes the centered structure is still clearly there; even a kind of awareness is still undeniable. A case from H. Bruhin's study of deer strikingly illustrates this:

The position of a male with high social standing is abruptly shattered as soon as he loses his antlers. In the spring of 1951, in the Basel Zoo, I was able to watch the moment when a fallowbuck sank to a lower level (α, β, γ, here describe the levels in social rank). On April 18th, at 3:45 P.M. the herd of five males and eight females were begging for food from the zoo visitors. Suddenly they were slightly startled by a playing child, so that some of them trotted off, including the α male. He happened to graze with the right side of his antlers the branch of a fir-tree lying in the enclosure. Immediately this half of the antlers fell clattering to the ground. Obviously upset, with tail raised, he sniffed at the piece he had just lost. Almost at the same moment the β buck realized what had happened, and attacked and pursued him vigorously. The other three yearling antler-less γ bucks took scarcely any notice of the occurrence nor did the does. After about half an hour both the α and β bucks had more or less calmed down and were again begging for food. But the former α buck was not tolerated at the fence by his rival, and therefore kept right at the back of the enclosure. There was only an indication of a social clash between α and γ bucks. Up to the evening the one-palmed animal carried out peculiar head movements, as were observed by Heck (1935) after the loss of antlers. On 23rd April the β buck also shed his first antlers. From this time on there was the same social ranking as had prevailed before the α animal shed his. In this case it can therefore be clearly established that the antlers, 'representing a particular social position', also lose their significance as representative organ when one half is lost, and this loss results in its owner going down the social scale.[13]

That the stag 'knows' about his antlers is confirmed by many other studies, notably, for example, those of Hediger on the psychology of animals both in captivity and in the wild.[14]

Now, admittedly as we move still further from styles of living like our own, it seems strange to speak at all of an animal's 'knowing' what it is doing, to speak of 'consciousness' or even in some more tenuous sense 'awareness'. Yet in some way even a hydra moving its tentacles or an amoeba its pseudopodia is showing in the broadest sense a kind of 'sentience' and through this, again, is controlling the patterns of its relation to the surrounding world. Looking downward across the continuum of life, we find universally some such principle, which is unlike anything in the inorganic world. Mechanists like Loeb, early in this century, hoped to reduce all living processes to 'tropisms'; Fraenkel and Gunn twenty-five years ago made a similar attempt to explain animal orientation in purely mechanical terms. But animals do not simply *move*; they *behave*. They do not simply display *reactions*, as acids and bases do, they perform – to borrow a psychologist's term – *transactions*. But transactions demand an agent, an individual center of giving and taking, of doing and ceasing to do. However strange to us the 'inner life' of animals, especially of invertebrates, remote as they are from our own patterns of living, we must in some extended sense admit the presence of such centers of experience where there is animal life at all. Even plants, although we should scarcely ascribed to them awareness in any sense at all, are still organized centers of growth and form.

This second pervasive and essential character of organisms Portmann describes by the phrase *Weltbeziehung durch Innerlichkeit*, again, a difficult phrase to make viable in English. 'Relation to the environment through inwardness': a hopelessly awkward expression! 'Inwardness' alone, on the other hand, is too exclusively subjective and fails to convey the *relatedness* that Portmann's concept entails. I shall use, therefore, a neologism, 'centricity', to convey what, as I understand it, Portmann means, and say that centricity, along with display, is a fundamental character of living things. Again, this statement is not meant to explain anything, but to acknowledge the existence of a range of phenomena too often neglected or even denied. But if we cannot, obviously, get inside the skins, so to speak, of other organic beings and feel how their way of living feels, we need not deny that such lives exist and have, in a way forever inaccessible to us, their own style of significance. In a passage reminiscent of our earlier reflections on the Galilean language of nature, Portmann writes:

The inwardness of these forms, widely different from our own organisation, speaks to us through its appearance. That we cannot translate this language into human words is no reason not to see the appearance itself. If, in a distant country, I attend a dramatic performance, of which I understand not a single word, I shall not on that account assert that nothing at all is being presented, that nothing is happening but a random noise.[15]

Not that centricity cannot be studied. It is indeed not only a quality of life acknowledged at the boundary of science but a domain of life accessible indirectly to analytical methods, as the experimental ingenuity of modern research into animal behavior has amply proved. Portmann has described some of these studies in his book *Animals as Social Beings*, from which I have already quoted, and there are numerous other places where one can read about them, like Tinbergen's *Social Behavior in Animals* or Lack's *Life of the Robin* or Lorenz's *King Solomon's Ring*.[16] What Portmann is emphasizing, however, is an implication of all this work which ethologists themselves, reared in the faith of 'pure objectivism', often fear to face. Thus he says, for example, in discussing the phenomenon of *territory* (in a chapter on 'The World of the Dragonflies'): 'The realization that territory is a biological fact has helped to obtain full attention for the subjective spheres of owning and defending it, and has made the most subjective thing of all, the individual's experience and social impulses, a field for new objective research.'[17] This frank avowal of the existence of the *subject*, not simply as a fringe phenomenon, but as the object of scientific research, may have, like the concept of display, farreaching consequences for our thinking about nature, about science, and about ourselves.

Revolutionary concepts, however, are hard to take seriously. Perhaps we may strengthen our grasp of this concept by considering some objections that might be made to it by more orthodox biologists.

Our most immediate and massive experience of centricity comes from our own experience of our own inner life; the inner core that holds together the threads of organism-environment relation in other species we can never directly know. What Portmann is doing in counting centricity a basic character of all life, therefore, is to extrapolate something we know in ourselves and apply it, if in diminished form, to all organisms. Yet Portmann hesitates, as we have seen, to follow the extrapolation of the neo-Darwinian's basic concepts, mutation and natural selection, from experiments on present-day selective change, where they undoubtedly do serve as adequate explanatory principles, to the whole panorama of phylogeny, where, taken by themselves, they do not seem adequate. How, then, it might be objected, is he justified in extrapolating what may well be a unique character of our own, the possession of consciousness, to the whole living world? To this objection there would be several answers.

First, Portmann is by no means alleging that 'consciousness' or 'mind' is to be predicated of all animals, let alone plants as well. Consciousness as we experience it is one expression, one style, of centricity. Even in the human individual it forms in fact only a narrow band in the wider spectrum of mental life. And so, since even our own awareness is by no means wholly focal, we need no great imaginative effort to extend a generalized concept of sentience of some sort at least to other animals. Second, consciousness in us, or sentience in a broader sense in animals generally, is again but the inner expression of centricity as such: of the fact that organisms are centers of metabolism and development, of ordered reaching out toward an environment and taking in from it, of birth and death. It is this centered dynamic, dependent as it is on the existence of *individuals*, that *is* characteristic of all life and is *not* characteristic of inorganic phenomena.

Third, one reason for approaching the general concept of centricity *from* our own consciousness as its characteristic human expression is that it is this expression, this dimension of centricity, which we know most intimately and indubitably. We also know, immediately and massively, that some similar structure characterizes other higher animals; and, finally, even as we try in imagination to lessen the intensity of centricity in its aspect of inwardness, we are still *describing*, and describing what is common to the living forms we see before us and around us here and now. We are not suggesting abstract explanatory theories nor extrapolating these to a vast and remote past, but only trying to pin down with a fitting phrase a description of a common quality of our present experienced world. Portmann's extrapolation, therefore, is less adventuresome than the Darwinian, since it is descriptive and contemporary, rather than explanatory of an inaccessible past.

To such an answer, however, it might be objected that, however plainly 'descriptive' the concepts Portmann is introducing, they exceed the scientist's brief in a way in which the Darwinian extrapolation from present to past does not do. For even if centricity is not equivalent to consciousness or sentience, it does involve a reference to inwardness – not just to a geometrical center or a midpoint of centripetal forces or the like, but to a literally *subjective* center. And such a reference, it would be alleged by the orthodox biologist, is obviously nonobjective and therefore nonscientific. The Darwinian account of evolution as the product of chance mutation and external selective control, on the other hand, invokes, its adherents believe, no concepts not compatible with strict scientific mechanism: it is 'life' in Galilean terms. And it is true, indeed, as we have seen in our discussion of display, that Portmann's reflections about living things cannot be contained within the frame of Galilean science. Rather they arise from an effort like that of the phenomenologists to overcome the abstractions of the Galilean tradition and return 'to the things themselves'.

Moreover, such a return, and in particular the acknowledgment of centricity as a pervasive character of living things, permits not only a truer account of nature in general but a reassimilation in reasonable terms of human nature to the living world. In the main tradition of modern naturalism, man must appear either as wholly alien to nature, like Galileo's 'living creature', or as reduced to meaninglessness, simply one more expression of the laws of matter in motion. The achievements of man, art, religion, legal and political institutions, science itself, *can* have no significance in a naturalistic one-level world, where there *is*, on principle, nothing but particles in a four-dimensional space-time continuum. Admittedly, if mechanism were true – if the book of the universe spread before us *were* Galilean – we should have to resign ourselves to this dismal fact: the only appropriate philosophy would be one of absurdism or of despair. Yet why should we so resign ourselves? In loyalty to the 'facts'? But the naturalist interpretation of man is itself in palpable contradiction of the 'facts' of our experience, even of living nature other than man, let alone of the massive human fact of consciousness, of the inner lives we do in fact lead. It seems intellectually justifiable, therefore, to try to revise our thinking about nature in such a way as to assimilate harmoniously to our basic view of things those aspects of our experience, so close to our deepest hopes and needs, which Galilean science must either deny or exile to some limbo of paradox and anomaly. And it is this more harmonious philosophy, this reintegration of man *into* nature, I believe, that Portmann's account of the characters of living things can help us to achieve.

One more warning, however. Portmann is not in the least denying, but only supplementing, the great storehouse of biological knowledge accumulated by a more mechanistically oriented science. In particular, he is not denying the importance of *self-maintenance* as a characteristic of organic nature. In fact, we may add self-maintenance here, parenthetically, as a third basic character of living things. For all the intricate mechanisms of reproduction, growth, metabolism, regeneration, learning behavior, and so on are, of course, rightly interpreted by biologists as devices for the preservation of the individual and the species. Many biologists, however, would make self-maintenance, or 'survival', the sole explanatory principle for all organic phenomena. And where they are in fact studying in detail particular physiological or genetic mechanisms, there is, again, nothing wrong with this. But it is in the philosophical implications of their work, when they think about them, that they go astray.

Animals must indeed be adapted to their environments if they are to survive, and such techniques for survival do form a very substantial, perhaps even the major, segment of the totality of living structure and process. But biologists, under the guiding influence of Darwinian theory, often treat living things as *nothing but* aggregates of such techniques. Life, they seem to believe, *is* adaptation. And adaptation is the fitting of the organism into its internal and external environment in such a way as to enable it to survive; in other words, it is the sum total of *means to survival.*

Now, the governing concept in this kind of thinking seems to be *utility*. Organisms are understood as aggregates of devices useful for survival. Thus a woodpecker's beak is useful for getting insects out of trees; the brooding habits of fishes and birds are useful for keeping the young alive; the colors of snails are useful in hiding them from hungry thrushes; and so on through the whole vast variety of living functions and forms. Utility, however, is a doubly relative concept. A use is a use not only *for* something – in this case, survival – but also to *someone*. Birds, fishes, snails: in every case there is an individual or a group of individuals, that is, a species, to whom the activity or pattern or process in the case is useful. How is the biologist to interpret this term of the adaptive relation, the individual woodpecker who gets his meals or the individual snail who does not get eaten by the thrush? There are two alternatives. Either the individual is itself simply useful for some further end, or it is intrinsically valuable and presupposed as such in the original predication of utility. In the former case, we have to say, as Dobzhansky has done on occasion, that the whole life of multicellular organisms is *nothing but* a device for the self-duplication and recombination of the only part of living things that can survive indefinitely, namely, the genes. In other words, a hen is an egg's way of making another egg[18] – an amusing, but also surely an absurd fashion of looking at all the achievements of all living things, including the whole of human culture as well as all the immense variety of plant and animal life. The other alternative is to acknowledge frankly that mechanisms for self-maintenance, however complex and fascinating in themselves as objects of study, are all of them instrumental in relation to the intrinsic value of the entities so maintained. And that is just what Portmann's criteria of display and centricity are meant to convey. Living things as they display themselves to the world and as they carry on, out of their own centered existence, ordered transactions with their environments, are significant in themselves. The criterion of self-maintenance, important though it is, must be counted only as a third character in subordination to the other two.

In this connection it is revealing, as Portmann points out, that the first great experimental embryologist, Wilhelm Roux, insisted on the need to refer to *Innerlichkeit*, 'inwardness', as a pervasive character of life. Roux's *Mechanics of Development* is one of the great landmarks in the advance of modern biology toward the ideal of exact experimental science. Yet this great founding father of objectivism asserted that for every characteristic living process, development, reproduction, metabolism, heredity and the like, one should really put the prefix 'self-' ahead of each such term.[19] So in self-maintenance generally, in all the processes through which the living individual achieves and preserves its being, the self that is so maintaining itself is necessarily presupposed as the significant entity, the center of action, experience, success, and failure, whose being all its manifold techniques for survival subserve.

The social life of animals, as Portmann interprets it, plainly exhibits the balance between self-maintenance and centricity as dominant features of organic life. He writes:

The observer of social life must time and again be shifting his attention between two different aspects. In one of these, society is an organization where the individual has value merely by being a member of it: it is considerably limited in possibilities of self-expression, but contributes towards achievements only realisable in society and is integrated with other individuals for purposes which can be seen only in judging the whole. In the other aspect, equally essential and indeed complementary, the individual finds complete fulfillment only in and through social life, which allows the greatest possible 'individuality' and vitality for individuals.[20]

In the first of these aspects, social life is geared to the maintenance, not, it is true, primarily of the individual, but of the species. In the selective process it is those characters which increase the probability of an organism's leaving descendants that are favored. Portmann mentions as one example of such a process the breeding habits of the Emperor penguin as compared with its near relations:

During the reproductive season, when these penguins live on land, the conditions of existence are more than forbidding. In Terre Adélie, where they were thoroughly investigated by the French Antarctic expedition, they lived in temperatures of from −13° to −31° (Fahrenheit) and endured snowstorms which rage at eighty miles an hour. According to the expedition's doctor and biologist, J. Sapin-Jaloustre (1952), 'a human being is blinded by the ice mask which forms on his face in a minute, whatever his protective clothing; his breathing is laboured, and he finds himself incapable of the slightest effort. A small area of naked human skin freezes in about 40 seconds. If he moves as much as 50 yards from his shelter, he will have lost all his physical faculties including seeing and hearing and any kind of bearing, and will never find his way back!'

In such conditions, in the dark winter of the Antarctic, the Emperor penguins breed. Besides many favourable physical dispositions, they are helped to survive by special modifications in behavior: e.g., their strikingly motionless waiting in one place, which avoids loss of energy; and perhaps also the production of sounds as means of communication instead of the very 'detailed' gesture-language used by other penguins. The conspicuous absence of many typically penguin ceremonials is bound to work in those border regions of life as a species-preserving economy in the metabolism.

These giant penguins are remarkably peaceable, without any aggressive urges, any need for a breeding couple's territory, or any hierarchical order. This is particularly useful at the beginning of a snowstorm, when the colony throng together in a narrow space, forming the so-called 'tortoise-shell', whereby they face the raging blizzard as a compact mass, exposing to it only the smallest possible area of their bodies and completely protecting within the shell the chicks which are specially endangered.

They only become combative where an egg or a chick is concerned, for in Sapin-Jaloustre's description, 'the drive to possess and nurse a chick is common to all the adults, and is so powerful and striking that Wilson called it "pathetic". Immediately a chick emerges from the skin-flap of the parent-bird's stomach, or is left by the parent bird, those penguins which have no chicks vie with each other to take charge of it. Pressing and pushing, the old birds peck dangerously, trying to thrust the chick on to their feet (on which the chicks are isolated from the ice surface of the ground). In the process the chicks are roughly handled, and their skin is often injured. Many are thus destroyed by love! Wilson tells us that they often try to escape and hide in cracks in the ice, to avoid this terrible devotion – that they prefer to starve or freeze to death!'

It thus happens that an egg has many hatchers, a chick many protectors, not only the actual parent birds. By and large, all eggs and chicks are cared for by the whole colony. 'This communal brooding and rearing of young by a succession of old birds means a considerable economy, compared with the family structure of the Adélie penguins (the smaller Antarctic species) at the same stage of reproduction. The old birds can thus afford to spend four or five times as long in the fishing which is so necessary in this climate – all the more necessary just now because they must assuage the chicks' hunger as well as their own.'[21]

We can see how such a species would have developed from a species like the King penguin, living in slightly less severe conditions. Any small mutations 'which weaken penguins' normal drive for a brooding territory and limit aggressiveness to the drive to possess an egg or a chick'[22] would clearly have selective value, and the accumulation of many such mutations would finally issue in a pattern of behavior adapted, as the Emperor penguin is, to the most severe Antarctic conditions. The advantage, finally, of the birds' adapting themselves to this rigorous habitat is the access it gives them to an abundance of food and at the same time the absence of predators in that forbidding environment.

In general, variants in territory-possession, associated as this phenomenon is with reproduction, are clearly functions of the need to preserve the species: or strictly speaking, in selective terms, they are phenomena in the absence of which the species in question would have become extinct. Yet at the same time, territory-possession is not only a device for species preservation but marks the enhancement of individual life as well.

Portmann cites, for example, R. A. Hinde's study of the social life of the great tit:

In winter the tits live in small flocks which roost and fly out foraging together. Among these flocks are the pairs which reared a brood together in the summer, sometimes keeping a little more 'on their own', though without cutting themselves off from flock life. In early spring the males especially often have moments of aggressiveness against each other. The flock slowly disbands, because its members become increasingly affected by new inner states. The males are the first to break loose, and soon they are seen pairing up in a chosen area: the flocking season is once again over. The male now looks for favourite spots in his territory where he can sing. There are a great many of them when the pair first separate from the flock, but soon he shows greater selection, and there are only a few such points, to which he is tied by habit. Finally he develops a clear preference for a particular place, which he defends fiercely. He also defends the area round it, though with less intensity the further he gets from the favourite haunt.[23]

The tits come to 'own' a brooding territory: ...

So there develops what biologists call 'ownership behavior', for it shows a striking resemblance to the human drive toward ownership: M. Meyer-Holzapfel (1952) has made a thorough comparative study of this drive in human beings and in animals. Such a comparison is far more than a piece of 'biologist's licence' to bring nearer home a bird's life. For in the case of these tits, for instance, their inner state, as revealed by external forms and behavior, can best be described by reference to familiar human experience, however uncertain the relation may be between that experience and the birds'. At any rate the tit pair 'owns' – in a real sense – a brooding territory, which can be expanded in the brooding time or shrink somewhat (as careful investigations have shown).

   This territory belongs to the pair: the favourite places make it into a specially attractive environment, in which very soon the nest will occupy a new favourite place. The individual tit is now a creature with its own private space, containing centres of particular importance. This space has its structure, like a magnetic field with lines of force which the physicist makes visible; indeed considering the progressive intensity with which it is defended, we might speak of the tension of the field and the intensity of the various points in it – as long as we manage to keep in mind the special, non-homogeneous character of a living 'magnetic field'.[24]

This kind of limited spatiality, Portmann continues, serves to enhance the individuality of the single animal:

The male tit singing in his look-out is recognized from a distance by other tits in the vicinity as a particular bird. The territory thus adds distinguishing marks to those of body and behavior, it becomes a part of the whole individual, also an expression of his capacity for self-assertion. It has its inner side in the bird's experience: he knows his territory exactly, recognises it again when he returns there. Even the cuckoo, laying her eggs in other birds' nests, is governed by familiarity with her territory: although in choosing the nests she shows a preference for those belonging to particular species of song-birds, she will still, if need be, lay eggs in any unfamiliar nest (as many observers have reported) rather than go outside the bounds of her territory.[25]

Associated with such individual ownership behavior, moreover, is the mutual value to each other of a mating pair:

In the roaming phases one bird will recognize and seek out his or her mate in their territory. Here as often, preservation of the species and enhancement of individuality go hand in hand.[26]

In a real sense, moreover, Portmann argues, a territory becomes *home*:

An inner attachment to this place develops, it becomes a 'home' associated with special values and feelings of familiarity and safety. Home is a place where through peace and security essential moods of every higher animal find most satisfaction. M. Holzapfel has shown the importance of such a home for a craterspider: if it has caught a fly in its web, its 'appetite for home' is greater than its appetite for the prey, and it does not start sucking the fly's blood till it reaches the 'soothing' atmosphere of its favourite haunt. We have seen earlier how 'home-like' are sand-wasps' sleeping places. All these examples point to the deliberate satisfaction of a drive, which fosters a positive mood within the animal.[27]

This is only one instance among many. The existence of social ranking, for example, as in the case of the stag quoted earlier, is a conspicuous example of individuality in social life. Or again, the greater ferocity of combative behavior in mammals as against birds, Portmann believes, also reflects their increased individuality.

Ownership behavior, the sense of home, focuses organisms at a center in space. But centricity is expressed in time as well as space, in rhythms of living as well as in the cherishing of a dwelling place. Portmann reminds us that, as zoo keepers know, animals can be bored, needing to 'fill time' as we do. Quite generally, he suggests, those aspects of social life which heighten individuality give meaning to life as it passes. So termites, for example, are more active, more 'lively', in the presence of their royal pair, or bees of their queen.

Nor is time simply a stretch, waiting to be filled. Ordered process, rhythm, is indeed the fundamental expression of centricity. In a paper on 'Time in the Life of the Organism', Portmann writes: 'Every form of life appears to us as a *Gestalt* with a specific development in time as well as space. Living things, like melodies, might be said to be configured time; life manifests itself as configured time'.[28] I can cite here only two examples from this remarkable paper. First, consider the duration of a minimal impression, or *moment*, in the life of a given species of organism. Experiments devised to measure such minimal durations indicate that the human moment lasts only about $\frac{1}{18}$ second; a dog's moment is roughly equivalent to ours, while the snail needs $\frac{1}{4}$ second, but the fighting fish only $\frac{1}{30}$ second to register an impression. Portmann comments :

It is hard to say what consequences such differences in the duration of the moment have for the animal's subjective experience and the structure of its world, and the scientist will do well not to say too much about the inner world of the snail or fighting fish – that is, the nature of their experience. Still, these animal experiments are of great value in that they show us once again how dependent our experience of the world is on our structure, and how significant a question it is whether this structure is adequate to the apperception of hidden reality. The infinite variety we may expect to find in this realm of 'relations with time' is suggested by motion pictures of hummingbirds in flight. In the space of $\frac{1}{8}$ second these tiny creatures, while remaining in one place, effect a complete revolution on their longitudinal axis and another on their transverse axis; at the same time they complete four wing beats. From this we may infer an amazing reaction velocity, which must in turn be reflected in the nature of the hummingbird's experience.[29]

Second, and obviously, the most dramatic integration of rhythms of living with larger temporal rhythms is the phenomenon of bird migration. Portmann, having warned of the extreme complexity of the processes involved, presents as one example of the correlation of bird life with the annual cycle the life of the arctic tern:

This delicately built white sea bird, related to the gulls, breeds in the entire Arctic Zone; its most southerly breeding grounds are situated on the German islands of the North Sea. The eggs are set down in the sand, without a nest, and after a hatching period of from 20 to 23 days the chicks slip out. Let us pause to consider for a moment that in this hidden period all the organs required for the bird's migratory habit are built up in the egg by inherited processes: the slender wings, the special character of the nerve centers, the special hormone apparatus – all are achieved by an unconscious process of development. It is not without reason that in this connection some biologists speak of developmental instincts. Just as the adult animal acts 'instinctively', so the plasm of the species 'instinctively' regulates the temporal sequence of formative processes. The species in its first state as pure plasm builds up all its organs.

For roughly one month the chick remains in the 'nest'. Here the young, intensively fed by their parents, grow quickly. At the end of this period the migratory drive awakens in consequence of hereditary processes of development. 'A hereditary drive awakens'. This is easy to say, but what riddles such a statement creates! It implies an almost inconceivably complex assembly in the organism. We conceive of this assembly primarily as occurring in the embryo of each individual bird, but we must also think of it as an assembly that came into being down through geologic ages in the process of the evolution of the species. This historical process has transformed a sedentary bird into a migratory bird. Concomitantly there must have been corresponding changes in the germ plasm, which in each new generation produce the time-responses of a migratory bird with its annual rhythm. This evolutionary process impresses us with the significance of temporal structures for the organism. And what a powerful drive it is that 'awakens' in our tern! Long before the older birds and uninfluenced by them, the fledglings start on their first journey, which is literally a world tour. The Northern European terns cross the Continent and follow the coasts of Africa. In the distant south of Africa they sometimes meet American birds of their species, which may have flown from Labrador, crossed North and Central America, and then traversed the Atlantic. Sometimes the journey continues on into the Antarctic for these terns have been definitely observed as far south as the sixty-sixth parallel. When they fly away from our latitudes, it is summer; and when they arrive in their 'winter quarters', it is again summer with its long days. For a few weeks the terns remain in the southern summer. In the southern autumn, just as our spring is coming on, the unknown urge drives them northward again, where the days have once more begun to grow longer. In the first part of May they are back in the North Sea territory, in their hereditary breeding grounds. Twice a year the little white bird effects this immense flight that carries him almost from pole to pole, from one summer to another, from a life in the long light days of the northern hemisphere to another with equally long and bright days in the southern half of our earth. Experiments with banded birds have shown for certain that this species has a longevity of at least twelve years, but for all we know they may live longer. An aging tern, then, has carried out at least twenty-four flights, each measuring almost the entire length of a meridian. We have little knowledge of the physical performance required by these journeys, and even the biologist surely reflects too little on the extraordinary inward processes of a bird engaged in migration. This transformation of many structures in the course of the interval from breeding time to migration period also belongs to the picture of the bird as a being in whom the dimension of time is extraordinarily filled with varying content, with transformations of structure and action – an extreme example of configured time.[30]

Nor is the wonder we feel in the face of such phenomena a mere 'subjective' addendum to what is 'really' a story merely of mechanisms of survival more complicated than most:

There is no doubt that bird migrations represent a solution to certain ecological problems, that they enable the bird to exchange unfavorable seasons for more propitious ones. But if we consider all the aspects of these migratory phenomena in their immense variety, it becomes increasingly plain that they surpass elementary practical needs, such as the preservation of the species. All necessity is transcended in these great formative processes, into which tellurian events are integrated as wonderfull alarm-signals for the awakening and enrichment of organic life in time. The passage of clock time, meaningless in itself, is employed for the enrichment of life. It need hardly be added that human life is a magnificent configuration of time in this same sense, offering in its successive ages ever new possibilities of development in time and hence of living riches.[31]

IV

As the passage just quoted indicates, Portmann's account of the characters of living things has important consequences also for our view of man's place in nature. Although he started from research on invertebrates, moreover, Portmann's comparative study of morphology and development led him, as long ago as 1937, to a comparative investigation of the ontogenesis of mammals and in particular of primates – an investigation which has produced important results for the theory of human nature and development.[32] Portmann's theory is closely allied to the arguments we shall be studying in later chapters; but, like them, it has, as against more widely current conceptions, revolutionary implications both for the philosophy of the biological sciences and for our philosophical self-knowledge. Again, we may see this more plainly if we approach Portmann's work in this field too by a brief historical detour.

Generalizations in the history of ideas are almost always oversimple, and doubtless the present one is no exception. Portmann's own account, for example, of men's changing views, in the past two hundred years, of our relation to the apes presents fascinating variations of which I can here take no account. But I think that speaking very roughly and generally we can distinguish four different views of man's place in the natural world. There was, first, the popular medieval conception of man as the microcosm set firmly within the macrocosm which he mirrored. Here we had our place, a little lower than the angels, in a neat, many-leveled, statically ordered cosmos. The seventeenth-century intellectual revolution, bursting open this tight little order, left us, in the terms Descartes handed down for our forebears, the only spiritual beings in an otherwise machine-like universe. There were mathematicians' minds, and the matter spread out in space for their contemplation, and, except for the God who made them both, no more. Man himself, of course, was a strange hybrid of the two finite substances, minds and bodies, a delicate and, as it turned out, unstable synthesis. For its time, however, Descartes's vision was definitive, and Western thinkers since then have had to start from the heritage he left them. But starting from the Cartesian concept of man and nature, there seem to be but two alternatives left. As men's confidence in God and in their own immortal souls dwindled, they could either deny the existence of anything mindlike in the universe – the issue accepted by the recurrent materialisms of the last three centuries – or take the works of human reason – as Hobbes did three hundred years ago – as *artifacts*. In other words, they could see nature, including all animal life, from the point of view of a pure materialism, but contrast with it the *cultural* life of man, which is not born but *made*. In the nineteenth century, the triumph of Darwinism seemed to many the triumph of materialism pure and simple. In its twentieth-century version, however, Darwinian theory has not only become, through its mating with genetics, much more subtle and statistical in its interpretation of natural change. In the writings of Huxley and Dobzhansky especially, it has also placed over against the mutation-selection model of all evolution up to Homo sapiens a new brand of evolution, *cultural* evolution, characteristic of human history (and prehistory) alone. This is change through tradition, through teaching and learning, through religion, arts, and science, through all the articulate devices of human imagination and inventiveness.

Now, of these four views, Portmann's position is in harmony with the last, but with two significant differences. First, as I have already pointed out, he is not convinced that the technique of evolution over the whole stretch of life's existence is explained by the concepts of micromutation and selection. While, of course, accepting evolution as a fact, and a fact of very great significance for biology and for the study of man, he prefers to leave the great question of ultimate origins as for the present unanswered and perhaps even in some ways unanswerable. Second – and this is the main point in the present context – his comparative study of ontogenesis and postembryonic development has led him to modify the dichotomy of nature and culture and to emphasize the rootedness of man's social life in his *biological* nature. It is this integration of man as a uniquely cultural animal into a solid biological foundation that enables us, through Portmann's eyes, to see ourselves once more *both* as human beings and as at home in the natural world: because we are *biologically* formed to be *cultural* animals.

The context within which Portmann develops this thesis is an investigation of the comparative development of mammals. Animals can be divided as to their postnatal development into two classes: nidicolous species, that is, those whose young are born relatively immature and remain for a period 'in the nest' before venturing independently into the world; and nidifugous species, that is, those whose young are relatively free-moving and even free-feeding at birth or shortly thereafter. Among birds the more primitive species are nidifugous; it is the more highly developed types that produce helpless and sightless young needing a protracted course of parental care to prepare them for the adventure of living. Among mammals, evolution has taken the opposite course: the young of relatively primitive species are nest-dependent, while the newly born offspring of higher mammals are ready with their first breath to disport themselves on their own four feet. One has only to compare newborn kittens with a newborn foal to see the difference. Associated with this striking difference in mobility are two other important contrasts: the young of nidicolous species are usually born in large litters and have their sense organs relatively undeveloped at birth, while nidifugous young are fewer in number per parent but as wide-awake to their surroundings at birth as they will be in maturity. A comparison of the ontogeny of the two types shows the evolutionary process involved in the last of these contrasts. In the embryos of nidicolous species, the eyelids are formed early, then close, and birth occurs before they reopen. In nidifugous species, the stage of lid-closing, now meaningless in the life of the organism, is retained, but birth is postponed till later when the eyes have once more opened.

Now, where does man fit into this classification? Nowhere, it seems. We are born helpless, but with our eyes and ears wide open. Even though a baby does not focus on objects for the first few weeks, he does respond to colors very soon indeed and to sounds as well. He takes from the first moment a powerful sight of notice. In other words, we have retained the sensory characteristics of the higher mammals (as well as the small number in our litters!), but have reverted in other respects to an earlier pattern in our infantile way of life. We have become *secondarily nidicolous*. Thus the evolution of mammals as a whole shows a change of course: from nidicolous to nidifugous, and then back again to a new kind of nidicolous state. That this is so is strikingly confirmed in comparative studies of postembryonic development by Portmann's associates at Basel. In nidicolous animals, as one might expect from the condition of their sense organs at birth, the brain has to develop during life to from 8 to 10 times its size at birth; while for nidifugous species this figure is on the average 1.5 to 2.5 – somewhere around 2. This holds even for our nearest kin, the anthropoid apes; they are still nidifugous. For even though the baby clings more closely to its mother than do other nidifugous animals, it does so, so to speak, by its own volition: it is free-moving. Moreover, the rate of growth of the brain from birth to maturity is still the 'normal' one of about 2. With man, however, this figure has sharply altered: his brain has to develop from birth to maturity to *four* times the natal size: a clear reversal of the evolutionary trend. Moreover, the proportions of a human infant are much further from those of the adult than are the proportions of a newborn ape from the mature individual. But at the same time the human infant is much heavier at birth than any newborn anthropoid, almost twice as heavy, even among peoples who are comparatively slight in build when full-grown. On all these counts, our pattern of development is unique among mammals.

What is the significance of all this? Let us look at the comparison a little differently. Instead of comparing the newborn young of each species, let us ask: when does man reach, in behavior and in physical proportions, a stage of development comparable to that of the newborn young of other highly evolved mammals? It is usually at about a year that the child takes his first steps and so assumes the upright posture characteristic of his kind, and at this age also his build has come much closer in its proportions to its eventual adult shape. At about twelve months, also, he begins to act like a person: that is, he begins to speak, and he begins to perform what we recognize as voluntary, responsible actions. Taking the pattern of nidifugous development as our model, therefore, we see that man is born prematurely and achieves the status of an ordinary nidifuge at about twelve months of age. This hypothesis is further confirmed by a comparison of the rate of growth of the human baby and the young of closely related species. Thus the apes show a steady development from birth to maturity, while the human baby changes with an amazing speed, in fact, a rate of growth comparable to that of the embryo, up to a year, then slows down to a rate similar to that of 'normal' postembryonic development. If we look at ourselves as mammals among other mammals, therefore, we see that we should be born twelve months later than we are: it is at that period that we emerge like other advanced mammals into the world of our peculiar kind, that we take on full human nature.

But what is this 'full human nature'? Portmann has mentioned its three chief characters: upright posture, speech, and rational action. Now, all these have to be *learned* by the infant in its first months through contact with human adults or in particular with one adult, his mother or foster mother. Of course parental care, and social life in general, are essential, in a host of intricate and marvelous ways, to a host of other species as well. But nowhere else in living nature on this planet does this pattern of premature sociality occur. Looking at it from the perspective of evolution and of comparative development, Portmann calls this unique period of postembryonic, yet still embryonic, growth the period of *social gestation*. To take its place alongside a newborn orang or chimp, at an analogous level of development toward the adult form, in other words, the human being demands not only nine months in the physical uterus of the mother but a further twelve months in the social uterus of maternal care. We must be careful also, Portmann warns us, not to take this lately developed pattern in evolution as a mere addendum to an otherwise 'ordinary' mammalian ontogenesis: our unique pattern of development is not an 'afterthought' tacked onto a standard embryogenesis. The human attitudes and endowments which we must acquire in infancy are prepared for very early indeed in embryonic growth: thus the first preparation for the upright posture, in the development of the pelvis, occurs in the second month of the foetus's growth. The preparation for the acquisition of speech, moreover, involves glottal structures very strikingly and thoroughly different from those of any other species. And the huge size of our infants relatively to the young of apes – born more 'mature' but very much smaller – is probably related, Portmann conjectures, to the immense development of the brain necessary for the achievement of human rationality – a development which begins, again, very early indeed in ontogenesis. In short, the whole biological development of a typical mammal has been rewritten in our case in a new key: the whole structure of the embryo, the whole rhythm of growth, is directed, from first to last, to the emergence of a culture-dwelling animal – an animal not bound within a predetermined ecological niche like the tern or the stag or the dragonfly or even the chimpanzee, but, in its very tissues and organs and aptitudes, born to be *open to its world*, to be able to accept responsibility, to make its own the traditions of a historical past and to remake them into an unforeseeable future.

Not only the first year of life, moreover, but the further pattern of human development testifies to the fact that, even in their physiological foundations, our lives are so ordered as to facilitate social learning, the perpetuation and modification from one generation to another of traditional lore. The postponement of sexual maturity to a date relatively later than is characteristic of other species permits a long period of apprenticeship and the gradual assumption of responsible adulthood. And at the other end of the time span, the prolonged and gentle slope of human senescence provides opportunities for rising generations to profit from the richer experience of the old and wise. Thus at every level and at every stage of our existence we live out the uniquely flexible, uniquely creative pattern of configured time that is man.

From this vantage point, then, we can see at one and the same time how wonderfully various are the modes of life of many, many kinds of living beings and also how wonderful is our own scheme of life, unique in its intricate order, as every specific life history is, yet unique among all these uniquenesses in its 'natural artificiality', in the dimension of cultural evolution which is not only added to, but enmeshed in and expressed through, every level of the many levels of order that converge to produce a human life.

This is a zoologist's contribution to our understanding of ourselves, founded, not on a mere wish to overcome our alienation from nature, but on careful morphological and developmental research over many years and with the help of many hands. But its implications for what is sometimes called 'philosophical anthropology' should be clear. Without suggesting a sentimental reversion to a once stable but now vanished 'scale of nature', Portmann's findings nevertheless allow us to accept a secure place *within* the natural world. Yet a place with a difference, too. We need no longer try, absurdly, to see all the achievements of men's minds – art, science, and religion – as epiphenomena to molecules in motion, or, alternatively, to set them, as mere conventions, over against the single level of change that is thought to constitute the balance of the organic world. Admitting from the start the intrinsic significance of life itself in a thousand thousand forms, we can acknowledge too the deep-reaching and far-flowing consequences of our own natural-and-cultural, biologically determinate, *and* traditional form of life.

There is one more point about Portmann's view of man which I ought finally to mention – a point which I have touched on already at least indirectly in introducing the concept of display.[33] Each of us lives, I pointed out there, in a primary life world, out of which and within which the world of science, or of any other highly articulate discipline, develops. Portmann has repeatedly emphasized the importance of giving due heed to *both* these aspects of our lives. By the world of primary experience he means, however, something more inclusive than the concept of a 'life world', as I introduced it earlier, may at first suggest. The world in which, from infancy, we come to live, and the human world shared by members of all cultures, does, of course, include the surface of experience, the colors, the sounds, the rhythms of movement that confront us on all sides. But it includes also our feelings, our desires, our dreams, the creative aspirations of artists, the vision of saints and prophets, even the delusions of the insane. No single term can adequately characterize this whole range of primary experience; perhaps we can still speak here of the 'life world' if we remember that it is more than the plain, open order of 'common sense' to which we are referring. Such a life world, then, with all its opacities and ambiguities, stands in contrast with the limited but lucid sphere governed by the operations of the intellect – and that means, in our culture, by the operations of science and technology. Human nature comprises both and can dispense with neither.[34]

I can perhaps point to the nub of Portmann's dichotomy by returning once more to Galileo's polemic in *The Assayer*. Galileo reviles the unhappy Sarsi for quoting poetry at him; and this was, indeed, on Sarsi's part, an irrelevant answer to a scientific argument. But it is the grounds of Galileo's objection that are revealing. It has been truly said, he remarks, 'that nature takes no delight in poetry'. With this truth, he goes on, Sarsi seems to be unacquainted: 'He seems not to know that fables and fictions are in a way essential to poetry, which could not exist without them, while any sort of falsehood is so abhorrent to nature that it is as absent there as darkness is in light.'[35] This is indeed the perspective of modern objectivists. As the 'mathematical language', the instrument of an impersonal reason, is seen as the sole medium of truth and light, so poetry has come to be mere taletelling, at best invention and entertainment, at worst obscurity and untruth. And it is not only poetry in the narrow sense, the craft of making verses, that is here exiled from reality but the whole work of the imagination: myth and metaphor, dream and prophecy. In the bare mathematical bones of nature there is truth; all else is illusion. Yet that 'all else' includes the very roots of our being, and we forget them at our peril. Indeed, even the scientist himself, no timeless, placeless spirit, derives from his aspirations and imaginings, his dreams and disappointments, the sustenance, the very existence, of his enterprise. Galileo himself, passionately evicting poetry from nature, has evoked the ancient metaphor of darkness and light. Portmann uses the same image to adjure us: 'In a world in which the apparatus of gleaming glass, the bright research laboratories and men and women in white have acquired an almost symbolic value, we must look again and see how great is the darkness out of which the light that fills the bright spaces of the intellect wells up'.[36] We must try, in other words, to achieve anew a whole vision of our nature – a revision which by its very character research alone is unable to provide. And Portmann is not speaking here, remember, as a writer or artist envious or ignorant of the achievements of science and technology. He is speaking as a scientist – a scientist looking beyond science to the wider, if obscurer, problems of our lives. So he says:

I myself work every day, through research and teaching, at the advancement of knowledge, and it is out of my own inner impulse that I have chosen this work. Hence perhaps my readers too will see the demand of the present hours more clearly if, out of the very passion for research, I emphasize the inevitable narrowness of every image of man that is formed through natural science alone, that does not draw its powers from all the sources of man's being.[37]

NOTES

[1] K. Z. Lorenz, in *Physiological Mechanisms in Animal Behavior, Symposia of the Society for Experimental Biology* IV, Cambridge University Press, Cambridge, 1950, p. 235.
[2] A. Portmann, *Animals as Social Beings* (trans. by O. Coburn), Hutchinson, London, 1961, pp. 108–109.
[3] A. Portmann, *New Paths in Biology* (trans. by A. Pomerans), Harper and Row, New York, 1964; *Neue Wege der Biologie*, R. Piper, Munich, 1960. The translation is inadequate at some crucial philosophical points, especially in the rendering of *eigentlich* and *uneigentlich* and also in the omission of some important passages.
[4] *Discoveries and Opinions of Galileo* (ed. and trans. by S. Drake), Doubleday, Garden City, New York, 1957, pp. 237–238. Galileo Galilei, *Il Saggiatore* VI, Edizione Nationale, Florence, 1965, p. 232.
[5] This thesis has been most clearly articulated and defended in Husserl's late work. Husserl's thesis is paralleled in Portmann's own work in the distinction he makes between *Welterleben* and *Weltwissen*, between our primary experience of the world and the intellectual understanding of it we acquire through education, and in particular through science.
[6] *Discoveries*, p. 272; *Il Saggiatore*, pp. 347–348.

[7] See A. Portmann, 'Der biologische Beitrag zu einem neuen Bild des Menschen', *Eranos Jahrbuch* **XXVIII** (1959), 459–492, especially 466–472.

T. Holtsmark, 'Goethe and the Phenomenon of Color', in *The Anatomy of Knowledge* (ed. by M. Grene), University of Massachusetts, Amherst, 1968, pp. 47–71.

E. P. Land, 'Color Vision and the Natural Image', *Proceedings of the National Academy of Sciences USA* **XLV** (1959), 115–129. 636–644; 'The Retinex', *American Scientist* **LII**, No. 2 (1964), 247–264; 'Color in the Natural Image', *Proceedings of the Royal Institution of Great Britain* **XXXIX**, No. 176 (1962), 1–15.

[8] A. Portmann, 'Gestaltung als Lebensvorgang', *Eranos Jahrbuch* **XXIX** (1960), 359. I shall return to this point again in connection with Portmann's account of the social life of animals.

[9] A. Portmann, *Animal Camouflage* (trans. by A. Pomerans), University of Michigan Press, Ann Arbor, 1959.

[10] *Neue Wege*, p. 148.

[11] See *Animal Camouflage*.

[12] Trans. by H. Lucas (unpublished).

[13] Quoted in *Animals as Social Beings*, pp. 182–183.

[14] See H. Hediger, *Wild Animals in Captivity*, Dover, New York, 1964.

[15] *Neue Wege*, p. 225.

[16] N. Tinbergen, *Social Behavior in Animals*, Wiley, New York, 1953. D. Lack, *The Life of the Robin*, H. F. and G. Witherby, London, 1946. K. Z. Lorenz, *King Solomon's Ring* (trans. by M. K. Wilson), Crowell, New York, 1952.

[17] *Animals as Social Beings*, p. 26.

[18] T. Dobzhansky, *The Biological Basis of Human Freedom*, Columbia University Press, New York, 1956, p. 17.

[19] See *Neue Wege*, p. 59.

[20] *Animals as Social Beings*, p. 160.

[21] *Ibid.*, pp. 170–171.

[22] *Ibid.*

[23] *Ibid.*, pp. 175–176.

[24] *Ibid.*, p. 176.

[25] *Ibid.*, p. 177.

[26] *Ibid.*

[27] *Ibid.*

[28] 'Time in the Life of the Organism' in *Man and Time III*, Princeton University Press, Princeton, 1957, p. 312.

[29] *Ibid.*, p. 311.

[30] *Ibid.*, pp. 317–319.

[31] *Ibid.*, p. 320.

[32] See especially A. Portmann, *Zoologie und das neue Bild des Menschen*, Rowohlt, Hamburg, 1956; Die Stellung des Menschen in der Natur'', in *Handbuch der Biologie*, IX, No. 19, Hachfeld, Constance, 1961, p. 437–460; *Vom Ursprung des Menschen*, Reinhardt, Basel, 1958. Cf. also A. Portmann, 'Beyond Darwinism', *Commentary*, **XL** (1965) 31–41.

[33] See, for example, 'Der biologische Beitrag', pp. 459–492, 'Welterleben und Weltwissen', in *Erziehung und Wirklichkeit*, R. Oldenbourg, Munich, 1959.

[34] Portmann has put this contrast sometimes as that of experiencing the world (*Welterleben*) and knowing the world (*Weltwissen*), sometimes as that of the Ptolemean and the Copernican in each of us. I have not been able to find two English phrases to

carry smoothly the connotations of *Welterleben* and *Weltwissen*, and I am not quite happy about the other pair. As earth-bound creatures, tied to the history of our species and the traditions of our community, we may be said to be Ptolemeans; as free-ranging analytical intellects, to be Copernicans. And perhaps, since Kant, 'Copernican revolution' has indeed come to mean any basic change of perspective which reverses a more naïve or natural point of view. Yet Copernicus himself was still so deeply imbued with neo-Pythagorean mysticism, and Ptolemy himself so sophisticated a mathematician, that I prefer to approach the contrast without using these particular names.

[35] *Discoveries*, p. 238.
[36] *Nationalzeitung*, Basel, February 16, 1964.
[37] *Ibid.*

CHAPTER XVII

THE CHARACTERS OF LIVING THINGS

II: *The Phenomenology of Erwin Straus*

I

Straus's *Vom Sinn der Sinne* was published in 1935, Kurt Goldstein's *Der Aufbau des Organismus* had been published the previous year, E. Minkowski's *Le Temps Vécu* in 1933, Helmuth Plessner's *Die Stufen des Organischen und der Mensch* in 1928. In the European literature of philosophical anthropology and, more broadly, of philosophical biology, all these works have exerted a profound influence. In particular, when one reads this literature, the phrase '*das schöne Buch von E. Straus*' becomes almost a fixed epithet like 'swift-foot Achilles' or 'the incomparable Mr. Newton'. Even Buytendijk, who takes issue with Straus's theory of sensing, pays tribute to – and borrows a great deal from – particular Strausian themes.

Yet the philosophical influence of all these writers outside Europe has been negligible (and this despite the fact that two of them, Goldstein and Straus, have been active in this country since the thirties). True, their work is now beginning to bear fruit, indirectly, through the influence of Merleau-Ponty, who – also in part indirectly – owed much to their conceptual reforms. But outside a very small circle their own work is still ignored. This is doubtless due in part to problems of semantics. Minkowski and Plessner are still untranslated; Goldstein's major work, which appeared in an execrable 'translation' in 1939, remains, understandably, unread. A similar fate seems so far to have overtaken Straus's book. A second German edition appeared in 1956, and this was published under the unfortunate (and ungrammatical) title, *The Primary World of Senses*, in 1963.[1] It is by no means an adequate translation, but that hardly justifies its total neglect. It is to be hoped that the more recent collection of essays, *Phenomenological Psychology*,[2] will find a readier reception and will thus call attention to the earlier, and basic, work.

Linguistic difficulties, however, are not the major obstacle to the acceptance of these authors in America and England. As Straus himself emphasizes, the chief hurdle is metaphysical. The scientific study of behavior has been hamstrung since its inception by its Cartesian heritage: a thesis which not only Plessner, Buytendijk, Merleau-Ponty, and others have argued, but which Straus himself defends in a detailed study of the presuppositions of and contradictions inherent in S-R theory in particular and objectivist psychology in general. Putting Ryle's 'ghost in the machine' in other words – and supporting his accusation with a detailed critique both of Pavlov's original work and of its later sequels – he refers to the 'one-and-a-halfism' of psychological theory, in which an impotent epiphenomenal consciousness floats over the allegedly effective, and purely mechanical, neurological processes which are thought to determine wholly the course of action.

There has been of late, admittedly, a widespread movement among philosophers of empiricist cast to rectify the Humean-Pavlovian 'causal'-associative theory of action. See, for example, among many others, such writers as Melden, MacIntyre, or Hart. They share with Straus the aim of diverting attention from a misleadingly abstract 'scientific' construct to the massive facts of action as we perform, and hence pragmatically 'know' it. But there are two major differences between their approach and that of Straus, which make their arguments, in my view, the close of a tradition and a rejection of philosophizing, a caution, like Hume's 'to live at ease ever after', and Straus's, in contrast, a beginning of a novel tradition and a *Wegweiser* to a new and more fruitful style of philosophical thought. The first is that Straus's reflections are founded, as I have already said, massively and concretely in a critique of the opposing theory, not only in a general and philosophical vein, but through a particular examination of the paradoxes to which experimental psychology itself gives rise.[3] Secondly, and much more fundamentally, Straus tries to lead us, not to complacent games of backgammon, but to critical and constructive reflection on the nature and presuppositions of action, and, more generally, of what it means to be, both in perception *and* motility, in sensing *and* performing, 'an experiencing being'. His aim, in short, is, as he says, to vindicate 'the unwritten constitution of everyday life'.[4]

Perhaps Melden, for example, would protest that in *Free Action* he has done the same. But look at the climax of his argument. Rejecting the causal theory and yet denying that one can seek the characteristics of a person or of an action as such, he remarks: 'One can say that one wants to know what these are, but one can also bark at the moon'.[5]

This is, as Austin, Ryle, and others also habitually do, laboriously to approach a philosophical problem, only to turn one's back on it when it comes plainly into view. 'Don't shoot when you do see the whites of their eyes'! seems to be these writers' tactic. However, not only do some of us driven by the motives to speculation which Kant has told us are inescapable, though hopeless, drives of the human mind, want to bark at the moon. We believe that, in view of the conceptual and moral inadequacies of the still powerful, alternative behaviorist position, we ought to undertake this (to empiricists) seemingly fruitless task: that we ought to seek, in fundamental reflection, to renew speculative daring and to justify the fundamental beliefs about man and nature which, outside the psychological laboratory – and, as Straus demonstrates, even inside it – we still irresistibly hold. We believe, in short, in line with Burtt's critique of Strawson, that metaphysics should set itself not only a descriptive, but a revisionary task.[6]

Straus's work furnishes, not such a metaphysic, but a solid, sound, and richly fruitful prolegomenon to it. There are, as we shall see, metaphysical (and epistemological) issues on which he takes an ambiguous and even inconsistent stand, but he does raise basic problems and suggests, at least, lines along which some of them may be answered.

The range of Straus's insights I cannot hope to exhaust in such an introductory essay. My principal hope is an exhortatory one: that I may persuade some already sympathetic readers, i.e., readers already unhappy about the philosophical implications of orthodox theories of mind and already open to the possibility of speculative thought, to read his work. I shall try to implement this aim here: first, by indicating the basic distinctions underlying Straus's phenomenology of experience; secondly, by referring to some of his detailed studies, where, as seldom happens, the experience of a clinical psychiatrist and the reflections of a philosophical mind meet to open new conceptual avenues to otherwise neglected problems; and thirdly, by suggesting, in the course of this exposition, some of the philosophical perspectives which, on the one hand, his analysis opens up, and which, on the other, it sometimes seems to blur.

II

As a psychiatrist, Straus starts from a critique of psychological theory, and in particular of Pavlov. In its underlying metaphysic, he argues, S-R theory is still Cartesian, and in particular it still rests on the 'time-atomism' which cuts off Cartesian thought from the original structure of experienced time. I shall return to some of these metaphysical implications of psychology later; but the distinction I want to start with is the one to which Straus frequently returns, between *stimuli* and *objects*. This distinction entails a fundamental change in epistemology and, through it, in metaphysics.

Within the tradition of empiricism it has been held that we must penetrate to the ultimate units of experience as presented: sensa, atomic facts, or whatever we call them. Everything else has been held to be super-structure upon these primary givens. Straus reverses this fundamental conception: it is *objects* that are primary, and 'stimuli', the 'scientific' equivalents of atomic facts, that are abstracted from them. True, the ur-empiricists, Berkeley and Hume, were looking for presented, not constructed, data. Berkeley really thought he could find a minimal visible, a minimal audible, and that it was judgment which, with the help of God, built objects out of them. Hume too, *pace* the missing shade of blue, thought he had found such minimal givens. But we can see now that this search was dictated, even for Berkeley, by an epistemological, if not a metaphysical, atomism: by the acceptance of a reduced and purely passive Cartesian *simple* which *must* be the isolable unit of knowledge, the building brick out of which an aggregate equivalent to 'experience' could be constructed. With the further reduction of mind to ghost or less, however, these singulars of 'experience' have become the presumably separate impacts of isolable physical events upon separate nerve denings. The physiological model of action takes over. So, says Boring, 'to understand man the doer, we must understand his nervous system, for upon it his actions depend.'[7] But if no one has ever discovered Berkeleyan minimum sensibles or Wittgensteinian atomic facts, so much the less can an observer discover in his own experience their presumed physiological equivalents.

To begin with, stimuli exist only as pure physical events, 'unstained by any secondary qualities': they are neither audible, tangible nor visible.[8] Indeed, they have no existence independent of the nervous system that has received them. But the objects with which we deal in our ordinary experience are not of this refined and reduced character. 'The wall over there, the writing pad, the pen and ink': these are things I confront, but they are not 'stimuli'. True, Straus admits, the light, once reflected from the wall or the paper, might be described as a stimulus, but only after it has reached the optical receptors, a process which I may postulate but have never experienced. And the experimenter, for all his alleged sophistication, is in no different situation in confronting his rats, Skinner boxes, or what you will. How can he set out to work, not with rats and apparatuses, but with the hidden, hypothetical events of which, on his own theory, his own behavior consists? Indeed, he cannot – for the temporal relation between objects and the experimenter's action upon them is precisely the converse of that between stimuli and responses:

Because stimulation precedes response, nobody can handle, nobody can manipulate stimuli; they are out of reach. I as an experiencing being may stretch out my hand toward the pen on my desk; a motor response cannot be directed to optical stimuli already received in the past.[9]

Thus 'stimuli' can be neither *observed* nor *manipulated*.[10] Nor, since they are, by definition, events *in* the receiving organism, can they be shared, in a psychological experiment, between 'subject' and experimenter:

Those stimuli which provoke responses in the experimental animal never reach the eyes or ears of the observer. Stimuli cannot be shared by two organisms.[11]

On all these grounds, then, it appears that 'stimulus' and 'object' are by no means synonymous terms: 'stimuli are constructs, never immediate objects of experience'.[12]

An experiencing being moves, on the contrary, not among stimuli, a life which even Pavlov's dogs sometimes resisted, but in a surrounding field within which things approach it and it approaches things. Its basic experience is of what Straus calls an 'I-Allon' relation – where 'Allon' means not just other persons but the organized totality of objects within which the living being moves. The stimuli with which the scientific observer operates, on the contrary, are not Alla to an experiencing being; they are not objects at all, but highly abstract constructs which split apart and render unintelligible this primary relation.

If he were consistent, therefore, the S-R theorist ought to legislate *himself* out of existence. If he too is a congeries of physical stimuli and physiological responses, he ought not to pretend that he can observe the behavior of his subjects or manipulate his appratus. But, 'spellbound by the magic of a venerated metaphysic' he neglects to notice his own inconsistency.[13] He continues, a necessary exception to his own alleged cosmology, to act as an experiencing being among objects which he can observe, manipulate and share. Thus experimental psychology, in its classical form, rests on a fundamental inconsistency.

That is not to say, however, that causal investigation in terms of S-R theory is useless; but it should be put into its due place *within* a theory of living beings and their experience. Causal analysis investigates the necessary, but not the sufficient, conditions of sensory experience. Thus 'the stimulus delimits what will be seen and determines that it will be seen.'[14] 'Intentionality of vision and causality of seeing', Straus insists, 'are fully reconcilable'.[15] But causal analysis can operate consistently only *within* the all-embracing I-Allon relation; it cannot put Humpty-Dumpty together again out of senseless fragments. Indeed, Straus argues, traditional theory has never even embarked upon its proper subject, which it has eliminated before it even begins:

Instead, traditional theory interprets sensory experience as the result of the interaction of two bodies. The sender produces in the recipient a phantom-like perception. Such sensory data do not belong to the outside world of objects; they represent them – it remains unclear how and to whom. In any case, while they represent an object, they are cut off from the outside world; they are said to be subjective; they belong to the subject. Since perceptions represent the outside world, but belong to the subject, they cannot be an object of action. Everyone, supposedly, carries in his consciousness a private gallery of such shadow images. The collector himself is not a part of his collection. He owns it, he has it; he does not belong to it. Sensory experience in classical theory does not include the experiencing being. The content of sensory experience is reduced to the appearance of a more or less distorted replica, a counterfeit of the outside world. Seeing is acknowledged as a physiological process, but not as the relation of seeing beings to things seen. The relation I-Allon is slashed. The Allon alone is left, but in a profoundly mutilated form. Perceptions are many; they follow one another in the order of objective time. They do not belong together in a meaningful context; they stick together through synaptic welding. Positivism from Hume to Skinner preaches the gospel that sense is repeated nonsense.[16]

In place of this absurd situation, Straus proposes that we abandon the myth of 'sensa' or 'impressions' and examine the process through which we do in fact find ourselves in contact, through sensory channels, with the world around us, a process which he calls sensing (*Empfinden*) in contrast to perception.

Sensing, Straus stated in an essay of 1930,[17] embraces both *gnostic* and *pathic* aspects – where 'gnostic' denotes the primitive forerunner of the cognitive, and 'pathic' 'the immediate communication we have with things on the basis of their changing mode of sensory givenness'.[18] But it is primarily pathic. For sensing is a way of receiving not merely atomic cues from a prompting environment (*Umgebung*), but moods, as well as lines, from fellow actors, from the stage set, from the whole ensemble within which we discover other things and persons, through cooperation (including rebellion against and revulsion from) with whom we develop our own roles as actors in a scene. My metaphor limps, for the *Umfeld* Straus is speaking of is no artifact. It is the living nature in which every animal is immersed and through contact with which it expresses its style of living and of being. From this encompassing Allon, we, and also to some degree other higher animals, have abstracted a perceptible world of stable, manipulable, and (for us at least) intelligible kinds of objects. But our primary, pathic sensing is the ground on which alone the chiefly gnostic achievements of perception can develop.

This, like every fundamental conceptual reform, is a difficult distinction to assimilate. I shall return later to the problems raised by Straus's cognitive theory of perception; but for the moment let me try to elucidate the concept of sensing by making a number of comparisons with more familiar philosophical theories of experience, and in particular of sensory experience.

Straus presents his theory as a reversion from Cartesianism to the open vision of an experienced world. Let us begin, however, not with the subtly ambiguous Cartesian, but with the more crudely ambivalent Lockean position, from which, through refinement and excision, traditional empiricism has developed. The real givens of sense for Locke were single separable ideas, pieces of mental content whose originals were resident in some material, but unknown, *X*, and which the mind could manipulate, abstract from, and return to for its intuitive knowledge, such as it was, of a 'real' world. But this was the real world of a good Newtonian, of a founding father of the Royal Society, for whom the 'corpuscular philosophy' had proved a liberation from the dead tags of scholasticism and useless Latin learning. It seemed common-sense because of the nonsense it had abandoned and because of the prospects for natural philosophy which it appeared, at first sight, to permit. Yet, except for our primary experience of motion, solidity, and weight, the experienced surface of the world, its colors, smells, and sounds, were held to be but secondary: mere expressions of as yet unknown processes in those underlying X's. But are our apprehensions of color, smell, or sound, in fact less experienced, less substantive in their shaping of our ongoing experience than our apprehensions of motions and shapes and of the solid resistance of bodies to touch?

Travel from the Sacramento valley an hour's drive into the snow country of the Sierras: this experience is not described, in its immediate quality as experience, by substituting for one congeries of 'ideas of sensation' another that includes more white bits and fewer gray and red and yellow, or more 'silences' – what are they in empiricist terms? – and fewer loud noises. The traveller finds himself *in* a different medium, his very being changes. Or compare, similarly, the difference between the impact on the ear – or rather on the person, through the ear – of a grating Bob Dylan ballad with the experienced effect of a well-performed baroque concerto. The first slaps at us; the second surrounds us: it places us, auditorily, in a different *landscape*. The summative plus representative plus hedonistic account of these experiences triply falsifies them. It forces us to reduce what is comprehensive to an alleged (not an experienced) atomic base, and to this it adds, on the one hand, an invented intellectual superstructure (like Helmholtz's 'unconscious inference') and, on the other, a 'merely subjective' feeling tone. But why must we insist that we infer, unconsciously or otherwise, the snow from the whiteness, the music from its discrete sounds? We have known since Ehrenfels that melodies are not experienced like that and since Wertheimer that neither are visual objects, let alone whole landscapes. Let us leave these cramped constructions and return, Straus exclaims with Husserl (though in a different spirit), to 'the things themselves'.

Not that that is easy or infallible. Straus performs no 'reduction' and hence possesses, or can claim to possess, no stringent method. On the one hand this reviewer agrees with Straus in suspecting the overintellectual, and, indeed, the presumptuous, method of phenomenological reduction. Yet admittedly, if we stay *in* the 'life-world', as Straus urges us to do, rather than bracketing its existence to seek its 'pure' structure, we risk substituting what are our own personally slanted descriptions, however universally we intend them, for what is truly universal.[19] But that is our condition, and we do better to face it than to substitute for the rich multidimensionality of our experience, both shared and single, some skeletal surrogate, whether in an abstract Lockean reconstruction of sensation or in the Husserlian highroad to 'transcendental subjectivity'. What Straus is seeking is a reinstatement of the life-world as lived, of that comprehensive horizon of earthbound experience which Descartes had distilled to a geometer's two-halved paradise: an improverishment which is still the starting point as much of Husserl's enterprise (whose *Cartesian Meditations* are not for nothing so entitled) as of Locke's *Essay*.

In its starting point, Straus's phenomenological enterprise most resembles that of Heidegger, and it may be useful, therefore, to compare his approach briefly with the *Daseinsanalyse* of *Sein und Zeit*. Straus's I-Allon relation is a variant, if you like, of Heidegger's being-in-the-world. A *Dasein* is an experiencing being, and the I-Allon description, like the first part of *Sein und Zeit*, turns its back on the divided world of *res cogitans* and *res extensa* to plunge directly into the inspection of human being in its entirety. The differences, however, are also significant for philosophy, as well, I should guess, as for psychiatry. Two points should be mentioned. First, despite one passing reference to a possibility of authentic *Fürsorge*,[20] *Mitsein* characterizes Heideggerian *Dasein* only on the level of forfeiture. The authentic existent who emerges in the second part of Heidegger's argument is, despite the bow to national destiny later developed in the chauvinistic vision of the *Introduction to Metaphysics*, the one, rare existential hero, utterly cut off in his true being from the contemptible '*das Man*' from whose distracting influence the rest of us never escape. In Heidegger's authentic existence there is no Allon. Not so for Straus. As a psychiatrist, and a humane psychiatrist, he looks with equal openness at the general character of all. One is reminded, in reading his psychiatric essays, of Jaspers' adjuration to the physician, in his *General Psychopathology*, that he must confront the patient as a person. This Straus accomplishes by seeing the pathology of the individual case as a rending in one way or another of the seamless whole that constitutes the norm of everyday life. Thus the I-Allon relation stands as the paradigm which becomes, in illness, split apart or deformed. For Heidegger, on the contrary, as for his hero Nietzsche, the norm is the deformity, and only the rare soul who hates and repels the norm can be said to live authentically. Indeed, although it is of course an *ad hominem* argument, it is perhaps not wholly irrelevant to point out that the two geniuses Heidegger most admires, Nietzsche and Hölderlin, both went mad. Not that Straus's philosophy of mind elevates philistinism or mediocrity at the expense of genius; he could interpret the I-Allon relation of an Einstein or a Goethe as well as of the rest of us; but he need not dismiss as despicable, as Heidegger contemptuously does, all that *is* ordinary.

And there is a second important difference. Heidegger's *Dasein*, like Sartre's *pour-soi*, or for that matter Jaspers' *Existenz*, is *only* human. Straus's 'experiencing being' is human *or* animal. It is the structure of all sentient living, not only of our relatively self-conscious living, that he wishes to reinstate as the foundation of knowledge and of action. And this aim – to reinstate man in nature – is an essential one for any philosophy that would finally exorcise the persistent Cartesian ghost.

For both these reasons, it seems to me, Straus's approach is more fruitful than that of other 'existential' psychiatrists, notably, for example, that of his friend Ludwig Binswanger (to whom his German collection of essays, *Psychologie der menschlichen Welt*, was dedicated).[21] Binswanger simply takes Heideggerian being-in-the-world, the very essence of which demands arrogance and hatred as the road from me to thee, and injects into it, with sublime incompatibility, a generous dose of love. But to restore a balanced vision of existence we need, both in our recognition of its positive rootedness in communion and in our recognition of the kinship of men and other animals, a broader and firmer foundation from the start.

One is tempted to compare Straus's proposed reform with one other effort to overcome the Cartesian tradition, in particular the Cartesian-empiricist theory of perception. Straus rejects, as we have seen, the empiricist concept of sensation in favor of a theory of sensing (*Empfinden*) as the fundamental sense-mediated road that links object with experiencing organism, and contrasts this pervasive process with the more sophisticated and at least primarily cognitive sensory awareness of objects in *perception*. This is obviously not the stock psychological distinction between sensation and perception,[22] but it is reminiscent, at least at first sight, of Whitehead's distinction between presentational immediacy and causal efficacy.[23] Yet it is also in some essential ways different. Whitehead is contrasting what is really present, what is 'enjoyed', in 'sensa' or 'pure' givens, with the opaque but powerful impact upon us of the processes beyond and around us in the world. The latter – causal efficacy – does in fact resemble Straus's 'sensing'. It is the constant interchange of experiencing being and surrounding field, the way in which an animal becomes the figure it is, expressing unity and contrast with its medium as ground, that they are both concerned to describe. And it is indeed this living dynamic of sensory awareness, a dynamic expressed, for example, in Aristotelian *aisthesis*, that the modern tradition has neglected and even denied. 'Presentational immediacy', however, is very different from Strausian 'perception'. It is the illusory surface of sense experience, while for Straus 'perception' is the sensory-interpretative process through which we know presented things as stable objects amenable to classification, definition, explanation, and the like.

On the other hand, if we liken Whitehead's distinction to Straus's distinction of stimuli and objects, we find that causal efficacy, which is another name for sensing, is indeed our everyday path to objects (in an ordinary, not a 'scientific' sense), while stimuli are by no means the givens of presentational immediacy, but intellectual artifacts constructed by abstraction and hypostatization to suit the demands of a physicalist metaphysic. So this distinction is not quite parallel to Whitehead's either. At the same time, we would notice that the data of presentational immediacy are after all the data, detached and delusive, from which Berkeley and Hume generalized to produce their theory of ideas (or of impressions), and they are the data which experimental psychologists have first fastened on and then forgotten in order to build beneath them the apparently more solid neurophysiological foundation with which they suppose themselves to work. The import of the two analyses seems to me, therefore, to be convergent, even though the distinctions used are not entirely congruent. In both cases it is the reinstatement of experience in its concrete significance that is at stake.

III

For the arbitrary models of empiricist psychology, then, Straus would substitute the conception of an experiencing being in its relation to a surrounding world. A host of philosophical themes are given new illumination by this change of ground. I have already touched on some of them but will mention more explicitly three: the concepts of time and space, the mind-body problem, and the problem of universals.

Philosophers in the empiricist tradition and experimental psychologists, who have depended on this tradition for their metaphysical nourishment, have taken, by and large, time and space as either objective and Newtonian or as subjective constructs equivalent to those uniform containers. But our experience of time and space is not thus uniform. Nor, in the case of time, does this assertion imply setting a Bergsonian *durée* or a literary stream of consciousness over against the uniform chronology of the 'real' world. Chronology is a product of culture, which we rely on to set our alarm clocks, to meet classes, or fry chicken, or catch planes; but like clocks, classrooms, frying pans, and Boeings, it is an artifact which we use in order to move about *within* the richer framework of lived time. And lived time is neither Cartesian-atomic nor Newtonian-continuous. It exhibits all, and more than, the modalities that Rosalind enumerated. It is not a measure, but a medium.

The concept of lived time, of course, is by no means original with Straus. His exposition is paralleled in Minkowski, in Merleau-Ponty, or, in a different style, in Heidegger, and it is reminiscent again, in its metaphysical implications, of Whitehead's philosophy of process. What distinguishes his work, however, is its linkage to specific psychological themes: as in his demonstration of the difficulties that follow from the time-atomism of Pavlovian theory, or in his treatment of the theory of memory-traces, or of time-disturbances in endogenic depression. Let us look at one example: his account of the phenomenon of infantile amnesia:

In organic and senile amnesias, earlier experiences prove more resistant than recent ones. Yet, infantile amnesia sets a barrier which prevents the recovery of memories from our first years.[24]

What can be done with this topic in Kantian, Lockean, or any other conventional philosophical terms, or in orthodox psychological theory? In terms of Straus's phenomenology, however, it makes perfectly good sense. 'The subject of remembering', Straus argues, 'is a human being who forms his life history within the temporal horizon of personal time'.[25] Thus within the medium of lived time I form a concept of my history, which is objective and chronological:

Human life evolves on two levels: on that of biological need and satisfaction in the circle of daily routine and on that of signal events, marked in the annals of [the] *curriculum vitae.* Corresponding to the two levels of existence, there are two modes of remembering and forgetting: one characterized by the familiar and the repeatable and the other by the new and unique.[26]

The infant still lacks this double structure; hence the baby's inability to remember. This insight permits a much simpler and more adequate explanation than psychologists have been able to devise. Straus compares his own explanation, for example, with Freud's:

In search of an explanation of this deficiency, Freud ... assumed – in line with tradition – that, since single, stimulus-bound impressions are preserved in memory, the earliest ones would also be available to us in later years, as long as they are not kept away from consciousness by the forces of repression. If Freud was right, one should expect that a child at the age of three would still remember well the events of his first and second years. This is not the case. Those early experiences are not preserved up to the advent of the Oedipal situation and then extinguished by repressive forces. Nothing needs to be repressed in this case, because nothing is preserved in its original form – for this reason: what is remembered is the *novum* in its particularity standing out from an invariant ground. This ground is built when the order of things (the world) is detached from the personal order. Before a child is able to remember, it has to fulfill the following conditions: (1) to detach the world from the moments of direct encountering, (2) to extend the temporal span beyond the moment, (3) to build an invariant framework into which single events can be entered, (4) to establish permanent and identifiable structures of particular things and events, and (5) to allow physiognomic changes no longer to interfere with the constancy of the invariant framework and with the identity of single events. But the baby lives from one moment to the other in the narrowness of his temporal horizon. A baby experiences the world basically in relation to himself. The early tendency to put things into his mouth is quite characteristic of his own attitude to the world. He lacks specification. There is an obvious lack of self-reflection; yet this is what is required to sever the order of one's own existence from the order of the environment. In short, there is a lack of a stabilized preserving order, of a schema in which events are to be registered in order to be recalled in later days. The conception of the historio-logical structure of memory not only states, concerning infantile amnesia, *that* it is so – which is a fact that everyone accepts – and not only explains *why* it is so – which Freud had tried do to – but, also, finally, makes evident that it *must* be so.[27]

Even more striking, because the topic is more habitually neglected, is Straus's treatment of space. The first essay in *Phenomenological Psychology*, 'The Forms of Spatiality', illustrates his approach to this problem.[28] If we look, without metaphysical prejudice, at the phenomenal difference between color and tone, we find that while colors are 'over there', 'on' an object, sounds act on us from a source: they approach us or move away from us:

Color clings (phenomenally) to the object while the tone produced by an object separates itself from it. Color is the mark of a thing, whereas tone is the effect of an activity.[29]

This is so, Straus argues, because while color and form constitute an object, sound essentially separates itself from its source. Sounds may indeed indicate objects, but they may also, as in music, separate themselves from objects to achieve an autonomous being. Nor do these differences between sound and color correspond to different judgments, unconscious inferences, or the like, superadded to differing sensory data. They express different pathic modalities, essentially diverse ways of receiving, or achieving, sensory experience. They express different forms of spatial being. Thus:

Tone has an activity all its own; it presses in on us, surrounds, seizes, and embraces us. Only in a later phase are we able to defend ourselves against sound, only after sound has already taken possession of us, while in the visual sphere we begin to take flight before we have been prehended. The acoustical pursues us; we are at its mercy, unable to get away. Once uttered, a word is there, entering and owning us. Nor can it be rendered unspoken through any pretence or apology.[30]

But color, on the other hand:

... not only presents itself from over there, opposite to us, but also is demarcated at the same time that it demarcates, articulating space as regions, laterally and in depth. In optical space, things stand out from one another with sharply defined boundaries; the articulation of the optical is governed by contour. The optical image appears as a representative of the concept, just as the melody serves as a natural representative of the unity of the Gestalt.[31]

Further:

Artistic activity is dependent on the pathic moment of the optical phenomenon – reactively dependent down to the tiniest details of technique. Baroque painters did everything to keep things from appearing simply juxtaposed; to reduce the contours through representation in the plane, they manipulated such devices as the density of paint on the canvas, the distribution of light, the picture's dimensions, its frame, *chiaroscuro* modelling as in etching, and indirect representation of contour. One need only call to mind the work of Rembrandt, especially his famous landscapes.[32]

Nor are these random observations, but recordings of the lawful patterns of the pathic aspect of sensory experiences. They are essential, moreover, to the understanding of movement in its relation to sensing. Thus in dance, which is dominated by the musical, we move, not *through* a space, as in practical purposive action, but *in* a space [33]. Dance, Straus infers, expresses '... the tendency of live body space to expand against surrounding space and to actualize itself symbolically'.[34] Dance transforms the oriented optical space of our routine practical activities into a 'presentic' acoustical space. Consider, for example, the experience of dizziness outside and inside the dance. Dizziness in climbing a ladder is purely unpleasant; dizziness in the dance exhilarates. The proprioceptive sensations in both cases are identical, yet 'they are embodied in different structures of immediate experience.'[35] It is worthwhile quoting Straus's description at some length:

The space in which we move on the merry-go-round or in dance – which we are discussing here – has lost its directional stability. Of course, it is still a space with extension and direction, but direction is no longer disposed in a certain way around a fixed axis; rather, direction moves and turns with us as it were. The dissolution of defined direction, and, correspondingly, of topical valences, homogenizes space. In a space of such modality, it is no longer possible to act; one can only enter into it as a participant. Actually we don't live in space but in spaces, spaces somehow demarcated and stabilized by a system of fixed axes. One need only imagine a room perfectly quadratic, without windows and indirectly illuminated, in the middle of each wall is a door, while furniture and pictures are arranged in a strictly symmetrical manner so that each wall appears as a mirror image of the opposite wall. If one were to spend some time in this room and then walk back and forth several times, one would become confused about the entrance, having lost his orientation to neighboring, surrounding areas; one would be bewildered and bewitched like a person in a magic maze. To enter such a space in fantasy is sufficient to show why we make our rooms rectangular rather than square, why we prefer asymmetry – a clearly and distinctly apprehended difference between length and width, as proportions in the ratio of the Golden Mean. Action demands a system of definite, distinct direction determining loci with valences varying in accord with their relationship to the directional system. When the spatial structure changes, as happens in dance, the immediate experience of confrontation also changes that tension between subject and object which, in ecstasy, completely dissolves. When we turn around while dancing, we are, from the very start, moving in a space completely at odds with oriented space. But this change of spatial structure occurs only in pathic participation, not in a gnostic act of thinking, contemplating, or imagining. That is to say, presentic experience actualizes itself *in* the movement; it does not produce itself *by means* of the movement. Even though a dance occupies a considerable interval in objective time, the entire movement is still integrally presentic. In itself, it does not produce any changes in immediate experience nor any changes in the external situation, as does action which must abandon its starting point to reach its goal. Every action demands that a particular condition or position be left behind in order to reach another condition, another position. This defines both direction and limits for action. When the new condition is reached, the old one belongs to the past; action is a historical process. Presentic movement, on the other hand, is free of direction or limits; it knows only waxing and waning, ebbing and flooding. It does not bring about this change; it is not a historical process. It is for just this reason that we term it 'presentic', despite its duration in objective time. The dissolution of the subject-object tension, cumulating in ecstasy, is not the aim of the dance; rather, the very experience of dancing originally arises within it.[36]

This general description is confirmed by a comparison of different styles of dancing, say a minuet and a waltz:

The dancer of the minuet performs his steps over the basic rhythm. The 'filling' of space is only figuratively represented by the formation of couples, the 'visits', etc. The dancer of the minuet senses the harmonizing influence of the music without yielding to it entirely; he remains an individual... The different ways of life of social classes and the changes of sentiment dominating different historical periods are directly reflected in their forms of dancing. The sequence: minuet, waltz, jazz, strikingly demonstrates the extent to which individual existence has been abandoned and swallowed by mass movements.[37]

This contrast, be it noted, moreover, does not indicate a contrast between the spatiality of dance as pathic and agnostic, Euclidean space as our everyday medium:

The space in which we live is as different from the schema of empty Euclidean space as the familiar world of colors differs from the concepts of physical optics... As immediately experienced, space is always a filled and articulated space; it is nature or world.[38]

Thus, the contrast between dancing and walking indicates, not a contrast between lived space and Euclidean space, but between two forms of lived spatiality. Our ordinary motility is the progressive one of purposive movement, and it is this that is reversed in the abandon of the dance. But purposive movement is not Euclidean, spread out in three dimensions indifferently to time, it is *historical*. It is the space of action.[39] In it we move ahead in preference to back, out from a stable *here* to the goal of our proposed action. This, not the infinite extension of geometry, is the ordinary lived spatiality which we forget in the self-abandonment of the dance. The contrast between acoustical and optical space is still a constrast within the pathic aspect of sensing.

Straus is often repetitive, often given to cryptic epigrams: e.g., 'repetition is possible in the acoustical sphere, while the optical sphere is limited to reduplication (*Vervielfältigung*)'.[40] (Cannot colors repeat themselves in a pattern or a costume?) Or: 'The space that extends before us is, thus, a metaphor of the approaching future: the space that lies behind is a metaphor of the past that has receded from us. When we hear something, we have already heard it.'[41] But in general his exposition is not only illuminating in itself: it can serve to rouse us from our dogmatic slumber. We take it for granted that non-Euclidean geometry has undermined the Kantian theory of spatial intuition. But why should we ever have thought that our everyday experience of space was that of the 'infinite container' Kant envisaged? We could find, if we seek philosophical precedent, a more faithful rendering of our experience in Aristotle's concept of place, or we could find a corrosive critique of the empiricist conventions about sensation in Hegel's argument, in the *Phenomenology of Mind*, on the here and now. In the main, however, philosophers have failed to build on these insights; but here in Straus we have an alternative approach which is tied, moreover, not to alternative philosophical systems, but to concrete psychological insights. Such an account may perhaps induce us to abandon the poverty of our usual school examples – the desk or the tree, the building across the quad – not, indeed, for irrational wallowing in 'situation', but for the structured descriptions of phenomenal realities, which may serve as the coping stones of a sounder metaphysic.

In studies like those of the forms of spatiality, or of lived movement, we have already – as I suggested earlier – emerged from the shadows of Cartesian dualism to look at human existence in its unitary, mental-and-embodied being. Straus has developed such non-Cartesian thinking in many areas, but nowhere more strikingly than in his essays on the upright posture.[42] Man's posture, he argues, exhibits, and conditions his nature both as animal and rational: neither has priority. Language, he points out, has long taken cognizance of this fact:

> The expression 'to be upright' has two connotations: first to rise, to get up, and to stand on one's own feet and, second, the moral implication, not to stoop to anything, to be honest and just, to be true to friends in danger, to stand by one's convictions, and to act accordingly, even at the risk of one's life. We praise an upright man; we admire someone who stands up for his ideas of rectitude. There are good reasons to assume that the term 'upright' in its moral connotation is more than a mere allegory.[43]

But that is not to say that the more 'basic' meaning is the moral one. Our posture, unique among animal species, is physiologically grounded in all the details of our nervous, muscular, and skeletal development:

> ... there is no doubt that the shape and function of the human body are determined in almost every detail by, and for, the upright posture. The skeleton of the foot; the structure of the ankle, knee, and hip; the curvature of the vertebral column; the proportions of the limbs – all serve the same purpose. This purpose could not be accomplished if the muscles and the nervous system were not built accordingly. While all parts contri-

bute to the upright posture, upright posture in turn permits the development of the forelimbs into the human shoulders, arms, and hands and of the skull into the human skull and face.[44]

Upright posture preestablishes our way of being-in-the-world. To begin with, it is an achievement, not of the species only, but of each individual, an achievement which every infant must realize – and glories in – for itself. 'Man must become what he is'.[45] He has to learn to walk and he has to learn to speak. Nor can his defeat of gravity ever be made definitive: we have to abandon our upright posture as part of our daily rhythm and assume it again, often reluctantly, next day. If, moreover, the young child enjoys his success in learning to stand and to walk, both child and adult enjoy the abandonment of the struggle against gravity in 'reclining'. Indeed, the rhythm of waking and sleeping, in its human form, is linked essentially to the rhythm of standing and lying down. To stand up is to rise to command the world in opposition to it; in sleeping, in contrast, we do not so much leave the world as give in to it, let it play upon us kaleidoscopically without giving it the order that in waking we are able to impose.

In its ever renewed attainment, then, the upright posture displays manifold significance. So, Straus argues, does the very fact of standing, once acquired. In standing, he points out, we put ourselves in three ways at a distance from the Allon: from the ground, from things, and from our fellows. The second of these brings us our confrontation with objects as objects and prepares our formal ways of handling them: 'spoon and fork', Straus remarks, 'do not create distance; tools can be invented and used only where distance already exists.'[46] Distance from our fellows, similarly, underlies both our formalities and our expressive relaxations of them. Such forms, of course, vary from culture to culture; yet everywhere the vertical is predominant, the vertical as presenting the aloof or the solemn, the deviation from it – 'inclination' – representing the abandonment at least in part, of such aloofness:

There is only one vertical but many deviations from it, each one carrying a specific, expressive meaning. The sailor puts his cap askew, and his girl understands well the cocky expression and his 'leanings'. King Comus at the Mardi Gras may lean backward and his crown may slip off-center. However, even the disciples of informality would be seriously concerned if, on his way to his inauguration, the President should wear his silk hat (the elongation and accentuation of the vertical) aslant. There are no teachers, no textbooks, that instruct in this field. There are no pupils, either, who need instruction. Without ever being taught, we understand the rules governing this and other areas of expression. We understand them not conceptually but, it seems, by intuition. This is true for the actor as well as the onlooker.[47]

The constancy of the vertical dimension, Straus adds, is exhibited in the fact that young children can draw a vertical or horizontal line or a square while still unable to copy the same square presented as a diamond.[48]

Walking, finally, depends, like standing, on a highly complex combination of physiological conditions, which predetermine the way in which we experience the world:

Human bipedal gait is a rhythmical movement whereby, in a sequence of steps, the whole weight of the body rests for a short time on one leg only. The center of gravity has to be swung forward. It has to be brought from a never stable equilibrium to a still less stable balance. Support will be denied to it for a moment until the leg brought forward prevents the threatening fall. Human gait is, in fact, a continuously arrested falling. Therefore, an unforeseen obstacle or a little unevenness of the ground may precipitate a fall. Human gait is an expansive motion, performed in the expectation that the leg brought forward will ultimately find solid ground. It is motion on credit. Confidence and timidity, elation and depression, and stability and insecurity are all expressed in gait.[49]

All this is entailed in the very fact of standing up and walking; but there is more. The development of hand and arm, which has so strikingly enlarged our body schema, and permitted the development of so many human skills, depends on upright posture. And so does the function of the head, which liberates the 'visage':

Eyes that lead jaws and fangs to the prey are always charmed and spellbound by nearness. To eyes looking straight forward – to the gaze of upright posture – things reveal themselves in their own nature. Sight penetrates depth; sight becomes insight.[50]

The contrast with animal orientation is evident:

Animals move in the direction of their digestive axis. Their bodies are expanded between mouth and anus as between an entrance and an exit, a beginning and an ending. The spatial orientation of the human body is different throughout. The mouth is still an inlet but no longer a beginning, the anus, an outlet, but no longer the tail end. Man in upright posture, his feet on the ground and his head uplifted, does not move in the line of his digestive axis; he moves in the direction of his vision. He is surrounded by a world panorama, by a space divided into world regions, joined together in the totality of the universe. Around him, the horizons retreat in an ever growing radius. Galaxy and diluvium, the infinite and the eternal, enter into the orbit of human interests.[51]

Among the many changes implicated in the transformations of the animal into the human head, that of the jaws into the mouth is of special significance as a prerequisite for the development of language. But it is only one of the many preconditions of human speech laid down in the achievements of upright posture:

In upright posture, the ear is no longer limited to the perception of noises – rustling, crackling, hissing, bellowing, roaring – as indicators of actual events, like warnings, threats, or lures. The external ear loses its mobility. While the ear muscles are preserved, their function of adapting the ear to actuality ceases. Detached from actuality, the ear can comprehend sounds in the sounds' own shape – in their musical or phonetic pattern. This capacity to separate the acoustical Gestalt from the acoustical material makes it possible to produce purposefully and to 're-produce' intentionally sounds articulated according to a preconceived scheme.[52]

In this 're-production' of sounds, moreover, we can already recognize the achievement of generalization or abstraction characteristic of human thinking: for 'the phoneme itself is universal'.[53] Speaking is already what some philosophers like to call 'rule-governed behavior':

A spontaneous cry can never be wrong. The pronunciation of a word or the production of the phoneme is either right or wrong. The virtuosity acquired by the average person in expressing himself personally and individually in the general medium of language hides the true character of linguistic communication. It is rediscovered by reflection when disturbances of any kind interfere with the easy and prompt use of language or when the immediateness of contact does not tolerate linguistic distance, and the word dies in an angry cry, in tender babble, or in gloomy silence.[54]

Indeed, language requires, in Straus's view, all three aspects of that distance from the world which we acquire through upright posture. Distance from the ground, with the disappearance of a mobile outer ear, leaves us to receive the phoneme in its pure acoustical character. Distance from things enables us to make them the objects of discourse mutually understood. And only a distance between speaker and listener could be overcome, as it is, through the mediation of speech. In short, as Herder observed, 'the upright gait of man is the organization for every perfection of his species and his distinguishing character'.[55]

Such studies may appear 'unphilosophical' – mere excursions into anthropology; but they are the kind of exercise we need if we are to pay more than lip-service to our alleged rejection of mind-body dualism and to the conception of a 'lived body' which we hope to set in its place.

At the same time, Straus's phenomenological insights are sometimes too easily won. In 'The Upright Posture', for example, he contrasts the horse, who can sleep on his feet, with man, who must relinquish his conquest of gravity and lie down to sleep. But surely this example makes the contrast look sharper than it is. If one thinks of a cat, domestic or not – indeed, a tiger or leopard comes most obviously to mind – the posture of repose in contrast to the active gait, with the aura of delight in both, is at least as striking as in our case. Yet cats do not stand on two legs; even moving as 'easily' as they do on four plainly demands rest as its contrary. Could it be that hunters show this contrast more clearly than grazers? Similarly, Straus contrasts 'earth-bound' animal with liberated man; but what of the air-borne vertebrates to whom we refer when we call some one 'as free as a bird'? Or again, Straus analyzes 'awakeness' as the mode in which persons act, and sleep, or dreaming, as the contrary state which can be judged as such only *from* the waking state. This is, I believe, a valid answer to the 'Am I dreaming?' question, but again it seems false to make the rhythm of sleep and waking a perquisite of human life alone. Other animals, too, 'act' only when waking and need the refreshment of sleep. In short, while Straus is correct both in rooting human cognition in a wider I-Allon (i.e. I-Other where Other is not necessarily other persons) relation of any experiencing being to its surrounding field and in seeing a deep-seated difference, inextricably linked to the physical basis of our being, between men and other animals, he sometimes makes this distinction in an aphoristic style that fails to stand up to closer analysis.

The same weakness infects Straus's treatment of the third problem I want to touch on, the problem of universals. Here too an illuminating account is confused by hasty, and indeed contrary, generalizations, and again just where the distinction between men and animals is concerned.

Both in ancient and modern philosophy, the problem of universals has arisen from the contrast between the mere particularity of sensory givens and the generality of language or of thought. Let us start, for example, from Plato's position in the *Theaetetus*.[55a] Knowledge, to be knowledge, Plato argued, must be both *infallible* and *real*. Sensation, as the particular, immediate, *given-to me*, is indeed infallible. But it fails the test of reality, since 'existence' can be grasped, not by sense, but only, *through* the senses, by the mind alone. Sensation, therefore, cannot qualify as knowledge. It presents us with the particular, meaningless *this*; but only through the comprehension of general concepts, like existence, can we *know* that the presented datum not only presents itself, but *is*.

Straus's position seems to stand in a peculiar relation of agreement and disagreement with this classic text. On the one hand, he is convinced that philosophers, from Plato and Democritus onward, have been, in the main, unfair to sensory experience. Sensing is not a delusive blooming buzz of meaningless particulars on which we must turn our backs in order to reach an intelligible world where the mind can feed on its proper objects. It is an all-inclusive road of access to the world, our means, over the varied spectrum of the five senses, of communicating with reality. It is thought, not sense, which cuts itself off, by a negation of which men alone among animals are capable, from immediate rootedness in the real, to spin out its gnostic constructions in separation – or at least in quasi-separation – from the more immediate immersion in reality of sensing in its pathic mode. Such sensing, moreover, far from being a congeries of meaningless bits, is itself already general. Though limited to, or at least ranging out from and returning to, the *here* and *now*, it nevertheless grasps the presented world in its generality. The experience of generality is intrinsic to sensing, and also common, therefore, Straus suggests, to men and animals:

I maintain that animals, too, experience the general – for example, sound. They have this experience not because they think in general terms, but because the relationship of an experiencing being to the world is a general relationship, whereas the singular moment is merely a constriction of this relationship. The content of each moment is determined in part by that from which it is distinct, that is, by what it no longer is, as well as by what it is to be. How, otherwise, could animals experience signals, which are midway between an undifferentiated and a differentiated situation and which announce the transition from the one to the other?[56]

This seems to me an important insight, and one which helps us to recognize a minimal continuity at least in the styles of being-in-the-world of all sentient beings. 'Generality' is not a human invention, which we have superadded to the merely particular data that make up, Hume-wise, the raw givens of animal experience. It is of the very fabric of sentience itself. And it is from *within* the world of an experiencing being that we extract, as it were, the more refined universals of language and of the articulate knowledge which it enables us to acquire.

So far, so good – or so it appears. But if Straus rejects the contrast between the sensed as merely particular and the known as general, and professes to find in sensory experience the full-bodied medium of all experience, even of the most refined cognition, he insists, at the same time, as strongly as did Plato, that sensory experience is *not* knowledge, is not even an 'inferior' brand of knowledge, but differs radically from it, and differs precisely in this matter of generality. Thus in contrast to perception – *Wahr*nehmung – which is cognitive, and grasps things in their objectivity as things, of one or another kind, sensing, *Empfinden*, after all, Straus asserts later in his argument, grasps only the here and now:

In sensory seeing the thing is for me, for me here and now in a passing moment. But after the step to the world of perception, this being-there-for-me is apprehended as a moment in a universal, general chain of events.[57]

It is 'only by the use of universals', he argues, 'that I can describe a thing as it is for me ... and for every one',[58] and such description, he seems to feel, is entailed in the very act of perceiving. Sensing, in contrast, permits no such generalization, no such enlargement to logic and taxonomy, to the rational use of 'all' and 'some'. This contrast is surely Platonic, but a strange transposition of Plato's dichotomy. In the *Theaetetus* we have the contrast of *aisthesis* with knowledge; here we have a different dichotomy: sensing (= aisthesis?) is indeed non-cognitive, and perception – which is not to be identified with sensory experience – is taken as at least the primordial level of, if not equivalent to, knowledge. 'Perceiving', Straus writes, 'and not sensing, is a knowing'.[59] Its theme, we are told, is the factual, which is constituted – made (*factum*) – by means of a breach in sensory experience, in the singling out of some*thing* against a neutral background of objective space and time. Thus by a fundamental negation man breaks through the sensory horizon, in which he like all animals originally experiences the world, to attain the geographical space of objectivity. He never does so totally, indeed, except in illness: the melancholic is precisely he who has lost touch with the landscape; 'frozen in unmoving time, ... he looks at the world ... in a bird's eye view'.[60] But normally we live in both spaces, that of landscape and that of geography. Routinely we sense, reflectively we perceive.

Indeed, as Straus introduces the distinction of sensing and perceiving, perception seems to be the sharpening of our sense-mediated attention that arises in response to language:

We see a thing a thousand times and yet have not really seen it. A question forces us to look at it properly for the first time. The first seeing was a sensing, a participation in expression; the second seeing, however, is a perception. Questions force us into a new order of understanding. We are asked about 'something' and wish to answer what and how that something is. We speak now of things or of a thing, we speak of its properties, its possible modifications. We speak of one thing which we see at this moment in front of us, or which we visualize in its particular place. We speak of one single thing, but we distinguish it with general words.[61]

So we have, if you like, generality in sensing, but 'true' universals, let us say, in language-mediated perception. Thus the perception *of* speech ('The phoneme is already universal' – see above, p. 313) becomes the model for the perception also of events and objects, which are seen or heard as such only within the universe of discourse that language has already shaped. Can it be that the 'generality' of sensing is that of the concrete universal, while perception, like all cognitive processes, depends upon the more abstract universals that language enables us to understand? It may be some such dialectical solution that will enable us to reconcile the seeming contradiction.

In that case, however, only language users perceive, and at one place at least Straus suggests that only language-users learn.[62] Thus, starting out to put our sensory experiencing on a par with that of other animals, and to exhibit the continuity on the ground of which our unique achievements have arisen, he seems, in his doctrine of perception, to outdo even the Cartesians in his relegation of animals to outer darkness. Yet surely that animals learn and in some sense acquire knowledge is as well attested as that human knowledge is in some way unique.

Moreover, in his account of perception, even Straus's analysis of human experience is strangely contradicted. In his account of the dance, for example, we have seen that he contrasts both 'presentic' or expressive and purposive or 'historical' space with the uniform, infinite space of Euclid or Newton. Here, however, he contrasts the presentic space of landscape directly with the uniform space of geography and geometry, which he identifies with that of history. Thus, he points out, a 'remote valley' is, geographically, on a uniform plane with my present location in this room. But the farmer living in the remote valley is immersed in it as his landscape. Granted, but the farmer ploughing a furrow, for example, is not moving like the dancer in a space, he is advancing purposefully – and historically – through a space to reach a goal. Both these forms of spatiality are to be contrasted – as Straus himself has contrasted them – with the geographical framework, say, of an agricultural survey. It seems, therefore, that we have here not a simple dichotomy, but a many-one relation. On the side of sensing there are diverse forms of lived spatiality;

on the side of perception, there is the one geographical space accessible to objective thought. And the concept of 'history' seems, if with different significance, to fall into both categories. Purposive action is historical; an expressive activity like the dance is not. But the universal framework of history entails 'history' in a more refined and critical sense.

Such ambiguities need ironing out if we are to work with Straus's basic concepts, and particularly, once more, when we are faced with the teasing question of the distinction between men and animals. As Straus argues in 'The Upright Posture', in agreement with the comparative studies, for example, of Adolf Portmann, men differ fundamentally from animals even in their anatomical and developmental endowments as animals. But is the division to be seen so sharply that perception and knowledge are wholly denied to other animals? There seems to be here a radical transmutation indeed, but of a common gift. Were not generality embedded in sentience itself, the power of language to mediate assertions with universal intent, and hence to aim, not at awareness only, but at truth, would remain, as it does for traditional empiricism, an unintelligible mystery.

NOTES

[1] E. W. Straus, *The Primary World of Senses* (trans. by J. Needleman), Free Press, New York, 1963.
[2] E. W. Straus, *Phenomenological Psychology* (trans., in part by E. Eng), Basic Books, New York, 1966.
[3] In this respect Charles Taylor's *Explanation of Behaviour*, New York 1964, which admittedly owes much to Merleau-Ponty, forms a striking exception to the general rule in recent Anglo-American thought.
[4] *Phenomenological Psychology*, p. xi.
[5] A. I. Melden, *Free Action*, London 1961, p. 198.
[6] E. A. Burtt, 'Descriptive Metaphysics', *Mind* 72 (1963), 18–39.
[7] Quoted in *Phenomenological Psychology*, p. vi.
[8] *Ibid.*, p. viii.
[9] *Loc. cit.*
[10] Cf. *Ibid.*, pp. 269–70.
[11] *Ibid.*, p. viii.
[12] *Loc. cit.*
[13] *Loc. cit.*
[14] *Ibid.*, p. 26.
[15] *Loc. cit.*
[16] *Ibid.*, p. 272.
[17] 'The Forms of Spatiality', in *Phenomenological Psychology*, pp. 3–37.
[18] *Ibid.*, p. 12.

[19] The concept of universal intent is derived from Michael Polanyi's *Personal Knowledge*, London 1958.

[20] M. Heidegger, *Sein und Zeit*, Halle 1927, p. 122.

[21] *Psychologie der menschlichen Welt*, Berlin 1960.

[22] See for example D. Hamlyn, *Sensation and Perception*, New York 1961.

[23] A. N. Whitehead, *Symbolism: Its Meaning and Effect*, Cambridge 1927.

[24] *Phenomenological Psychology*, p. 72.

[25] *Ibid.*, p. 73.

[26] *Loc. cit.*

[27] *Ibid.*, pp. 72–3.

[28] 'The Forms of Spatiality', *Ibid.*, pp. 3–37. Cf. also Ch. II, 'Lived Movement' and the account of action and space in 'The Upright Posture.'

[29] *Ibid.*, p.8.

[30] *Ibid.*, p. 16.

[31] *Ibid.*, pp. 16–17.

[32] *Ibid.*, p. 17.

[33] Cf. *ibid.*, p. 23.

[34] *Ibid.*, p. 28.

[35] *Ibid.*, p. 30.

[36] *Ibid.*, pp. 30–31.

[37] *Ibid.*, pp. 31–32.

[38] *Ibid.*, p. 32.

[39] Cf. *ibid.*, p. 150.

[40] *Ibid.*, p. 16.

[41] *Loc. cit.* The second passage is rendered even more difficult by a change in paragraphing in the translation.

[42] 'The Upright Posture', in *Phenomenological Psychology*, pp. 137–165; cf. 'Born to See, Bound to Behold', *Tijdschrift voor Filosofie* 27 (1965), 659–688.

[43] *Ibid.*, p. 137.

[44] *Ibid.*, p. 138.

[45] *Ibid.*, p. 141.

[46] *Ibid.*, p. 145.

[47] *Ibid.*, p. 146.

[48] *Ibid.*, p. 147.

[49] *Ibid.*, p. 148.

[50] *Ibid.*, p. 162.

[51] *Loc. cit.*

[52] *Ibid.*, p. 163.

[53] *Loc. cit.*

[54] *Loc. cit.*

[55] Quoted in *ibid.*, p. 164.

[55a] On this interpretation of the Theaetetus, however, cf. note to Chapter II above.

[56] *Primary World of Senses*, p. 96.

[57] *Ibid.*, p. 317.

[58] *Loc. cit.*

[59] *Ibid.*, p. 329.

[60] *Ibid.*, p. 328.

[61] *Ibid.*, p. 317.

[62] *Ibid.*, p. 147.

CHAPTER XVIII

THE CHARACTERS OF LIVING THINGS

III: *Helmuth Plessner's Theory of Organic Modals*

I

Our understanding of ourselves and our place in nature constitutes, if not *the* central, at least *a* central problem of metaphysics. Yet, faced with this question, modern philosophical thought has for the most part swung helplessly between an empty idealism and an absurd reductivism. It is time we overcame our narrow factionalism and learned not only to think more independently ourselves about persons, minds, and living nature, but to profit from the efforts of those who have already given us concepts and arguments which could help us along this road. Among such writings, Helmuth Plessner's major work, *Die Stufen des Organischen und der Mensch*,[1] seems to me outstanding, in that it provides both a firm rational basis for the biological sciences, in their many-levelled structure, and for the sciences of man.

What I propose to do here is to offer an introduction to the central portion of Plessner's argument. I must perforce omit much: both in the background of his work and in its detailed application to philosophical anthropology. Of the complex of problems from which *Stufen* takes its origin, let me, however, very briefly enumerate three points. First, Plessner was dissatisfied with the alleged solution of the problem of the foundations of social science, which placed *verstehende Wissenschaft* outside and over against the 'exact' sciences; resolution of this conflict, he believed, must be made at a higher and inclusive level. Secondly, he considered such a mediation to be possible only in terms of a philosophy of the organic, which alone could find a well-articulated place for man within the whole of the living world. Here the problem of the social sciences converged with that of biology as he had met it in the conflict of the young Köhler's reductive isomorphism with Driesch's vitalism. To overcome the last-named issue within the philosophy of biology, thirdly, demanded, he believed, a fundamental shift of outlook from the ceaseless pendulum of classical mechanism *versus* classical teleology to a biological philosophy which takes as fundamental neither means nor ends but *significance* or *form* as the fundamental category of life. In another context (only indirectly influenced by Plessner) this proposal will be familiar to students of Merleau-Ponty's *Structure du Comportement*. It resembles also, with the crucial difference that we are dealing here with form in a world of process, a return to an Aristotelian foundation for biology. (The argument of Collingwood's *Idea of Nature* is apposite here.)

From these brief hints at historical context and contemporary affinities, let me now plunge into the central argument of *Stufen*.[2]

II

Plessner's exposition begins, as any argument must which attempts to come to grips with the foundations problem in biology, with the sharp alternative set by Cartesian dualism. Is there, Plessner asks, in truth the strong disjunction envisaged by Cartesian thought between the inner and the outer aspect of living things? For Descartes himself, of course, such a dichotomy obtains in fact only in the case of human beings. Animals, let alone plants, are simply machines; in our case an 'inner' non-material soul has been mysteriously added. But the point is – and it is for the philosophy of biology, and, in the context of Plessner's investigation, for the social sciences, a point of fateful importance – that within the Cartesian heritage, which is the principal heritage of the modern intellect as such, there is no alternative for any form of life except to be either a mere body spread out in space, completely 'external', or a bit of subjectivity, completely and secretly 'within', or an unintelligible combination of the two. It is this conceptual framework which has up to our time made a rational foundation for the biological and social sciences impossible.

The first step then is to look and see whether living things do in fact display a two-aspect character which does not fit the rigid Cartesian alternative. If we examine carefully the things we perceive and the way we perceive them, we notice that there is in this respect an experienced difference between inanimate and animate objects. We do of course always perceive only a given aspect of any single thing; we cannot see, for example, all around a tennis ball, nor can we feel its texture simply by seeing it. But the multiplicity of aspects in such a case does add up to the total object. There is nothing that resists a total grasp, through several senses, of the object as a totality. Perceiving a cat is different: the cat is present to me in a fashion which cannot be wholly specified in terms of particular views from particular positions or through different senses. The cat confronts me in a way in which the tennis ball does not. It is just this resistance to specification, of course, just this confrontation with life, which has long made people suspicious of the Cartesian theory of the *bête-machine*. But the statement of a disturbing fact is not enough; a philosophical theory must provide us with concepts through which to understand such facts.

Plessner's procedure in this situation is as follows. He asks, in Chapter Three of *Stufen*, what would be the distinguishing character of living things as perceived objects which would make them resist the Cartesian alternative of outer and inner? This distinguishing criterion, call it for the moment *A*, he then takes as a postulate for the development of his argument. He asks: given, hypothetically, *A*, do the acknowledged characters of living things, such as development, self-regulation, organization, and so on, follow logically from *A*? Such characters are usually listed, empirically, as ultimate qualities of life. They are, if any, those characters which cannot be reduced by analysis to other qualities. In other words, they are organic *modals*. And what Plessner will have done if he succeeds in deducing them from his initial postulate is to produce a *theory of organic modals*.

What we need, then, is a property directly present to us in our immediate awareness of living things, not an abstract character or one which needs to be gleaned from a complicated chain of experiences – like the characters of metabolism or reproduction. We can find this, quite simply, by considering under what circumstances a shape 'looks alive'. Plessner refers in this connection to Buytendijk's systematic study of such judgments. Proceeding from the presentation of simpler to more complex shapes, he found that a certain irregular regularity both of shape and motion makes us react to forms as living or like living. Every one will know this experience from watching animated cartoons. It is, to put it briefly, a certain plasticity of form, 'freedom of form within form', that constitutes the immediate criterion of life. In the case of animated cartoons, of course, life is simulated; one's judgment here may be, humorously, or even seriously, in error. But the problem is to establish the minimal criterion by which we judge a body to be alive – whether or no in a given case it 'really' is so. It is a question here of the intuitive perception quite naively and ordinarily involved in such judgments. Consider again the example of the cat contrasted with the tennis ball. A cat playing with a ball, for example, confronts the ball in a fashion different, to our perceptive apprehension, from the way in which the ball confronts the cat. There is a difference here analogous to Merleau-Ponty's distinction between lived, bodily spatiality and objective spatiality. An inanimate object, a house or a mountain or a pebble, simply fills space. In Whitehead's terms, it is 'simply located'. But a living thing *takes* its place. It not only has a position in the coordinate system of space and time; it has *its* place. The venerable concept of natural place, so hampering to the perspectives of physics, is essential to biological thought.

The concept we need here, however, is not *only* spatial. It is a question of the whole way in which an organism 'takes its place' in an environment – arises in it, is dependent on it, yet opposes itself to it. Plessner tries to sum up what is intended here in a single term: *positionality*. This is the fundamental concept which he puts, hypothetically, at the head of his argument, and from which he proceeds to elicit, first, the other characters of living things in general, and then the distinctions between different styles of living – notably between plants and animals, animals and men.

Before I try to explain what Plessner means by this term, however, I must ask the reader's patience. This is a difficult concept to grasp at first meeting, but it does, I believe, gain in intelligibility as the argument proceeds. It may help at this point to anticipate some of the difficulties.

For one thing, positionality, if we take it, as I have been doing, as identified with 'natural place', seems to apply, as in my feline example, more readily to animals than to plants. Plants too have natural places – or, in modern language, ecological niches – but they do not seem, as animals do, to *take* their place. They can exist only in certain environments and disintegrate outside them: but so can crystals, for instance, keep their shape only out of solution. If positionality is to serve as a basic concept for philosophical biology, it must mean, therefore, more than *just* natural place.

Would Buytendijk's formula 'freedom of form within form' do better? It does apply to plants – to the shape of a leaf, for example, as against a crystal. But positionality is not, or at any rate is not *only*, a concept of shape, or even, more generally, of physical form. Plessner has argued recently that his theory is confirmed by modern evolutionists who hold that the origin of life can be located as the first establishment of an entity enclosed in a semipermeable membrane.[3] And the concept of positionality does indeed have to do, as we shall see, with the way in which an organism establishes its boundary over against a surrounding medium. Yet positionality is *not* the boundary, or the shape described by the boundary: it is the *way an organism bounds itself* that is essential. It is a question not only of a *Grenze*, but of *Begrenzung*.

To put it this way, however, again seems to lead us back to a distinction involving an active aspect inappropriate to the description of plants. Yet, as I hope to show in what follows, Plessner's basic concept, and the theory he builds upon it, are neither arbitrary nor meaningless. Perhaps the best crutch I can ask the reader to lean on at this stage is yet another example from modern biology. Sir MacFarlane Burnet, in his work on immunology argues that the ability of a living body to 'know' its own cells from invading material is a deep-seated property of life.[4] Now of course a cell, or indeed an organism, with few exceptions, cannot be said to 'know' anything in a conscious or intentional sense. Yet it is some such power of distinguishing itself from the environment of which not only immunological phenomena, but the very existence of semi-permeable membranes, are the biochemical expression. I think that is the sort of thing Plessner is after: natural place, plasticity of shape, semi-permeable membrane, immunological activity all express facets of the one very general, and essential, character of *positionality*, which he is taking as common to all living things.

With this warning, then, let me return to Plessner's argument and see, first, how he introduces his basic concept, and then what he proceeds to do with it.

The hypothesis which Plessner puts at the head of his argument is a statement about the nature of the *boundary* between a living thing and its medium. A body, he argues, can have one of two relations to its boundary. Either its boundary is merely the point where it stops, and identical with its contour or outline, or its boundary is a part of itself, not merely a virtual in-between between body and medium, but an actual boundary belonging to the body and setting it over against the medium, and indeed over against the body itself whose boundary it is. Plessner puts these two possibilities diagrammatically:

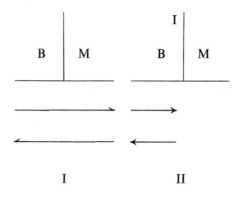

The first is characteristic of inanimate objects. A pebble or a box simply is where it is and stops where it stops. Living things, by contrast, have their boundaries as part of themselves, or, strictly, as part of the bodies which they are. In the one case we have the relation: $B \leftarrow I \rightarrow M$ (Body ← In-Between → Medium). In the other: $B \leftarrow B \rightarrow M$, where the relation of the body to its boundary is a relation of *itself* to *itself*. In this situation the boundary is placed over against the body and beyond it and at the same time directed in *to* it. This relation of a body to its boundary is, physically speaking, paradoxical, but nevertheless it does not contradict the physical existence of the boundary as such. It appears rather, Plessner says, as a *marginal phenomenon* of a physical system. Moreover, it is a phenomenon which cannot be specified even in an exhaustive listing of the explicit sensed characters of the body, but which is nevertheless presented directly to perceptual awareness, even though when we try to specify the character in question all we manage to state are the characters of the system as a gestalt. Again, if this seems mysterious, I must ask the reader to be patient, and to keep in mind the various expressions of positionality through which I have already tried to exemplify its meaning.

Of course, if we can accept this concept, and work with it, as I hope we shall succeed in doing, we do have here exactly that double-aspect character we were in search of as a way out of the Cartesian dichotomy. For a body with this kind of boundary structure is not simply divisible into inner and outer; it is, through its relation to its boundary, *both* directed *out* beyond the body that it is *and* back *into* it again.

III

Having established positionality, then, as the fundamental character from which his theory takes its start, Plessner proceeds to consider what follows for a body possessing this property. First, positionality entails process. The double-directedness of a living body beyond the body and back into the body can be realized only in change. But this is not change simply, as a rock wears away or the tide ebbs. A living body becomes something, and *has* this process of becoming something as a property of *its own*. Ebbtide is ebbtide, not an entity which ebbs. A living body, in order to take on the two-directional relation to its boundary by which we have defined it, must change, but it is at the same time an entity undergoing change, itself exhibiting the process which it undergoes, and being the end-product toward which the process tends.

Organic process, in other words, is change governed by *type*. Inorganic forms change, but they change simply. There are 'typical' rock formations and 'typical' cloud formations, but they are 'typical' simply in an inductive sense: because that is how they often occur. If we 'explain' them, moreover, it is by showing that, given certain conditions, they follow necessarily (by logical necessity) from the laws of physics or chemistry. Organic process, however, displays an essential tension between the series of physico-chemical changes through which it is necessarily exhibited and the type toward the realization of which it moves. Inorganic forms are gestalten simply. Organic form stands in an essential relation to a 'gestalt-idea', a type which is its norm.

From the very shape which organic bodies exhibit, therefore, or, better, from the way in which they exhibit it, from their positionality, it follows that they exist on more than one level. They exist essentially in relation to the types they realize. Given the concept of type, or gestalt-idea, moreover, such types are themselves comparable in terms of near or far likeness, of 'kinship' – quite apart from, and prior to, any theory of factual descent. I have tried elsewhere to show how theories of evolution which attempt to make phylogeny prior to morphology import surreptitiously a concept of type into their arguments.[5] Plessner here shows why this is so: the very way in which living bodies enter into our experience, as bodies having the property of positionality, already entails their essential relation to morphological principles. The particular type or level attained by a particular living thing and its relation to others, both morphologically and phylo-genetically, is of course a matter for the empirical scientist to determine. Plessner is not, one must say explicitly, suggesting a return to the 'Natur-philosophie' of the early nineteenth century. But he *is* saying that the possibility of systematics is a logical consequence of the perceptible character of living bodies – not a fortunate accident which just happens to suit the convenience of biological research. Anyone who is dealing with living things at all *is* thinking in terms of their trueness to type, and if he is thinking in terms of trueness to type, he is thinking, Plessner argues, in terms of levels which can be hierarchically ordered. There is no need to be ashamed of such activities, which are part and parcel of our ordinary grasp of the things around us: as much so as recognizing the difference between a toy snake and a real one or between a doll and a baby.

Correlative to the notion of type, moreover, is that of the *individual*, which in its existence does and does not conform to its type. Individuality does indeed empirically characterize living things, and again, in terms of Plessner's analysis, it follows necessarily from the nature of positionality that it should do so. The fact that the boundary of a living thing belongs to the thing and cuts it off from the medium while at the same time setting it in relation to it clearly entails the marking out of that living thing as an individual. At the same time, individuality is the correlative of the gestalt-idea, of the norm in relation to which it is the kind of individual it is.

Individuality expressed through process, moreover, is *development*, again an empirically listed character of living things which in this context is seen to be entailed by the concept of positionality. Development is just the orderly *process* in which a living thing transcends the body it is, to become the body it is. And finally, the dynamical character of living things, as developmental, necessarily entails, Plessner argues, its own rounding off, its own cessation, in ageing and death. Dying is sometimes interpreted as an aspect of living: since to live is to change irreversibly and to change irreversibly is to pass away, all living is dying. This is incorrect, Plessner says, since death is not part of, but the radical other of life, its absolute negation. But it is true, he insists, that the very nature of living things as positional entails this radical negation. The structure $B \leftarrow B \rightarrow M$ exhibits a hiatus, a cut between body and medium. The tension between individuality and type already displays this kind of dividedness. But as development proceeds – always to a new 'stage', but a new stage of the same being – it approaches the final hiatus which cuts it off irrevocably in time, just as its own individuality, for all its reaching out beyond itself, has always, from its beginning, cut it out in place.

The inevitability of death, the approach of death, taken together with the whole spiral-like process of development that has preceded it, show us further, that living things, unlike inanimate objects, have a destiny. One can write a geological 'history' of a mountain or a valley, for example; yet rocks and rivers, nevertheless, just are where they happen to be, even though they gradually accumulate or are thrown up by cataclysms of the earth or erode or fill up and disappear. Living individuals and only living individuals, with their janus-like direction to and from the world around them, to and from the bodies that they both are and have, are *destined* to live as they do – and to die.

Plessner's argument on death is closely tied to his critical examination of German idealist theories, and perhaps for this reason his reasoning here seems rather murky. Does the necessity of death really follow logically from the premiss of positionality? The hiatus exhibited by the boundary of a living thing does indeed generate an irresoluble tension between individual and type. But why *must* growth grow toward total cessation? There is an analogy, I suppose, between the contrast: living body/ medium and the contrast of life with death. But we have started, concretely, from the model of a perceptible living thing – and simply to say: as it stops in place, so it must stop in time, seems to be, so far, stretching the model just that bit too far. And there is an empirical difficulty: if we look at the whole range of the things we do in fact call living, we find that not all of them die. Only those that exhibit sexual reproduction undergo death in the proper sense. This fact, though not a philosophical argument, confirms the reader's uneasiness, at this point, about the argument itself. We may leave this question open for the moment.

At the same time Plessner's introduction of the concepts of death and destiny at this point, in a book published in 1928, presents an interesting contrast with Heidegger's treatment of the same themes in *Sein und Zeit*. Whether or not Plessner's argument at this stage is formally valid, what is important is that he sees these concepts as essential over a much broader range than Heidegger does. For Heidegger only the rare authentic human individual has a destiny; for Plessner destiny is coterminous with life itself. And Being-to-Death, though not, as for Heidegger, the only meaning of human life, is one essential theme, in Plessner's view, of organic, not only of human, development. Thanks to this broader basis, Plessner's book lacks the dramatic urgency which made the publication of Heidegger's magnum opus so overwhelming a philosophical event. But it can also lead to more fruitful, and more balanced, philosophical consequences than have emerged, in the later pronouncements of Heidegger's 'quest for being', from *Sein und Zeit*. Indeed, where Heidegger's concept of being-in-the-world has proved most fruitful – as in Merleau-Ponty's work, for example – it has been, if only indirectly, fused with Plessner's more broadly based argument. *Human* being and Being can be harmoniously and truly interpreted only through the mediation of an adequate philosophy of *life*.

IV

The treatment of death and destiny completes Plessner's argument on the dynamics of living things. Now for their 'statics'. Let us start once more from the character of positionality. The body in the relation $B \leftarrow B \rightarrow M$ to its surroundings has its boundary as part of itself. There is, as there is not for inanimate objects, a cut between body and medium, but more than this: the body takes up a relation to its own configuration, to its boundary. It stops, as Plessner puts it, before it comes to an end, and sets its boundary beyond itself as its own limiting area. It is tempting to translate this rather enigmatic statement into visible and zoological terms and say, the living body has a skin between itself and its environment, a skin which is one of its organs and therefore both a part of it, yet not 'it'. But this, Plessner warns, would be misleading. It is true, of course, that the development of a patterned outer layer, especially in higher animals, becomes an important organ not only of self-maintenance and survival but of expression, of what Adolf Portmann calls display.[6] And it is also true that Portmann's 'display' – which he takes as an ultimate and irreducible character of living things – is by Plessner shown to follow from positionality as a fundamental concept. But what he is arguing at this stage is not that display follows from positionality – though it does do so – but rather that Portmann's other fundamental criterion of life, 'centricity', is entailed by the positional character of organisms.[7]

The body which has the property of positionality, we have seen, assumes with this property a two-way relation to that body: it is directed both into the body and away from the body. In his argument up to this point Plessner has avoided the reflexive form of statement: he talks of the relation of the body to 'it', not to it*self*. But he now makes explicit the reflexive character which we do indeed, both in common sense and science, constantly attribute to organic beings. Taken in terms of structure rather than process, positionality means that the body has an inner *core* out of which it *has* its parts, and an outer aspect which is the aspect *of* that core. Its parts are peripheral relative to its center, and the center is the point of reference that makes the parts its parts. But neither core nor periphery in this context are simply spatial. Spatial depth and externality can be studied by the instrumentalities of physics and precise measurement, but the distinction Plessner is making is not touched by such anallysis – though on the other hand he is not suggesting, as Driesch had done, that physical analysis of living things must somewhere call a halt. Positionality and the core-periphery relation it carries with it are irrelevant to and unaffected by physical and microbiological techniques, which can, however, proceed indefinitely to illuminate the conditions necessary to the existence of the entity exhibiting that quality.

This seems paradoxical. Plessner started from the *perceptible* difference between living things and inanimate objects and took positionality, hypothetically, as a clue to that perceptible difference. Now he says, the relation of inner and outer in the living body is not a measurable, spatial relation. How then is it perceptible or entailed in what is perceptible? The answer (which Plessner had elaborated in great detail in an earlier book)[8] is that perceptibility is not by any means exhausted by what can be precisely specified in physically intelligible spatial terms. The aspect of our perceptions amenable to precise measurement and hence to physicochemical analysis is only one dimension, so to speak, of perception in a richer sense. We perceive, directly and immediately, much that is not reducible to such precise terms. All qualities resist such reduction: the physicist tells us how colors are produced but he does not thereby exhaust or destroy the qualities of color as perceived. Similarly, the neurophysiologist tells us about the deep-lying structures of the central nervous system which is the inner controlling agent of the organism's outward behavior – but he has not thereby explained, let alone explained away, the quality of tension between inner and outer, the quality of being peripheral expressions of an inner core, that characterizes life. A living body is indeed a body, it occupies space; but its center is nevertheless not a spatial center, it is a core which transcends spatiality and at the same time controls the spatiality of the body whose core it is. Relative to the body which is its body, it is nowhere and everywhere. And it is this relation of core to periphery which we perceive when we perceive a living body, even though it is not a 'perception' that we can exhaustively specify in the spatio-temporal terms of classical, or even of modern physics. It is this relation, in short, that we are referring to when we speak of living things as *subjects having* such and such properties or parts.

It is extremely important that we understand this statement in the full generality with which it is intended. To be a subject having properties or parts is *not* to be a conscious subject. The quality of consciousness simply does not come into the present argument at all. There are biologists and philosophers who resist the allure of a one-level ontology, the temptation to reduce organic phenomena to 'nothing but' physics and chemistry, because and only because they feel that the quality of consciousness which they know from their experience, and on which the very existence of their scientific disciplines as knowledge depends, is left unexplained by this kind of one-level ontology.[9] So far they are right; physics cannot 'explain' consciousness and if it were so expanded that it could do so, it would no longer be what we call physics. And they are right also to this extent: that we do feel a kinship with anything we perceive to be alive – especially anything that we feel exhibits self-motion and animal life, and since our way of being alive is best expressed through reference to consciousness we feel also that 'something like' consciousness is present here. It is this road that Portmann follows in his exposition of centricity: extrapolating gently down a slope of diminishing awareness from our self-conscious awareness of our own awareness to more distant, and, we cannot help believing, 'dimmer' centers of experience.

But consciousness alone is too narrow a concept to carry either the problem here or its solution. Even our own mental life is, as we well know, by no means wholly conscious; consciousness is one, specialized expression of a much more broadly occurring phenomenon. To see the problem solely in terms of physics versus introspection, space-time coordinates versus conscious feeling, is to see it still in terms of the Cartesian alternative which is just what we are trying to overcome – which we *must* overcome if we are to do justice to our perceived, direct, undeniable experience of the quality of life. True, consciousness is one of the forms of life, one of the expressions of the general subject-having character Plessner is describing, and indeed an adequate theory of consciousness can be developed only in these or similar terms. But at the present stage of his argument Plessner is a long way from any reference to consciousness. He is not introducing a ghost into the living machine, because he has not started from the living body as a machine. He has started from the living body as positional, in its unique relation to its own boundary, and this relation is now seen to entail the property of being a subject which both is the body that it is and has that body as its body. There is absolutely no reference here to an 'inner' something in a ghostly, secret, Cartesian sense; what Plessner is describing is in its fullest range the quality that perceptibly characterizes all the bodies we call 'living', from the simplest to the most complex.

It is this quality of being a subject having properties or parts which is, at the same time, the body had by that subject: it is this quality which makes the living thing in fact a 'harmonious equipotential system'. This phrase, introduced by Driesch in conjunction with his concept of entelechy, in fact needs no such anti-mechanical concept for its support. What it refers to is covered by the character of positionality and its structural consequences.

The relation of subject-having to body-had, moreover, carries with it that most striking character of living things: *organization*. The relation of organs to organism is close to, indeed, is to be equated with, the relation of periphery to core. This statement has again at first sight a paradoxical air. For even if we take the relation of core to periphery conceptually rather than spatially, as we must do, some organs seem to be 'central' not only spatially but in the sense that they are indispensable to life, while others are peripheral not only in being at the physical boundary of the organism – limbs for instance as against heart or liver – but in so far as life can go on without them. This distribution, however, Plessner insists, is an empirical one, which does not affect the fundamental relation which holds for all organs. An animal *has* a heart; it is not to be identified with its heart even though it cannot live without it. The organism is, universally, the subject which possesses all its organs, however much it depends on their totality for being the organism it is.

At the same time each organ is what it is in relation to the whole, not simply as whole but as subject. Each organ is not only a part – as the windows or roof are parts of a house – each organ is part of the whole as *representing* the whole. It stands *for* the whole in relation to the 'outside' and to the organism itself. *Mediation* is essential to organization. Mediation here is a concept clearly derived from the perceptible boundary structure from which we began. A body having positionality is a body delimited from its medium by a boundary which it *has*, a boundary which turns back on itself as part of itself and which it nevertheless transcends – goes beyond – in its relation to its medium, and in its relation to itself. Its relation to its medium, its living in its medium, is necessarily mediated by and through the boundary, and out of this arises an essential mediation of its parts to itself. Plessner has generalized, as we have seen, the intuited relation of body to boundary to the relation of core to periphery, and it is this relation in turn which shows us the essential role of mediation in the structure of living things. To be a living thing is on one level to be a thing like any other, but at the same time to be a living thing is to be more than the physical thing it is. It is to be through the mediation of the parts – the periphery – which, as core, it *has*, and yet which, as body, it *is*. That is why a principle of life separate from body is inconceivable, yet at the same time life is not reducible to physical terms. A living body is, precisely, a body which as a subject is represented by the parts which in their totality are itself as object, and its relation to its medium – in biological terms, to its environment – is in turn necessarily mediated by this internal relation of subject-aspect to object-aspect of the living thing itself.

It is in this aspect of organization that the traditional means-end character of living things becomes apparent. The teleology that is involved here, however, must be carefully distinguished from the kind of teleology that seems to demand a reference to a supernatural 'maker' of organic beings. It is an inner teleology, where each part or organ possessed by the subject is a means to the end of the subject itself. Nor, *a fortiori*, is there here a question of a conscious subject, 'wanting' to see, or digest, or cool off, or whatever it 'does with' the organ or organ-system in question. It is still a question of an organized body which has as parts the organs which in their totality it is. The dialectic of materialism versus supernaturalism which seems to follow from a means-end treatment of living things results from the failure to see the teleological character of organism in subordination to their intrinsic significance. This acknowledgement of the intrinsic meaningfulness of life as a unique resonance of being and having, in the sense in which Plessner has been describing it, and only this acknowledgement, can rescue our conception of living things from the twin unintelligibilities of mechanism and the will of God.

The inner dialectic of organic means-end relations, on the other hand, can be better understood for what it is if it is described, as Plessner proceeds to do, in terms of *powers*. Once more an Aristotelian concept, 'dynamis', potency or power, proves itself – within limits – naturally suited to biological philosophy. An organ is an instrument, that is, something by means of which something *can* be done; a hand is the power to grasp, an ear the possibility of hearing, and so on. And the subject itself – this is an un-Aristotelian insight – is the central power of all these performances. Inanimate objects, even the most sophisticated machines, do not 'have' their powers in the same sense as organisms, for their specific 'capacities' are not held together by the core-capacity which is the living subject.

Now that Plessner has explicitly introduced the concept of organization we can look back for a moment and see more clearly how in his view the possibility of systematics arises. For, given the relation of organs and organ-systems to an organism, we can extrapolate to arrange organisms themselves according to the levels of complexity of their organization.

Plessner himself turns the argument back at this point to reinforce an earlier stage. It is in their temporal structure, he reiterates here, that organisms most centrally display their uniqueness. This reference back to the dynamics of living things arises inevitably from a consideration of organs and potencies. For a potency is essentially in reference to the future. Just as Heidegger did for existential time, so Plessner here shows that organic time is essentially future-directed. Only this forward pull transforms the simply flowing passage of change into process and produces a true present which *is now* its possibility of becoming other, and anticipates *in posse* what it is not yet. In this context, moreover, the inevitability of death becomes more intelligible than in its earlier introduction. The living thing is a nexus of finite possibilities. As each develops it is gone, and, when all the possibilities, which in their totality are the organism, have been exhausted, that is natural death. This is the limit of natural time, its absolute measure, just as natural place, i.e., its place in its environment, is, for the living thing, its limit, its measure in relation to space. Only in living things, moreover, are place and time wholly at one: development is the organism's way of taking its place in nature, and natural place is the inevitable expression of the rhythm of organic process, of organic time.

V

So far the theory of organic modals has demonstrated that, if we take positionality as our starting-point, a number of the properties which are in fact ascribed to living bodies follow from this basic quality. All these properties have to do primarily with the internal structure and individual destiny of living things. The distinction from which Plessner had started, however, was a distinction between the ways in which bodies can be delimited from their surrounding media. What is most striking, both in experience, and as the principle of the present argument, is the way living bodies set themselves over against a surrounding medium – cut off in independence of it yet living in and through communication with it. The living body is directed both beyond itself to the environment and from it back into itself. Positionality thus generates a relation between the organism on the one hand and, on the other, a *positional field* which is the environment in the broadest sense – internal as well as external, organic as well as inorganic. It is indeed the whole opposition to itself, otherness of itself, which positionality entails. Organism and positional field in fact appear as contrary poles of a structured rhythm of life, of what Plessner calls the *Lebenskreis*, or *biocycle*. Thus we have the rhythm of assimilation and dissimilation, that is, of metabolism, through which the organism turns the environment into itself, and breaks down again into the stuff of the environment. Second, there is the rhythm of adaptation and adaptability. Plessner stresses the need to take into account the organism itself which *can* adapt itself as one pole of this balanced rhythm. The active and passive aspects of adaptive phenomena, emphasized in mutual exclusion by Lamarckian and Darwinian theories respectively, are in fact equally and inextricably involved in the dynamic interaction of organism and environment. When we take the biocycle to its logical conclusion, further, we have again the opposition of death to life, and, through it, the balance back toward life again in heredity and reproduction.

Again, however, it must be emphasized, Plessner is not arguing that the forms empirically discovered to express these general properties of life can be deduced a priori from the concept of positionality. On the contrary, positionality *entails* contingency: entails the possibility of one or other type of itself as a general principle. For it is just the flexibility of form – freedom of form – as we saw at the outset, that this concept is meant to express. Life, Plessner argues further, entails *selection*: the exclusion of some possibilities in favor of others. And this leads us to discriminate, within the general pattern of positonality, certain essential variants: the forms of positionality represented, first, by plants as against animals, then by animals in general as against men in particular.

The two possibilities realized in plant and animal life Plessner characterizes as those of *open* and *closed* form respectively. In plant life the biocycle flows smoothly in the same direction as the environment. Growth is additive: stage *n* may be simply tacked on to stages *a-m*, which still remain visibly present. From this relatively simple and directly integrated relation to its medium the basic properties of plant life follow. First, most plants are autotrophic; they can live directly from the non-living. Secondly, they do not, for the most part, possess true locomotion, nor, thirdly, sensori-motor functions. Indeed, so little does a plant exhibit the individuality characteristic of animal life, that it might be called, a botanist has suggested, not an 'individuum', but a 'dividuum'.

Animal form, in contrast, is *closed*. It displays, in the relation between body and medium, two contrary rhythms: one unidirectional with, the other opposed to, the environment. By this opposition it cuts itself off more sharply from the environment, stands over against it, yet is also, as an individual, exposed to it. From this variant of positionality, again, there follow the basic properties of animal life. First, animals are heterotrophic: they can live only from life. Secondly, animals have developed sensori-motor functions. There is here a more complex biocycle, a rhythm which displays a gap between the receptive and the motor phases. Thirdly, the closed form of animal existence generates an *open* positional field. In the light of the concept of positionality, that is the meaning of *drives*. Animal life is closed back in itself, but as *against* the positional field, always *wanting*, always essentially unfulfilled.

If it is objected that the distinction between plants and animals is not always so sharp as Plessner makes it, he answers: empirically, of course not. Why should it be? It is a question of two *types* which are ideally opposed to one another. Whether a given organism conforms or fails to conform, and to what extent, to one pattern or the other, is a matter for the empirical investigator to ponder. But just as positionality marks off the type of living thing as such from the inorganic – even though we may be in doubt, or indeed mistaken, about the precise place to draw the line in a given case – so the ideas of open and closed form divide off, ideally, plant from animal as forms of life.

This distinction once given, moreover, much that is the case about the animal kingdom can be seen to follow logically from the principle of closed form. Positionality, we have seen, is a boundary structure such that the body both has its body and is it. This principle is relatively submerged in the open-formed life of plants. In animals, thanks to their closed form, it is much sharper and more explicit. Indeed, the difference is not just one of degree, but of *level*. Where there is closed form, the body becomes what Merleau-Ponty would call a lived body, a *Leib*, possessed by a subject – as distinct from the physical body which it nevertheless is. We have already noticed, in dealing with organization, this subject-having relation. It is stabilized and made explicit in animal life through the *mediation* of a central organ which carries the sensori-motor functions, and which thus makes *action* possible. Not that the central nervous system, or its more primitive equivalent in lower animals, *is* the subject, or is identical with the core which we have already contrasted with the periphery of the organism. But the nervous system *represents* the center, the subject, and mediates between it, its body *as* its body, and the environment as external, as all that is *not* 'itself'.[10] Through this new level of mediation – which is animal organization, sensori-motor organization, as well as organization simply – the animal achieves *self*hood. It becomes a *me* – though not, Plessner is careful to emphasize, an *I*. It *has* a body, and *is* a self, but does not yet *have* a self, to which in turn it can take a stand in reflective awareness. Indeed, there is no question here, Plessner insists, of introducing into the argument a foreign and mysterious factor of 'subjectivity'. It is simply a question of a variant of positionality, but one which by its structure brings to the argument a new level of existence: a level which we might describe in language other than Plessner's as that of true individuality, but not yet that of personhood.[11]

This level in turn can take two forms: either decentralized existence, where the subject is in effect eliminated, or the centralized form which is exemplified in a series of ascending levels of animal life. For both, but most clearly for the centralized variant, some of the most striking properties of animal life once more follow logically from the characteristic form. First, the rhythm of development, in its animal form, is drawn, as we have seen, into a center, but a center which is wholly absorbed into the *here and now*. Animal life, however much directed by its future potentialities and transformed by its past, is assimilated without residue to its present. There is for it its own body and the open field of the external world: of all that it needs and notices – and into this presence it is wholly absorbed. That is why – and how – animal individuality is the individuality of the *me* but not the *I*.

In its absolute here-now, however, as the body it has over against the environment which both threatens it and presents to it opportunities for living, in this situation, the animal displays *spontaneity*: it *acts*. It has, as against plants, a new dimension of freedom. And as agent it confronts the world: it exhibits what Plessner calls *frontality*. It takes not only a place, but a *stand*.

Characteristic of this new level of existence, Plessner has already said, is the *empty center* which mediates between the sensory and motor phases of the functional cycle distinctive for animal life. In higher animals, the sensory phase becomes more marked, and in particular we can distinguish higher from lower animals by the degree to which the body schema (their awareness of their own body) is developed. The gap is wider, the *having* of the body more clearly over against the subject.

Never, however, Plessner insists, do animals in their perception evade the intrinsic relation of perception to movement. Animals perceive *things*, he says, but not *objects*. 'Things' are what they are relative to drives, to uses: they are threats, or lures, playthings, friends, enemies, not objectified entities understood in and for themselves. There is indeed a gap between thing and organism, a hiatus represented by the self, by the power of perception which both exposes the animal to reality and protects it against reality. But this hiatus is never, radically never, adequate to the understanding of objects as such. Why not? Once more, because of the way in which the animal self is assimilated to the here and now. Animals, Plessner points out, have no sense of *negativity*. Higher animals do make generalizations of a sort, they abstract similarities from similar situations;

learning experiments, as well as everyday experience of animal behavior, make this quite plain. And this kind of abstraction does even involve in a way a grasp of generalities. Do animals then understand 'universals'? Yes and no: they are, many of them, capable of experience, ordering their perceptions along rather general lines, but only because for them particulars are no more fully particular than universals are fully universal. The distinction arises in its full significance only for us who have put, as we shall see, a second hiatus, a new kind of distance, between ourselves and the world. It is this radical distancing of which animals other than man are wholly, structurally incapable. Negativity is the price we pay for being each of us an I, not only a 'me'.

To be sure, higher animals are capable also of insight, but again only into concrete situations. Moreover, they retain their past, not only as the product of growth and development, but as a living residue in memory. And they retain such residues, further, only in relation to drives, that is, not automatically, but as the correlate of anticipations, of needs. All living things are related to their past through the mediation of development, but for animals, especially higher animals, the past can be present, yet present only insofar as the anticipated future has made it memorable, has built the past into a pattern of experience. With memory and learning Plessner contrasts instinct, which, prior to memory, belongs to the cycle of adaptation. Both instinct and learning, however, he insists, are always, in all animals, relentlessly oriented to the concrete needs of *action*.

The importance of Plessner's account of animal existence should, I hope, be obvious. In the philosophical tradition we have had, on the problem of animal intelligence, as on so many questions, only a few equally unpalatable alternatives before us. In this case there have appeared to be three possibilities. Either sentience, memory, experience, learning characterize men only (Descartes), or there is no such thing (behaviorism), or there is a kind of brute, irrational animal inference and nothing else (Hume). The second alternative is nonsense (though still potent nonsense), denying the plain facts of our own experience. The first denies the equally plain facts of animal existence; no one who has ever known 'personally' a single animal could seriously accept the Cartesian theory. And the third, though it allows us a kind of bastard reasoning – a 'wonderful and unintelligible instinct in our souls' – can make nothing of the problem of standards, of truth and falsity, right and wrong, beauty and ugliness. Plessner's schema for animal life, his account of the closed form of positionality with its corresponding open positional field, his deduction of the me-body-environment structure of animal experience, brilliantly and with absolute cogency, it seems to me, sets off animal from plant life on the one hand, and on the other human from animal. There is here, as he rightly says, *no* introduction of a secret subjective factor, no '*metabasis eis allo genos*', but simply a re-structuring of positionality, an introduction of a second level of opposition and mediation between core and periphery, inner and outer. And again, be it noted, these are not 'inner' and 'outer', separable in Cartesian fashion from one another, but an ineradicably two-sided pair. The body, though beyond the center, is nevertheless within, as against the outside world. And the self, the center, is wholly absorbed in its directedness to what lies beyond. But at the same time this ambiguity, mediated through the sensori-motor cycle and its representatives, the organs of the nervous system, is a more intricate ambiguity, a more deeply levelled complex, than that of positionality simply and in general. Yet it is still lacking the further complication which, as we shall see, is constitutive for the achievement of humanity.

VI

There remains, then, the next and last level, which is our own. Animals live *into* a center and out *from* a center but not *as* a center: the center of their experience, as we have seen, is absorbed without residue into the here and now. Men are animals, they still live, and experience, out of and into the center of their bodily lives. But they *are* also the center itself: or better, life out of the center has, in the human case, become reflexive, set itself to itself as its own. Man cannot free himself from his own centred, animal existence, yet he has placed himself over against it. This structure Plessner calls *eccentric positionality*. A living thing exhibiting this new level of positionality is still bound by its animal nature, yet detached from it, free of it. Its life has its natural place as all animal existence has, yet is at the very same time detached from locality, is everywhere and nowhere. Nor is there any new entity which comes from somewhere – like Aristotelian *nous* – to create this situation: there is only a new hiatus, a break in nature, which produces a new unity. Plessner writes:

Positionally there is a threefold situation: the living thing is body, is in its body (as inner life...) and outside the body as the point of view from which it is both (body and inner life). An individual which is characterized positionally by this threefold structure is called a *person*. It is the subject of its experience, of its perceptions and its actions, of its initiative. It knows and it wills.

And he concludes: 'Seine Existenz ist wahrhaft auf Nichts gestellt'.[12] 'Its existence is literally based on nothing'. For what produces humanity is not a new type of organization – as the closed form over against the open constitutes animal life as against plants. We are still animals, but animals at a double distance from their own bodies. We have not only an inner life distinct from – though not separable from – our physical existence; we stand over against both these, holding them apart from one another and yet together. It is our eccentric positionality that gives to our existence the ambiguity — of necessity and freedom, brute contingency and significance – which it characteristically displays.

The world constituted by this structure (*Welt*, as against the *Umwelt* of other animals) has three aspects. It is an outer world, an inner world and a shared world (*Mitwelt*). The nature of the person is constituted by his relation to all three. Each of these aspects displays its own variety of ambiguity, its own double aspect – we have come far indeed from the simple Cartesian alternative.

First, the external world. We live as animals in an environment of which our body is the center. But at the same time, in setting that center for ourselves, we place ourselves as physical bodies, as things, at a measurable point in a uniform space and time. We place ourselves, and everything, in the empty forms of space and time through which we constitute our world as external. Plessner is not here taking any special stand on the metaphysical problem of space and time, but outlining the consequences of our special type of positionality: interpret them as you will.

The inner world, secondly, has its own ambiguity. There is the flow of experience, of sensations, feelings, 'mental' events which go on whether we are aware of them or not, whether we like it or not, and which are ourselves despite ourselves. And there are the acts of perceiving, noticing, recalling which we can – but by no means always do – enact in respect to our experience as passive. Again, the ultimate and indissoluble two-in-one character of these two aspects of our inner world follows logically from the structure of eccentric positionality as such. I am the center of my lived experience, yet, as an I, I stand apart from it. Plessner's analysis not only leaves room, as neither empiricist nor rationalist philosophies can, for the vast range of unconscious mental life. It exhibits at one stroke the inevitability, and the significance, of that strange but central fact of our inner lives which most philosophers neglect: the fact that, for all my actions, even the most considered or the most self-consciously responsible of them, I never really know, is it *I* who perform them or something in me that is not 'really' *I*. Many writers, especially in the existentialist tradition, have argued *that* we are always both passive and active, both bound and free. Plessner has shown *why* this is so. This is just the ambiguity produced by a centered positionality that has turned back on itself, that holds its bodily and experienced aspects together by splitting them apart. The 'problem' of freedom and determinism is not a pseudo-problem, but it is an insoluble problem because, as living things exhibiting the structure of eccentric positionality, this is the problem that we *are*.

But there is more to come. Psychologists, sociologists, philosophers all have told us that each of us needs the others: that we cannot become fully human except in and through human fellowship. Plessner's theory of organic modals derives this patent truth from the nature of our positionality. To be an I or a person is to stand at a distance from one's own physical existence and one's own passive experience. All my experience is mine only, as an animal's is; my own position in space and time is non-transferable as any body's is. But as an I, I am both irrevocably myself, not you, *and* universally an I as such, an I in general. The detachment, the very nothingness, that constitutes a person *is* the power he has of putting himself in the place of any other person, indeed, of any living thing[13]. Where there is one person, Plessner says, there is every person. And conversely, there can be one person only where there is the possibility of every person: where there is a shared world (a *Mitwelt*). Animals too have social lives, which serve in intricate ways to maintain and to enrich their existence. And we too, of course, being animals, have social lives rooted in a biological foundation. But beyond our animal needs and satisfactions we are persons. And it is, Plessner argues, constitutive of a person to be both I and we, to be I through being we. Nor is the 'we' entailed here any particular empirical we, not even that of mother and child or husband and wife. It is communion as such that constitutes the person: 'as a member of the shared world, each man stands where the other stands'. Only as the product of such sharing, moreover, is there *mind* (*Geist*), and its product, objectivity. The opposition of I to it, the understanding that what confronts me is capable of being elsewhere or anywhere, in other words, the grasp of true universals which makes scientific knowledge possible: this understanding is not something on its own over against our personal lives. It is constituted, as the sphere of mind, of *im*personal judgment, by the radical togetherness that is the very being of the person as such.

Again, in his description of the *Mitwelt*, Plessner has taken a simple yet revolutionary step. What existentialists call the problem of the 'other', what empiricists call the problem of the 'knowledge of other minds' may be, as a matter of fact, a genuine puzzle to particular people in particular situations of isolation and misunderstanding. Philosophically, however, it is a pseudoproblem. For persons are persons, *by definition*, only insofar as they take a stand, in its nature general or generalizable, over against themselves as living individuals. To be an *I*, no matter how egocentrically, is to be able, in principle, to take such a stand. My own particular being, in my own limited, parochial situation, is a concretion, as every particular human being is, of this generality. That is what is true in *both* rationalist and empiricist ethics. Kant's categorical imperative, the maxim of my will generalizable to a universal law, is an abstract statement of this truth, which neglects its necessary embodiment in a particular living thing. On the other hand, Hume's moral sense, 'a general calm determination of the passions founded on some distant view or reflection', represents empirically the condition of the concrete self engaged in moral judgment, exercising personhood. But Hume's empirical description is philosophically self-denying since for him there is no self. Its factual correctness can be justified only if it is shown, as Plessner shows it, to follow from, to be the concretion of, the power of universalization which *is* the Kantian moral law. Not that this is an 'ethic': it is much more. For the 'person' Plessner is describing, the person that we are, is not a merely moral agent. The being of the person unites all three Kantian questions: he is the agent as such, whether of knowing, of doing or of hoping. Only as a member of the shared world, the *Mitwelt*, am I a person; only as a person can I submit myself to universal standards, whether for the objectification of perceptual experience that makes the external world knowable, or for the universalization of inner experience that generates moral law.

On the foundation of his concept of eccentric positionality Plessner develops, in the concluding chapters of *Stufen*, a trio of 'basic laws of anthropology'. I must omit here all account of these, or of their application in special studies, such as his well known book *Lachen und Weinen*.[14] I hope, however, that I have reported enough to indicate the relevance of his thought for those who are concerned with the ontology both of nature and of man.

NOTES

[1] Plessner's book was first published by de Gruyter (Berlin) in 1928, and re-issued in 1965. In the preface to the second edition, Plessner speculates on the possible reasons for its having been so long ignored, or nearly ignored (though its indirect influence has been deep if not wide). As far as the German philosophical audience goes, there were two important reasons. For one thing, the publication in 1927 of Heidegger's *Sein und Zeit*, had, as it were, cornered the market for philosophical surprises. And for another, the influence of phenomenology in many quarters amounted at that period to a resuscitation of idealism; the down-to-earth realism of much of Plessner's argument was from this point of view also uncongenial to the *Zeitgeist*. It was an unlucky turn of fate that this was so: for Plessner does himself, it seems to me, profit from the phenomenological revolution – though without the heavy emphasis on the new 'method' and its new certainty which makes much phenomenological philosophy so difficult for the outsider to penetrate. Moreover, as we shall see, he, like Heidegger, has something to say about such concepts as time, death and destiny, and not, like Heidegger, only in terms of human existence cut off from living nature, but in terms of the significance of those concepts for organic being as a whole. But this is just the corrective that *Sein und Zeit* needs: the analysis of human existence is necessarily distorted unless it is grounded in an adequate philosophy of living nature.

[2] I shall not attempt to convey this argument in the fullness of its philosophical background. German idealism, in particular, forms for Plessner a tradition to be reckoned with, if chiefly in criticism, while for the mid-twentieth century English-speaking reader his references to it provide chiefly stumbling blocks for the understanding of his own position. I shall therefore ignore much of this aspect of the book and try simply to reconstruct the core of the argument. Nor can I render literally much of Plessner's terminology; I can only follow the argument in outline and hope to remain faithful, on the whole, to its general theme.

[3] *Stufen* (second edition), introduction and postscript; also 'A Newton of a Blade of Grass?', in *Toward a Unity of Knowledge* (*Proceedings of the Study Group on Foundations of Cultural Unity*), *Psychological Issues*, 1969.

[4] Sir MacFarlane Burnet, *The Integrity of the Body*, Cambridge, Massachusetts, 1962.

[5] See my 'Two Evolutionary Theories', this volume, Chapter VII.

[6] M. Grene, 'Portmann's Thought', *Commentary* (November, 1965) and this volume, Chapter XVI.

[7] I am reversing history here, for Portmann's treatment of display, through the past twenty years, has in fact been influenced by Plessner's argument of 1928.

[8] Helmuth Plessner, *Die Einheit der Sinne*, Bonn, 1923 and 1965.

[9] See for example C. H. Waddington, *The Ethical Animal*, New York 1961; E. P. Wigner, 'The Probability of the Existence of a Self-Reproducing Unit', in *The Logic of Personal Knowledge*, London 1961.

10 It would be worth comparing Plessner's exposition with F. S. Rothschild's theory of *biosemiotic*. See for example his article 'Laws of Symbolic Mediation in the Dynamics of Self and Personality', *Annals of the New York Academy of Science* **96** (1962), 774–783.

[11] See Michael Polanyi, *Personal Knowledge*, Chicago, 1958 [Torchbook edition: New York, 1964], Part Four. Cf. also M. Grene, *The Knower and the Known*, New York 1966, Chapter VIII.

[12] *Stufen*, p. 293.

[13] The first step in primitive, and in childish, thought is to people the world with personalities. Only a more 'mature' humanity learns to use the category of the *in*animate; the first essential step beyond the positionality of animal life is to see eccentricity, to see personhood, everywhere.

[14] Helmuth Plessner, *Lachen und Weinen*, Bern [third edition], 1961.

CHAPTER XIX

PEOPLE AND OTHER ANIMALS

Late in the twentieth century, we still need to think through anew the basic principles of our view of nature and of man and especially of the relation between nature and man. The trouble we are still, and more urgently than ever, faced with in this endeavor, began, if one may say so yet again, in the seventeenth century when Descartes split apart the inner life of consciousness from the external, material world. For despite the ingenuity of some of his remarks about the interaction of mind and body, the radical disconnection of these two kinds of finite substances made a coherent understanding of the world in terms of them impossible. And as men's knowledge of physical nature increased by leaps and bounds, the mind understood as disembodied spirit shrivelled to a mere 'ghost in the machine', gibbering but impotent – or else vanished altogether into the software of the brain. Man was displaced from the center of nature and become no longer his Maker's image, but *only* a handful of dust. Paradigmatic of this displacement, and even debasement, of man, was Laplace's famous boast of 1814: that if an omniscient observer knew the position of all the particles in the universe at time t_0, he would be able to predict their position at time t_1, and so would know everything there was to know. Now without returning, or even longing to return, to the tidy theocentric cosmos of earlier centuries, we do need, it's clear, to repopulate the Laplacean desert, to find a way of understanding the world more coherent and more adequate to our own experience than the Laplacean conception: our own experience whether as ordinary mortals, as scholars or as scientists. This is the task that Wilfrid Sellars calls reconciling the manifest and the scientific image of man.

In very general terms, this reconciliation could take four forms. It could try to rehabilitate in some fashion the dualism from which the modern view began. Or secondly, it could embrace, as Sellars himself has done, some variant of the so-called identity theory, that is, a theory of identity between mind and brain, which thus transforms the manifest image into the scientific one. Or it could, thirdly, develop some kind of idealist solution, according to which everything is ultimately of the nature of mind: thus the scientific reduction of mind to brain would be turned upside down and the material world retranslated into terms of the mental. This is in fact the solution adopted by the German evolutionist Bernhard Rensch. Or, fourthly and finally, one can introduce as basic a new categorization of the human condition which tries to go between the horns of the traditional dilemma and to espouse neither matter nor mind, nor both of them, as its fundamental concepts.

What I want to do here is to present very schematically two recent attempts to effect the reconciliation in question. One is a version of the first possibility I mentioned, that is, in effect, a modification of the traditional dualistic view. The other is a version of the fourth possibility: it entails a radical rejection of dualism and attempts to start afresh on a more comprehensive foundation. I shall ignore the second and third possibilities, identity theory or idealism. Idealism, I must confess, I find just too unlikely a conjecture to worry about. Minds seem too rare in nature to count as the fundamental sort of thing there is. The identity theory, or materialism, in its cruder form also seems unreasonable and has indeed been repeatedly refuted. In its subtler form, on the other hand, it amounts to the other possibility I *am* going to discuss: that is, it amounts to a rethinking of the problem in terms of new categories, primarily in terms of the category of the 'person' rather than of either body or mind.

The first schema I shall be describing, then, is basically a dualistic one. It was elaborated by the distinguished neurologist Sir John Eccles in a lecture delivered at Berkeley in 1969 and is based on the so-called three world theory of Sir Karl Popper.[1] The other is a schematisation of my own derived from several sources, but chiefly from the work of the German philosopher and sociologist Helmuth Plessner. Plessner's philosophy of man, and other theories like it, are sometimes described as 'philosophical anthropology', although, I should add, any resemblance to the social science called anthropology is purely coincidental. To avoid confusion, therefore, let me call Plessner's view *the theory of the eccentric position*; I hope to clarify this eccentric nomenclature shortly.[2]

First, then, let us look at Eccles. Till recently he had followed his great predecessor Sherrington in espousing a two-world theory. There is the external, physical world, including my central nervous system, which has

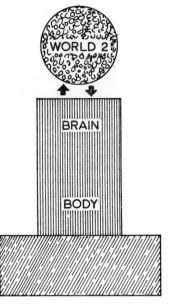

Fig. 1.

one kind of reality, and then there is the subjective world of my conscious experience, which constitutes another kind of reality. World 1 includes the necessary conditions for the existence of world 2; and world 2 seems in some way (though no one can say how) to effect, sometimes, some changes in world 1 (see Figure 1). There are at least three major difficulties about this view. First, there is the puzzle of psycho-physical causality already suggested. Secondly, there is the fact that each individual world 2 is isolated from every other. Eccles seems to take for granted the common 'argument from analogy' for the existence of other minds. Yet he talks, for example, about scientific research as a 'shared enterprise' in a way in which on this view no enterprise *can* be shared. For on the two-world view, all the achievements of man, including neurological science as represented by the knowledge of Eccles or Sherrington himself, belong to a detached, private, subjective world. And that produces a further difficulty. What guarantee is there that this inner world of subjective experience has any bearing at all on the 'real' nature of the physical world with which it purports to deal?

One possible answer is that our scientific theories if systematically mistaken would mislead us so seriously that natural selection would quickly wipe us out – and of course as far as *applied* science goes it may do just that. But philosophically this theory of knowledge, sometimes called evolutionary epistemology, is rather less than satisfying. It tells us nothing about the truth, or even the probable truth, of our theories, only about their practical utility. For of course as far as selection knows or cares a systematic set of illusions might well permit us to orient ourselves successfully in our environment and so to survive. We may be happily deceived and our theories nothing but fortunate fictions, including the theory of evolutionary epistemology. Given the disparity between worlds 1 and 2, there is no rational way to find out whether or not this is the case. Pure science for all we know may be just a kind of mumbo-jumbo, or, as Whitehead described this conception, a mystic chant over an unintelligible universe. It's an amusing game, useful careerwise, given the status of this kind of magic in our culture, but no more. I find this notion of science as magic depressingly prevalent among science students, especially in fact among gifted students, and in terms of any version of the two-world theory there is every reason why this should be so. (Also in terms of a classical identity theory, for that matter; but that's another story.)

Yet there is something plainly wrong with this conception, not only of science, but of other human activities as well, literature, the visual and auditory arts, religion, moral reflection. How can we correct it?

Eccles claims that he has found a new solution, an escape from the mere subjectivity of knowledge or of other seemingly significant aspects of conscious experience. He has found it, he says, in the conception of a third and objective world elaborated by Sir Karl Popper in two papers published in 1968. The 'third world', it turns out, is a modification of Plato's world of forms: it contains, not conscious experience itself, which is world 2, but everything that *could be* an objective content of thought: propositions, theories, arguments, even (and for Popper especially) problems. Eccles lists the contents of these worlds as he understands them (Figure 2). Actually the list is somewhat confused, for the records of intellectual (and artistic, moral or religious) effort belong in world 1, and, contrariwise, artifacts as *intelligible* belong in world 3. For world 3 *is* just Plato's intelligible world deprived of its metaphysical status and existing as the infinite class of potential objects of thought or imagination.

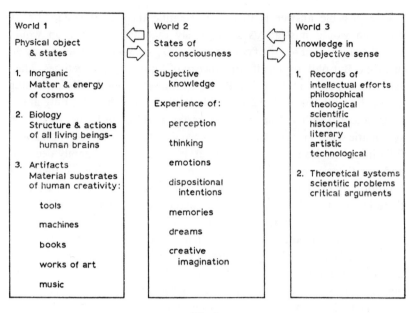

Fig. 2.

But never mind these details. What I want chiefly to ask is: how do these three worlds interact? World 2 of course is private; each individual is cut off in his own inner consciousness. Admittedly, world 3 has supposedly been created by our fellows, but one can only gather indirectly from one's own contemplation of it that there have probably been – and probably are – other world 2's who have produced the problems and problem solutions of science, art, philosophy to which my own world 2 happens to be attending just now. Still, there is no communication of any direct or even intelligible sort between one world 2 and any other. Thus the information flow of four different individuals is represented as in Figure 3.

Moreover, despite the arrows connecting worlds 3 and 2 on this diagram, the objective world – the world of problems, theories and the like – does not really act on world 2 or vice versa, except through world 1. This becomes clear when the information flow diagram of a single individual is represented in more detail, as in Figure 4. Here it is clear that a piece of world 3 has to be made into world 1 in order to react with the

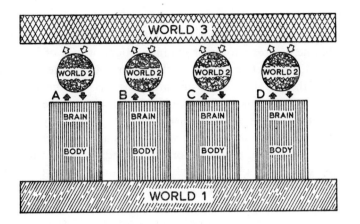

Fig. 3.

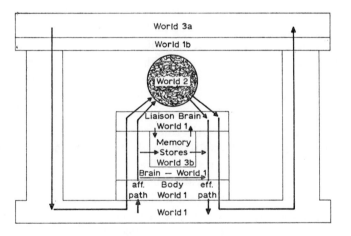

Fig. 4.

other little piece of world 1 that is my brain. There is no direct connection between worlds 3 and 2. Indeed, except for the mysterious transformation of neural events into thoughts and thoughts into neural events, my consciousness is an isolated entity, isolated in a way that no other entity in nature seems to be. Thus psycho-physical causality still remains a mystery; communication still remains a mystery; the epistemic claims of science

– let alone of art, religion or morality – still remain a mystery. Popper, incidentally, doesn't have to worry about these difficulties, for the simple reason that he ignores altogether the interaction, or lack of it, between worlds 2 and 3. In fact he ignores world 2. For him it is the structure of world 3 – of theories, hypotheses and problems – as such that constitutes science and the development of science (or of art). Scientists, he insists, and this should hold, he suggests, for artists too, do not *believe* anything; their aspirations, hunches, disappointments, don't count in the structure, even the historical structure, of science or art. Indeed, he entitles his theory 'epistemology without a knowing subject'. But, on the other hand, if one seriously tries to envisage what's going on in the Popperian three worlds, one does come back to Eccles' picture and its incoherences. What Eccles has really shown is that the alleged three-world theory still embodies a dualistic concept of nature and mind. When you really think about it, the third world, quite disconnected from the second, is a piece of the first world, and all the traditional problems of dualism remain.

What can we do about it? Let us look at Eccles' three pictures and see how we ought to revise them. First, his table of contents, so to speak. And first in that, world 1. To begin with, I want to rechristen this the *natural* rather than the physical world, to make it plain that the classes of living things, though included as subclasses in the class of physical systems, are nevertheless so organised as to differ from other physical systems in important ways. To move straight from *res extensa* to something called the mind is to miss the myriad styles of life cast up by the processes of evolution, of which the human style of life is one – with a difference. So let's call world 1 the *natural* world and subdivide it to bring out its stratified character (Figure 5). We start at the lowest and most inclusive level with world 1a, the class of all physical systems. Then we come to 1b, which is included in 1a but differentiated from it, that is, the class of all living things. The difference between 1a and 1b consists, not of course in any new 'stuff' or 'entity' superadded to 1a: living things, like any other things, are made of molecules. But they are made of molecules so ordered as to generate systems that function in some new and interesting fashion, and these new and interesting fashions constitute higher levels of organization within the physical system in question. To characterize this situation in summary fashion, I would like to adopt here G. L. Stebbins' concept of *relational order* as the unique characteristic of life. Given, as I have just

I. Natural World

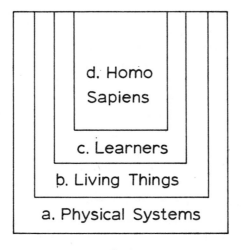

d. Homo
Sapiens

c. Learners

b. Living Things

a. Physical Systems

Fig. 5.

suggested, that living things are hierarchically organized physical systems, this basic concept is defined as follows: 'In living organisms, the ordered arrangement of the basic parts or units of any compound structure is related to similar orders in other comparable structures of the same rank in the hierarchy permitting the structures to cooperate in performing one or more specific functions.'[3]

So we have, then, worlds 1a and b. And next, inside world 1b we have a further subclass, 1c, which we may characterize (following David Hawkins) as the class of learners. I don't know quite where to start this class on the evolutionary scale. H. S. Jennings would have started with paramecia; perhaps that's too lowly. In any case it's the class of organisms not wholly dependent throughout their lives on their packet of genetic information, but able to increase and alter information through experience. Finally, inside the class of learners, but distinguished from it again, we have 1d, the class of human organisms: all the human babies that ever have been or will be born or are being born this minute. All these classes and subclasses, a–d, with their several levels of organization, are necessary before we can even begin to think about the nature of world 2.

On the other hand, however, to talk about the unique organization of world 1d is precisely to talk about the nature of worlds 2 and 3. One cannot really separate them in this enumerative way; but let me try. World 2, to begin with, is not subjective experience as such, but the world of *persons*. Every normal human organism at birth possesses by genetic endowment the capacity for becoming a person. To be a person, in other words, is an *achievement* of a human organism. It is not a kind of *thing*, it is something human organisms can *do*, and it takes the first year of life in particular, what Adolf Portmann calls the year of the social uterus, to do it: the year in which the human infant learns to stand upright, to speak and to perform responsible actions. And that's also where world 3 comes in, or has always already come in: for human infants become persons by *participation* in a culture or a social world. And by that I mean not just the world of knowledge, but the whole human world. This world is indeed *objective* in the sense that it is what the human beginner finds himself *in*; it surrounds him as surely and necessarily as the natural environment surrounds any organism. But it is also *conventional* in that it is man-made. Persons become persons through sharing in it, but persons past and present have made and are making it what it is. This social or cultural world then includes: first and most fundamentally languages, then other symbol-systems – religions, art, cognitive systems, including theories, problems, arguments, systems of morality, as well as all the institutions, political, social and economic, in which we live, move and have our being.

Now what about the relations between these systems as compared with Eccles' list? First, persons *are* human organisms, they are members of world 1, but world 1 *personalised*. They *have* subjective experience, but they *are animals*, subjectively experiencing animals, yet as surely embodied as any other animals – and like other animals they are also physical systems as well. But they achieve the personalisation of their bodily, and physical, nature through *participation* in a culture, that is in world 3. In other words, *to be a person is an achievement of a human organism mediated by participation in a culture.* Or to put it still another way, a developed human being is at one and the same time a *personalisation* of nature and an *embodiment* of culture. Conversely, his existence as a person *expresses* his culture *through* his participation in it. All this is schematized in Figure 6.

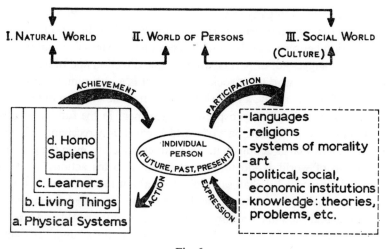

Fig. 6.

What happens now to Eccles' second illustration? Our equivalent is given in Figure 7. Note first that the person with his inner life is an embodied being; there is no separation here of two kinds of reality, only

WORLD III. SOCIAL WORLD: HISTORY, COMMUNITY

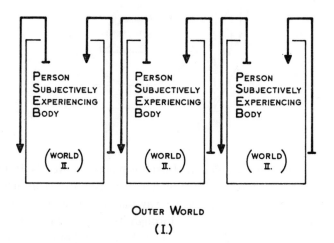

OUTER WORLD

(I.)

Fig. 7.

an inescapable ambiguity between two aspects of one reality. Second: the person is, or becomes a person, through sharing in a social world. The relation between worlds 2 and 3 is not one of external causality, but, as I've just suggested, of participation in one direction and of expression in the other. For world 3 is social. That is, it is a communal world, and also, and fundamentally, a historical world. It is the ongoing outcome of a tradition, as well as its prospective engagement in the future. Finally, a person is also, of course, part of world 1, constantly receiving information from it as well as acting on it. Yet characteristically, even the person's perception of world 1 and to a great extent also his actions on it are mediated by world 3. I shall say a bit more about this later; what I want to say for now is that although, as Plessner puts it, every human person has at one and the same time an inner world, an outer world and a social world, the inner and outer worlds are dependent in their human character on the structure of the social world. This will be clearer when we look at our equivalent to Eccles' third diagram, in Figure 8.

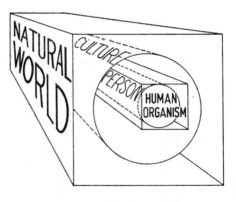

Fig. 8.

And this one is very different. First we put down worlds 1a to c as an outer container. World 1d, the human organism, forms an inner area which is related to 1a–c in the same way that any living thing is related to its environment: it is what it is through its adaptation (or sometimes its maladaptation) to its ecological niche. And all this of course is a process, not a static system. Both the environment and the organism are constantly interacting, and, through those interactions, ceasing to be what they have been and becoming what they are not yet. But now we have to ask: what, if anything, has happened to 1d that makes it different in its particular form of organization from that of other learners? What new kind of relational order have we here? And this is where worlds 2 and 3 come in. Every animal (here I am paraphrasing Plessner) both *is* a body, that is, a physical system, and *has* a body, that is, it lives as center of perceiving and acting, in and out of its given biological endowment and into and out of its given biological environment. But a human person not only *is* a body, that is, a physical system, and *has* a body, that is, a living nexus of tissues, organs, organ systems by which he lives. He also learns, as he assumes humanity, to *take a position* with respect to this bodily being and to his biological and physical environment. This is what Plessner calls the *eccentric position of man*. The existence of other animals is wholly absorbed into their medium. They have no power of detachment from it, of criticism of it. Man has acquired such powers; he can stand apart, to one side, 'eccentrically', from his biological and physical being and consider himself in relation to them. How can he do this? *Not* by possessing some new entity called soul or mind; but simply through the *achievement* of personhood as the *embodiment* of a culture. The achievement of the eccentric position of man, of *each man*, is dependent on the artifacts of culture through participation in which and in expression *of* which he achieves that position. In other words, just as much as a normally functioning central nervous system, an ongoing culture is a necessary, though not a sufficient condition, for the achievement of humanity. In terms of the concept of relational order, then, what we have here is, over and above the novel and interrelated anatomico-physiological structures that make personhood possible, such as cerebralization, the capacity for upright posture, the eye-hand field and so on: what we have over and above this is *the ordered relation of an achievement – personhood – to an artifact – the social world of a culture.*

I shall not try here even to begin to articulate in any detail this (so far as we yet know) specifically human version of relational order. If we had this worked out we would have a completed philosophical anthropology. All I want to suggest (or *have* to suggest) is the taking-off point for one. I think, though, that such work as that of Alfred Schutz on multiple worlds, or the extensive research that has been done both in Europe and America on the nature and function of social roles would bear out in detail the consequences of such a view.

Short of such applications, however, I would like, in conclusion, to mention rather cryptically some of the principles that Plessner has stated in connection with his concept of the eccentric position. They follow clearly from the basic situation as I have tried to sketch it. First – and I hope this is obvious from what I have been saying – is the principle of the *natural artificiality* of man. We become human, not just by being born *homo sapiens*, but by relying on a complex network of artifacts: language and other symbolic systems, social conventions, tools in the context of their use – artifacts which are in a way extensions of ourselves, but which in turn we actualize in our personal lives. It is our nature to need the artificial, art in the broadest sense of that term, or, indeed, poetry in the broadest sense of that term: making and the made. We cannot become human beings without this.

So profound is this need, secondly, that even our relation to the *natural* world is often, if not, at least in some degree, always, mediated by the world of culture. Even our bodily lives, our perceptions, our performances are experienced and enacted *through* the man-made sphere in which we dwell. Plessner calls this the principle of *mediated immediacy*. Take a very simple example. Sometimes in Davis when the smog over Sacramento lifts and the rain is gone one can suddenly see the Sierra from the ninth floor of our tallest office building. Now I suspect that no other animal, however lynx eyed, could see the Sierra. Certainly the dogs who accompany our visitors on those occasions don't even try to look at the view. The contrast I want to suggest is analogous to that made by Wittgenstein when he remarks: 'A dog can expect his master; but can he expect his master tomorrow?' A dog can see, but can he see the Sierra? What is this kind of seeing? It *is* a perceptual experience and, in its intrinsic quality, entirely immediate, not, as some philosophers would say, the product of an inference or a set of associations: suddenly those mountains are there. Yet it is, somehow, a *human* perception, not just a biological event. Nor do I mean by this that the gleaming whiteness of the mountains forms an added titillation to an otherwise dull day. To perceive them alters radically the world one is *in*. Admittedly, of course, any animal's world is changed by changes in its surroundings, by being caged instead of in the wild, by becoming an alpha rather than a beta animal, and so on. And we, being animals, retain these dependencies, too. Yet, again, with a difference. The structure of our given, communal and historical world 3, the man-made world through which we have developed as persons, permeates our very seeing – as well as our hearing and feeling. The natural world is present to us directly, as it is to any sentient organism, but present *through* the medium of the social world, by sharing in which we have become the persons we are, the seeing, hearing, feeling persons we are. Lifting up one's eyes unto the hills, in short, is a human experience, perceptual and in this sense direct, yet made the perception it is by the mediation of a culture, of the interests, concepts, values that define a given world 3, or some well-formed aspect of a given world 3. Merleau-Ponty has likened perception to communion: both are direct, unmediated, yet at the same time wholly mediated by human meanings.

Finally, Plessner adds a third principle which I found at first harder to understand, but can perhaps reinterpret. This is the principle of the Utopian Standpoint, as he calls it. Now of course I know that Utopias, messianic visions, cargo cults and so on are frequent occurrences in many societies, including our own. Yet it is hard to understand at first sight that this *must* be so: that somehow this kind of seeming illusion belongs to the very nature of the human condition. But if one calls it a principle of negativity, I think I can see better what it means. We rely for becoming persons on a received nexus of human artifacts, on a tradition, but we are *makers* of tradition also. We make ourselves what we are through the way we actively assimilate our received culture, and in so doing we remake it, and that is also to unmake it. The power of criticism (stressed by Popper), the power of denying what is in favor of what we believe ought to be, of demanding what is not in defiance of what is: this does indeed form the root, not only of religion – witness the tradition of negative theology – but even of science – witness the role of revision and self-correction in its growth. Man's so-called rationality resides, not in his grasp of some transcendent truths, but in his power to doubt, to criticize, to ignore or deny the actual in favor of the barely possible, if not the impossible. That is, it seems to me, what the Utopian Standpoint means. We *are* such stuff as dreams are made on. High abstractions can engage us more urgently than the hardest fact. To whatever is, we can say 'no' for the sake of what is not.

In short, our nature demands for its completion the unnatural, the indirect, and also the unreal. We have to make ourselves, and our making is also an unmaking. We *are*, like other things, physico-chemical systems; we live, like other animals, bodily lives dependent on bodily needs and functions, but we *exist* as human beings on the edge between nature and art, reality and its denial. That is both our peril and our opportunity.

NOTES

[1] J. C. Eccles, 'The Brain and the Soul', in *Facing Reality*, Springer, New York, 1970, pp. 151–175. Cf. K. R. Popper, 'Epistemology Without a Knowing Subject', *Proc. 3rd Int. Congr. for Logic, Methodology and Philosophy of Science*, Amsterdam, 1968, 225–277, and Sir Karl Popper, 'On the Theory of the Objective Mind', *Akten des XIV Internationalen Kongresses für Philosophie*, I, Vienna, 1968, 25–53.
[2] See H. Plessner, *Die Stufen des Organischen und der Mensch*, de Gruyter, Berlin, 1928, 1965; for a discussion in English, see my *Approaches to a Philosophical Biology*, Basic Books, New York, 1969, Ch. II and Ch. XVIII of this volume; also the translation of Plessner's *Lachen und Weinen, Laughing and Crying*, Northwestern University Press, Evanston, Illinois, 1970.
[3] G. L. Stebbins, *The Basis of Progressive Evolution*, University of North Carolina Press, Chapel Hill 1969, pp. 5–6.

INDEX

SYNTHESE LIBRARY

Monographs on Epistemology, Logic, Methodology,
Philosophy of Science, Sociology of Science and of Knowledge, and on the
Mathematical Methods of Social and Behavioral Sciences

Editors:

Donald Davidson (The Rockefeller University and Princeton University)

Jaakko Hintikka (Academy of Finland and Stanford University)

Gabriël Nuchelmans (University of Leyden)

Wesley C. Salmon (University of Arizona)

1. J. M. Bocheński, *A Precis of Mathematical Logic.* 1959, X + 100 pp.
2. P. L. Guiraud, *Problèmes et méthodes de la statistique linguistique.* 1960, VI + 146 pp.
3. Hans Freudenthal (ed.), *The Concept and the Role of the Model in Mathematics and Natural and Social Sciences, Proceedings of a Colloquium held at Utrecht, The Netherlands, January 1960.* 1961, VI + 194 pp.
4. Evert W. Beth, *Formal Methods. An Introduction to Symbolic Logic and the Study of Effective Operations in Arithmetic and Logic.* 1962, XIV + 170 pp.
5. B. H. Kazemier and D. Vuysje (eds.), *Logic and Language. Studies dedicated to Professor Rudolf Carnap on the Occasion of his Seventieth Birthday.* 1962, VI + 256 pp.
6. Marx W. Wartofsky (ed.), *Proceedings of the Boston Colloquium for the Philosophy of Science, 1961–1962,* Boston Studies in the Philosophy of Science (ed. by Robert S. Cohen and Marx W. Wartofsky), Volume I. 1963, VIII + 212 pp.
7. A. A. Zinov'ev, *Philosophical Problems of Many-Valued Logic.* 1963, XIV + 155 pp.
8. Georges Gurvitch, *The Spectrum of Social Time.* 1964, XXVI + 152 pp.
9. Paul Lorenzen, *Formal Logic.* 1965, VIII + 123 pp.
10. Robert S. Cohen and Marx W. Wartofsky (eds.), *In Honor of Philipp Frank,* Boston Studies in the Philosophy of Science (ed. by Robert S. Cohen and Marx W. Wartofsky), Volume II. 1965, XXXIV + 475 pp.
11. Evert W. Beth, *Mathematical Thought. An Introduction to the Philosophy of Mathematics.* 1965, XII + 208 pp.
12. Evert W. Beth and Jean Piaget, *Mathematical Epistemology and Psychology.* 1966, XXII + 326 pp.
13. Guido Küng, *Ontology and the Logistic Analysis of Language. An Enquiry into the Contemporary Views on Universals.* 1967, XI + 210 pp.
14. Robert S. Cohen and Marx W. Wartofsky (eds.), *Proceedings of the Boston Colloquium for the Philosophy of Science 1964–1966, in Memory of Norwood Russell Hanson,* Boston Studies in the Philosophy of Science (ed. by Robert S. Cohen and Marx W. Wartofsky), Volume III. 1967, XLIX + 489 pp.

15. C. D. Broad, *Induction, Probability, and Causation. Selected Papers.* 1968, XI + 296 pp.

16. Günther Patzig, *Aristotle's Theory of the Syllogism. A Logical-Philosophical Study of Book A of the Prior Analytics.* 1968, XVII + 215 pp.

17. Nicholas Rescher, *Topics in Philosophical Logic.* 1968, XIV + 347 pp.

18. Robert S. Cohen and Marx W. Wartofsky (eds.), *Proceedings of the Boston Colloquium for the Philosophy of Science 1966–1968,* Boston Studies in the Philosophy of Science (ed. by Robert S. Cohen and Marx W. Wartofsky), Volume IV. 1969, VIII + 537 pp.

19. Robert S. Cohen and Marx W. Wartofsky (eds.), *Proceedings of the Boston Colloquium for the Philosophy of Science 1966–1968,* Boston Studies in the Philosophy of Science (ed. by Robert S. Cohen and Marx W. Wartofsky), Volume V. 1969, VIII + 482 pp.

20. J. W. Davis, D. J. Hockney, and W. K. Wilson (eds.), *Philosophical Logic.* 1969, VIII + 277 pp.

21. D. Davidson and J. Hintikka (eds.), *Words and Objections: Essays on the Work of W. V. Quine.* 1969, VIII + 366 pp.

22. Patrick Suppes, *Studies in the Methodology and Foundations of Science. Selected. Papers from 1911 to 1969,* XII + 473 pp.

23. Jaakko Hintikka, *Models for Modalities. Selected Essays.* 1969, IX + 220 pp.

24. Nicholas Rescher et al. (eds.), *Essay in Honor of Carl G. Hempel. A Tribute on the Occasion of his Sixty-Fifth Birthday.* 1969, VII + 272 pp.

25. P. V. Tavanec (ed.), *Problems of the Logic of Scientific Knowledge.* 1969, XII + 429 pp.

26. Marshall Swain (ed.), *Induction, Acceptance, and Rational Belief.* 1970. VII + 232 pp.

27. Robert S. Cohen and Raymond J. Seeger (eds.), *Ernst Mach; Physicist and Philosopher,* Boston Studies in the Philosophy of Science (ed. by Robert S. Cohen and Marx W. Wartofsky), Volume VI. 1970, VIII + 295 pp.

28. Jaakko Hintikka and Patrick Suppes, *Information and Inference.* 1970, X + 336 pp.

29. Karel Lambert, *Philosophical Problems in Logic. Some Recent Developments.* 1970, VII + 176 pp.

30. Rolf A. Eberle, *Nominalistic Systems.* 1970, IX + 217 pp.

31. Paul Weingartner and Gerhard Zecha (eds.), *Induction, Physics, and Ethics, Proceedings and Discussions of the 1968 Salzburg Colloquium in the Philosophy of Science.* 1970, X + 382 pp.

32. Evert W. Beth, *Aspects of Modern Logic.* 1970, XI + 176 pp.

33. Risto Hilpinen (ed.), *Deontic Logic: Introductory and Systematic Readings.* 1971, VII + 182 pp.

34. Jean-Louis Krivine, *Introduction to Axiomatic Set Theory.* 1971, VII + 98 pp.

35. Joseph D. Sneed, *The Logical Structure of Mathematical Physics.* 1971, XV + 311 pp.

36. Carl R. Kordig, *The Justification of Scientific Change.* 1971, XIV + 119 pp.

37. Milič Čapek, *Bergson and Modern Physics,* Boston Studies in the Philosophy of Science (ed. by Robert S. Cohen and Marx W. Wartofsky), Volume VII. 1971, XV + 414 pp.

38. Norwood Russell Hanson, *What I do not Believe, and other Essays,* ed. by Stephen Toulmin and Harry Woolf. 1971, XII + 390 pp.

39. ROGER C. BUCK and ROBERT S. COHEN (eds.), *PSA 1970. In Memory of Rudolf Carnap*, Boston Studies in the Philosophy of Science (ed. by Robert S. Cohen and Marx W. Wartofsky), Volume VIII. 1971, LXVI + 615 pp. Also available as a paperback.

40. DONALD DAVIDSON and GILBERT HARMAN (eds.), *Semantics of Natural Language*. 1972, X + 769 pp. Also available as a paperback.

41. YEHOSUA BAR-HILLEL (ed.), *Pragmatics of Natural Languages*. 1971, VII + 231 pp.

42. SÖREN STENLUND, *Combinators, λ-Terms and Proof Theory*. 1972, 184 pp.

43. MARTIN STRAUSS, *Modern Physics and Its Philosophy. Selected Papers in the Logic, History, and Philosophy of Science*. 1972, X + 297 pp.

44. MARIO BUNGE, *Method, Model and Matter*. 1973, VII + 196 pp.

45. MARIO BUNGE, *Philosophy of Physics*. 1973, IX + 248 pp.

46. A. A. ZINOV'EV, *Foundations of the Logical Theory of Scientific Knowledge (Complex Logic)*, Boston Studies in the Philosophy of Science (ed. by Robert S. Cohen and Marx W. Wartofsky), Volume IX. Revised and enlarged English edition with an appendix, by G. A. Smirnov, E. A. Sidorenka, A. M. Fedina, and L. A. Bobrova. 1973, XXII + 301 pp. Also available as a paperback.

47. LADISLAV TONDL, *Scientific Procedures*, Boston Studies in the Philosophy of Science (ed. by Robert S. Cohen and Marx W. Wartofsky), Volume X. 1973, XII + 268 pp. Also available as a paperback.

48. NORWOOD RUSSELL HANSON, *Constellations and Conjectures*, ed. by Willard C. Humphreys, Jr. 1973, X + 282 pp.

49. K. J. J. HINTIKKA, J. M. E. MORAVCSIK, and P. SUPPES (eds.), *Approaches to Natural Language. Proceedings of the 1970 Stanford Workshop on Grammar and Semantics*. 1973, VIII + 526 pp. Also available as a paperback.

50. MARIO BUNGE (ed.), *Exact Philosophy – Problems, Tools, and Goals*. 1973, X + 214 pp.

51. RADU J. BOGDAN and ILKKA NIINILUOTO (eds.), *Logic, Language, and Probability*. A selection of papers contributed to Sections IV, VI, and XI of the Fourth International Congress for Logic, Methodology, and Philosophy of Science, Bucharest, September 1971. 1973, X + 323 pp.

52. GLENN PEARCE and PATRICK MAYNARD (eds.), *Conceptual Chance*. 1973, XII + 282 pp.

53. ILKKA NIINILUOTO and RAIMO TUOMELA, *Theoretical Concepts and Hypothetico-Inductive Inference*. 1973, VII + 264 pp.

54. ROLAND FRAÏSSÉ, *Course of Mathematical Logic – Volume I: Relation and Logical Formula*. 1973, XVI + 186 pp. Also available as a paperback.

√ 55. ADOLF GRÜNBAUM, *Philosophical Problems of Space and Time*. Second, enlarged edition, Boston Studies in the Philosophy of Science (ed. by Robert S. Cohen and Marx W. Wartofsky), Volume XII. 1973, XXIII + 884 pp. Also available as a paperback.

56. PATRICK SUPPES (ed.), *Space, Time, and Geometry*. 1973, XI + 424 pp.

57. HANS KELSEN, *Essays in Legal and Moral Philosophy*, selected and introduced by Ota Weinberger. 1973, XXVIII + 300 pp.

58. R. J. SEEGER and ROBERT S. COHEN (eds.), *Philosophical Foundations of Science. Proceedings of an AAAS Program, 1969*. Boston Studies in the Philosophy of Science (ed. by Robert S. Cohen and Marx W. Wartofsky), Volume XI. 1974, X + 545 pp. Also available as a paperback.

59. ROBERT S. COHEN and MARX W. WARTOFSKY (eds.), *Logical and Epistemological Studies in Contemporary Physics*, Boston Studies in the Philosophy of Science (ed. by Robert S. Cohen and Marx W. Wartofsky), Volume XIII. 1973, VIII + 462 pp. Also available as a paperback.

60. ROBERT S. COHEN and MARX W. WARTOFSKY (eds.), *Methodological and Historical Essays in the Natural and Social Sciences. Proceedings of the Boston Colloquium for the Philosophy of Science, 1969–1972*, Boston Studies in the Philosophy of Science (ed. by Robert S. Cohen and Marx. W Wartofsky), Volume XIV. 1974, VIII + 405 pp. Also available as a paperback.

63. SÖREN STENLUND (ed.), *Logical Theory and Semantic Analysis. Essays Dedicated to Stig Kanger on His Fiftieth Birthday*. 1974, V + 217 pp.

65. HENRY E. KYBURG, JR., *The Logical Foundations of Statistical Inference*. 1974, IX + 421 pp.

66. MARJORIE GRENE, *The Understanding of Nature: Essays in the Philosophy of Biology*, Boston Studies in the Philosophy of Science (ed. by Robert S. Cohen and Marx W. Wartofsky), Volume XXIII. 1974, XII + 360 pp. Also available as a paperback.

In Preparation

61. ROBERT S. COHEN and MARX W. WARTOFSKY (eds.), *For Dirk Struik. Scientific, Historical and Political Essays in Honor of Dirk J. Struik*, Boston Studies in the Philosophy of Science (ed. by Robert S. Cohen and Marx W. Wartofsky), Volume XV. Also available as a paperback.

62. KAZIMIERZ AJDUKIEWICZ, *Pragmatic Logic*, transl. from the Polish by Olgierd Wojtasiewicz.

64. KENNETH SCHAFFNER and ROBERT S. COHEN (eds.), *Proceedings of the 1972 Biennial Meeting, Philosophy of Science Association*, Boston Studies in the Philosophy of Science (ed. by Robert S. Cohen and Marx W. Wartofsky), Volume XX. Also available as a paperback.

67. JAN M. BROEKMAN, *Structuralism: Moscow, Prague, Paris*.

68. NORMAN GESCHWIND, *Selected Papers on Language and the Brain*, Boston Studies in the Philosophy of Science (ed. by Robert S. Cohen and Marx W. Wartofsky), Volume XVI. Also available as a paperback.

69. ROLAND FRAÏSSÉ. *Course of Mathematical Logic – Volume II: Model Theory*.

SYNTHESE HISTORICAL LIBRARY

Texts and Studies
in the History of Logic and Philosophy

Editors:

N. KRETZMANN (Cornell University)
G. NUCHELMANS (University of Leyden)
L. M. DE RIJK (University of Leyden)

1. M. T. BEONIO-BROCCHIERI FUMAGALLI, *The Logic of Abelard*. Translated from the Italian. 1969, IX + 101 pp.

2. GOTTFRIED WILHELM LEIBNITZ, *Philosophical Papers and Letters*. A selection translated and edited, with an introduction, by Leroy E. Loemker. 1969, XII + 736 pp.

3. ERNST MALLY, *Logische Schriften*, ed. by Karl Wolf and Paul Weingartner. 1971, X + 340 pp.

4. LEWIS WHITE BECK (ed.), *Proceedings of the Third International Kant Congress*. 1972, XI + 718 pp.

5. BERNARD BOLZANO, *Theory of Science*, ed. by Jan Berg. 1973, XV + 398 pp.

6. J. M. E. MORAVCSIK (ed.), *Patterns in Plato's Thought. Papers arising out of the 1971 West Coast Greek Philosophy Conference*. 1973, VIII + 212 pp.

7. NABIL SHEHABY, *The Propositional Logic of Avicenna: A Translation from al-Shifā':al-Qiyās*, with Introduction, Commentary and Glossary. 1973, XIII + 296 pp.

8. DESMOND PAUL HENRY, *Commentary on De Grammatico: The Historical-Logical Dimensions of a Dialogue of St. Anselm's*. 1974, IX + 345 pp.

9. JOHN CORCORAN, *Ancient Logic and Its Modern Interpretations*. 1974, X + 208 pp.